Practical Heat Treating
Second Edition

Jon L. Dossett
Howard E. Boyer

ASM International®
Materials Park, Ohio 44073-0002
www.asminternational.org

Copyright © 2006
by
ASM International®
All rights reserved

No part of this book may be reproduced, stored in a retrieval system, or transmitted, in any form or by any means, electronic, mechanical, photocopying, recording, or otherwise, without the written permission of the copyright owner.

First printing, March 2006
Digital printing, October 2009

Great care is taken in the compilation and production of this Volume, but it should be made clear that NO WARRANTIES, EXPRESS OR IMPLIED, INCLUDING, WITHOUT LIMITATION, WARRANTIES OF MERCHANTABILITY OR FITNESS FOR A PARTICULAR PURPOSE, ARE GIVEN IN CONNECTION WITH THIS PUBLICATION. Although this information is believed to be accurate by ASM, ASM cannot guarantee that favorable results will be obtained from the use of this publication alone. This publication is intended for use by persons having technical skill, at their sole discretion and risk. Since the conditions of product or material use are outside of ASM's control, ASM assumes no liability or obligation in connection with any use of this information. No claim of any kind, whether as to products or information in this publication, and whether or not based on negligence, shall be greater in amount than the purchase price of this product or publication in respect of which damages are claimed. THE REMEDY HEREBY PROVIDED SHALL BE THE EXCLUSIVE AND SOLE REMEDY OF BUYER, AND IN NO EVENT SHALL EITHER PARTY BE LIABLE FOR SPECIAL, INDIRECT OR CONSEQUENTIAL DAMAGES WHETHER OR NOT CAUSED BY OR RESULTING FROM THE NEGLIGENCE OF SUCH PARTY. As with any material, evaluation of the material under end-use conditions prior to specification is essential. Therefore, specific testing under actual conditions is recommended.

Nothing contained in this book shall be construed as a grant of any right of manufacture, sale, use, or reproduction, in connection with any method, process, apparatus, product, composition, or system, whether or not covered by letters patent, copyright, or trademark, and nothing contained in this book shall be construed as a defense against any alleged infringement of letters patent, copyright, or trademark, or as a defense against liability for such infringement.

Comments, criticisms, and suggestions are invited, and should be forwarded to ASM International.

Prepared under the direction of the ASM International Technical Books Committee (2005–2006), Yip-Wah Chung, FASM, Chair.

ASM International staff who worked on this project include Scott Henry, Senior Manager of Product and Service Development; Bonnie Sanders, Manager of Production; Madrid Tramble, Senior Production Coordinator; and Kathryn Muldoon, Production Assistant.

Library of Congress Control Number: 2005055552
ISBN-13: 978-0-87170-829-8
ISBN-10: 0-87170-829-9
SAN: 204-7586

ASM International®
Materials Park, OH 44073-0002
www.asminternational.org

Printed in the United States of America

Cover photograph courtesy of MT Heat Treating Inc., Mentor, Ohio

Contents

Preface v

CHAPTER 1 What is Heat Treating? Importance and Classifications 1

CHAPTER 2 Fundamentals of the Heat Treating of Steel 9

CHAPTER 3 Hardness and Hardenability 27

CHAPTER 4 Furnaces and Related Equipment for Heat Treating 55

CHAPTER 5 Instrumentation and Control of Heat Treating Processes 85

CHAPTER 6 Heat Treating of Carbon Steels 97

CHAPTER 7 Heat Treating of Alloy Steels 125

CHAPTER 8 Case Hardening of Steel 141

CHAPTER 9 Flame and Induction Hardening 159

CHAPTER 10 Heat Treating of Stainless Steels 175

CHAPTER 11 Heat Treating of Tool Steels 191

CHAPTER 12 Heat Treating of Cast Irons 207

CHAPTER 13 Heat Treating of Nonferrous Alloys 231

CHAPTER 14 Assuring the Quality of Heat Treated Product ...243

APPENDIX A   Glossary of Heat Treating Terms 257

APPENDIX B   Decarburization of Steels 275

APPENDIX C   Boost/Diffuse Cycles for Carburizing 279

APPENDIX D   Use of Test Coupons for Process Verification 283

Index 287

Preface

Nearly 44 years ago, I started into practical heat treating by operating furnaces and induction equipment at Ross Gear in Lafayette, Indiana, while studying for my metallurgical degree at Purdue University. After graduation, I spent the next 17 years involved in heat treating and foundry operations of major manufacturing companies. In 1976, I became involved in commercial heat treating as Division Manager of the Melrose Park (IL) plant for Lindberg Heat Treating Company and continued in commercial heat treating until I sold Midland Metal Treating, Inc. (Franklin, WI) in 2004.

During my career, with the encouragement of my lovely wife, Gwendolyn, I have remained involved with ASM International and the Heat Treating Society. That work involved preparation and dissemination of practical heat treating knowledge by working on technical papers and Handbooks, teaching, serving on technical committees and boards, reviewing papers, organizing heat treating conferences, and helping to organize and then later serving as Editor of ASM's *Journal of Heat Treating*.

It is an honor to update *Practical Heat Treating*. The first edition was compiled by Howard Boyer, who is now deceased. I had the pleasure of knowing Howard and working with him on several *ASM Handbook* sections. *Practical Heat Treating* covers the fundamentals and practical aspects of the broad field of heat treating. Since many of the fundamentals have not changed in the past 30 years, in this updated edition we concentrated on adding information about the new processes and process control techniques that have been developed or refined during the intervening period.

Jon L. Dossett, P.E.

CHAPTER 1

What Is Heat Treating? Importance and Classifications

THE GENERALLY ACCEPTED TERM for heat treating metals and metal alloys is "heating and cooling a solid metal or alloy in such a way so as to obtain specific conditions and/or properties." Heating for the sole purpose of hot working (as in forging operations) is excluded from this definition. Heat treatments sometimes used for nonmetallic products are also excluded from coverage by this definition.

Importance of Heat Treatment

It is difficult to imagine how our lives would be changed if the properties of metals could not be altered in a variety of ways through the use of heat treatment. With those benefits derived from heat treating, many of the products manufactured by the transportation, aerospace, construction, agricultural, mining, and consumer goods sector of our economy would not be available for use.

The village blacksmith performed crude heat treatments in years past, which improved the standard of living for society at that time. However, the understanding of the science and underlying principles of various metal and metal alloy systems and their related heat treatments has been significantly developed only during the past 75 to 100 years.

Almost all metals and alloys respond to some type of heat treatment, in the broadest sense of the definition. The response of various metals and alloys, however, is by no means equal. Almost any pure metal or alloy can be softened (annealed) by means of a suitable heating and cooling cycle; however, the number of alloys that can be strengthened or hardened by heat treatment is far more restricted.

Practically all steels respond to one or more type of heat treatment. This is the main reason that steels have been so extensively used in the manufacturing sector of our economy during the twentieth century. The underlying principles of the heat treatment of steel are discussed in Chapter 2, "Fundamentals of the Heat Treating of Steel."

Many nonferrous alloys—namely aluminum, copper, nickel, magnesium, and titanium alloys—can be strengthened to various degrees by specially designed heat treatments, but not to the same degree and not by the same techniques as steel. The heat treatment of nonferrous metals is covered in Chapter 13, "Heat Treating of Nonferrous Alloys."

Classification of Heat Treating Processes

In some instances, heat treatment procedures are clear cut in terms of technique and application, whereas in other instances, descriptions or simple explanations are insufficient because the same technique frequently may be used to obtain different objectives. For example, stress relieving and tempering are often accomplished with the same equipment and by use of identical time and temperature cycles. The objectives, however, are different for the two processes.

The principal heat treating processes are described subsequently. Additional information on these processes as well as the terminology associated with them can be found in Appendix A, "Glossary of Heat Treating Terms."

Normalizing

The term normalize does not characterize the nature of this process. More accurately, it is a homogenizing or grain refining treatment, with the aim being uniformity in composition throughout a part. In the thermal sense, normalizing is an austenitizing heating cycle followed by cooling in still or slightly agitated air. Typically, work is heated to a temperature of approximately 55 °C (100 °F) above the upper critical line of the iron-iron carbide phase diagram, and the heating portion of the process must produce a homogeneous austenitic phase. The actual temperature used depends on the composition of the steel, but the usual temperature is approximately 870 °C (1600 °F). Because of characteristics inherent in cast steel, normalizing is commonly applied to ingots prior to working, and to steel castings and forgings prior to hardening. Air-hardening steels are not classified as normalized steels because they do not have the normal pearlitic microstructure typical of normalized steels.

Annealing

Annealing is a generic term denoting a treatment consisting of heating to and holding at a suitable temperature, followed by cooling at a suitable rate; the process is used primarily to soften metals and to simultaneously produce desired changes in other properties or in microstructures. Reasons for annealing include improvement of machinability, facilitation of cold work, improvement in mechanical or electrical properties, and to increase dimensional stability. In ferrous alloys, annealing usually is done above the upper critical temperature, but time-temperature cycles vary widely in maximum temperature and in cooling rate, depending on composition of the steel, condition of the steel, and results desired. When the term is used without qualification, full annealing is implied. When the only purpose is relief of stresses, the process is called *stress relieving* or *stress-relief annealing*.

In full annealing, steel is heated 90 to 180 °C (160 to 325 °F) above the A_3 for hypoeutectoid steels and above the A_1 for hypereutectoid steels, and slow cooled, making the material easier to cut and to bend. In full annealing, the rate of cooling must be very slow to allow the formation of coarse pearlite. In process annealing, slow cooling is not essential because any cooling rate from temperatures below A_1 results in the same microstructure and hardness.

In nonferrous alloys, annealing cycles are designed to: remove part or all of the effects of cold working (recrystallization may or may not be involved); cause substantially complete coalescence of precipitates from solid solution in relatively coarse form; or both, depending on composition and material condition. Specific process names in commercial use are final annealing, full annealing, intermediate annealing, partial annealing, recrystallization annealing, and stress-relief annealing. Some of these are "inshop" terms that do not have precise definitions.

Stress Relieving

Residual stresses can be created in a number of ways, ranging from ingot processing in the mill to the manufacture of the finished product. Sources include rolling, casting, forging, bending, quenching, grinding, and welding. In the stress-relief process, steel is heated to approximately 595 °C (1105 °F), ensuring that the entire part is heated uniformly, then cooled slowly back to room temperature. The procedure is called stress-relief annealing, or simply stress relieving. Care must be taken to ensure uniform cooling, especially when a part has varying section sizes. If the cooling rate is not constant and uniform, new residual stresses, equal to or greater than existing originally, can be the result. Residual stresses in ferritic steel cause significant reduction in resistance to brittle fracture. If a steel, such as austenitic stainless steel, is not prone to brittle fracture, residual stresses can cause stress-corrosion cracking (SCC). Warping is the common problem.

Surface Hardening

These treatments, numbering more than a dozen, impart a hard, wear-resistant surface to parts, while maintaining a softer, tough interior, which gives resistance to breakage due to impacts. Hardness is obtained through quenching, which provides rapid cooling above a steel's transformation temperature. Parts in this condition can crack if dropped. Ductility is obtained via tempering. The hardened surface of the part is referred to as the *case,* and its softer interior is known as the *core.* Four of the more popular surface hardening treatments are carburizing, carbonitriding, nitriding, and ferritic nitrocarburizing.

Carburizing consists of absorption and diffusion of carbon into solid ferrous alloys by heating to some temperature above the upper transformation temperature of the specific alloy. Temperatures used for carburizing are generally in the range of 900 to 1040 °C (1650 to 1900 °F). Heating is done in a carbonaceous environment (liquid, solid, or gas). This produces a carbon gradient extending inward from the surface, enabling the surface layers to be hardened to a high degree either by quenching from the carburizing temperature or by cooling to room temperature followed by reaustenitizing and quenching. Carburizing is discussed in greater detail in Chapter 8, "Case Hardening of Steel."

Carbonitriding is a case-hardening process in which a ferrous material (most often a low-carbon grade of steel) is heated above the transformation temperature in a gaseous atmosphere of such composition as to cause simultaneous absorption of carbon and nitrogen by the surface and, by diffusion, create a concentration gradient. The process is completed by cooling at a rate that produces the desired properties in the workpiece.

Carbonitriding is most widely used for producing thin, hard, wear-resistant cases on numerous hardware items. For more details on carbonitriding, see Chapter 8.

Nitriding. The introduction of nitrogen into the surface layers of certain ferrous alloys by holding at a suitable temperature below the lower transformation temperature, Ac_1, in contact with a nitrogenous environment is known as nitriding. Processing temperature is generally in the range of 525 to 565 °C (975 to 1050 °F), and the nascent nitrogen may be generated by cracking of anhydrous ammonia (NH_3) or from molten salts that contain cyanide. Quenching is not required to create a hard, wear-resistant and heat-resistant case (see Chapter 8 for more details on the nitriding process).

Nitrocarburizing is used to describe an entire family of processes by which both nitrogen and carbon are absorbed into the surface layers of a wide variety of carbon and alloy steels. The sources for carbon and nitrogen may be either molten salt or gas, and temperatures are generally below the lower transformation temperature of the alloy; that is, below Ac_1. Nitrocarburizing not only provides a wear-resistant surface, but also increases fatigue strength. Retention of these properties depends largely on the avoidance of finishing operations after the heat treatment because nitrocarburized cases are extremely thin. There are a number of proprietary processes that comprise the family of nitrocarburizing processes (see Chapter 8 for further details on and applications of nitrocarburizing).

Quenching/Quenchants

Steel parts are rapidly cooled from the austenitizing or solution treating temperature, typically from within the range of 815 to 870 °C (1500 to 1600 °F). Stainless and high-alloy steels may be quenched to minimize the presence of grain-boundary carbides or to improve the ferrite distribution, but most steels, including carbon, low-alloy, and tool steels, are quenched to produce controlled amounts of martensite in the microstructure. Objectives are to obtain a required microstructure, hardness, strength, or toughness, while minimizing residual stresses, distortion, and the possibility of cracking. The ability of a quenchant to harden steel depends on the cooling characteristics of the quenching medium. Quenching effectiveness is dependent on steel composition, type of quenchant, or quenchant use conditions. The design of a quenching system and its maintenance are also key to success.

Quenching Media. Selection here depends on the hardenability of the steel, the section thickness and shape involved, and the cooling rates needed to get the desired microstructure. Typically, quenchants are liquids or gases.

Common liquid quenchants are:

- Oil that may contain a variety of additives
- Water
- Aqueous polymer solutions
- Water that may contain salt or caustic additives

Most common gaseous quenchants are inert gases, including helium, argon, and nitrogen. They are sometimes used after austenitizing in a vacuum.

A number of other quenching media and methods are available, including fogs, sprays, quenching in dry dies, and fluidized beds. In addition, some processes, such as electron-beam hardening and high-frequency pulse hardening, are self-quenching. Very high temperatures are reached in the fraction of a second, and metal adjoining the small, localized heating area acts as a heat sink, resulting in ultrarapid cooling.

Tempering

In this process, a previously hardened or normalized steel is usually heated to a temperature below the lower critical temperature and cooled at a suitable rate, primarily to increase ductility and toughness, but also to increase grain size of the matrix. Steels are tempered by reheating after hardening to obtain specific values of mechanical properties and to relieve quenching stresses and ensure dimensional stability. Tempering usually follows quenching from above the upper critical temperature.

Most steels are heated to a temperature of 205 to 595 °C (400 to 1105 °F) and held at that temperature for an hour or more. Higher temperatures increase toughness and resistance to shock, but reductions in hardness and strength are trade-offs. Hardened steels have a fully martensitic structure, which is produced in quenching. A steel containing 100% martensite is in its strongest possible condition, but freshly quenched martensite is brittle. The microstructure of quenched and tempered steel is referred to as tempered martensite.

Martempering of Steel. The term describes an interrupted quench from the austenitizing temperature of certain alloy, cast, tool, and stainless steels. The concept is to delay cooling just above martensitic transformation for a period of time to equalize the temperature throughout the piece. Minimizing distortion, cracking, and residual stress is the payoff. The term is not descriptive of the process and is better described as marquenching. The microstructure after martempering is essentially primary martensite that is untempered and brittle.

Austempering of Steel. Ferrous alloys are isothermally transformed at a temperature below that for pearlite formation and above that of martensite formation. Steel is heated to a temperature within the austenitizing range, usually 790 to 915 °C (1455 to 1680 °F); then quenched in a bath maintained at a constant temperature, usually in the range of 260 to 400 °C (500 to 750 °F); allowed to transform isothermally to bainite in this bath; then cooled to room temperature. Benefits of the process are increased ductility, toughness, and strength at a given hardness; plus reduced distortion that lessens subsequent machining time, stock removal, sorting, inspection, and scrap. Austempering also provides the shortest possible overall time cycle to through harden within the hardness range of 35 to 55 HRC. Savings in energy and capital investment are realized.

Maraging Steels. These highly alloyed, low-carbon, iron-nickel martensites have an excellent combination of strength and toughness that is superior to that of most carbon hardened steels, and are alternatives to hardened carbon steels in critical applications where high strength and good toughness and ductility are required. Hardened carbon steels derive their strength from transformation hardening mechanisms, such as martensite and bainite formation, and the subsequent precipitation of carbides during tempering. Maraging steels, by contrast, get their strength from the formation of a very low-carbon, tough, and ductile iron-nickel martensite, which can be further strengthened by subsequent precipitation of intermetallic compounds during age hardening. The term *marage* was suggested by the age hardening of the martensitic structure.

Cold and Cryogenic Treatment of Steel

Cold treatment can be used to enhance the transformation of austenite to martensite in case hardening and to improve the stress relief of castings and machined parts. Practice identifies −84 °C (−120 °F) as the optimum cold treatment temperature. By comparison, cryogenic treatment at a temperature of approximately −190 °C (−310 °F) improves certain properties beyond the capability of cold treatment.

SELECTED REFERENCES

- *Heat Treating,* Vol 4, *ASM Handbook,* ASM International, 1991
- *Heat Treater's Guide: Practices and Procedures for Irons and Steels,* 2nd ed., ASM International, 1995
- *Heat Treater's Guide: Practices and Procedures for Nonferrous Alloys,* ASM International, 1996

CHAPTER 2

Fundamentals of the Heat Treating of Steel

BEFORE CONSIDERATION can be given to the heat treatment of steel or other iron-base alloys, it is helpful to explain what steel is. The common dictionary definition is "a hard, tough metal composed of iron, alloyed with various small percentage of carbon and often variously with other metals such as nickel, chromium, manganese, etc." Although this definition is not untrue, it is hardly adequate.

In the glossary of this book (see Appendix A, "Glossary of Heat Treating Terms") the principal portion of the definition for steel is "an iron-base alloy, malleable in some temperature range as initially cast, containing manganese, usually carbon, and often other alloying elements. In carbon steel and low-alloy steel, the maximum carbon is about 2.0%; in high-alloy steel, about 2.5%. The dividing line between low-alloy and high-alloy steels is generally regarded as being at about 5% metallic alloying elements" (Ref 1).

Fundamentally, all steels are mixtures, or more properly, alloys of iron and carbon. However, even the so-called plain-carbon steels have small, but specified, amounts of manganese and silicon plus small and generally unavoidable amounts of phosphorus and sulfur. The carbon content of plain-carbon steels may be as high as 2.0%, but such an alloy is rarely found. Carbon content of commercial steels usually ranges from 0.05 to about 1.0%.

The alloying mechanism for iron and carbon is different from the more common and numerous other alloy systems in that the alloying of iron and carbon occurs as a two-step process. In the initial step, iron combines with 6.67% C, forming iron carbide, which is called *cementite*. Thus, at room temperature, conventional steels consist of a mixture of cementite and ferrite (essentially iron). Each of these is known as a phase (defined as a physically homogeneous and distinct portion of a material system). When a steel is heated above 725 °C (1340 °F), cementite dissolves in the matrix, and a new phase is formed, which is called *austenite*. Note that phases of steel should not be confused with structures. There are only three phases involved in any steel—ferrite, carbide (cementite), and austenite, whereas there are several structures or mixtures of structures.

Classification of Steels

It is impossible to determine the precise number of steel compositions and other variations that presently exist, although the total number probably exceeds 1000; thus, any rigid classification is impossible. However, steels are arbitrarily divided into five groups, which has proved generally satisfactory to the metalworking community.

These five classes are:

- Carbon steels
- Alloy steels (sometimes referred to as low-alloy steels)
- Stainless steels
- Tools steels
- Special-purpose steels

The first four of these groups are well defined by designation systems developed by the Society of Automotive Engineers (SAE) and the American Iron and Steel Institute (AISI). Each general class is subdivided into numerous groups, with each grade indentified. The fifth group comprises several hundred different compositions; most of them are proprietary. Many of these special steels are similar to specific steels in the first four groups but vary sufficiently to be marked as separate compositions. For example, the SAE-AISI designation system lists nearly 60 stainless steels in four different general subdivisions. In addition to these steels (generally referred to as "standard grades"), there are well over 100 other compositions that are nonstandard. Each steel was developed for a specific application.

It should also be noted that both standard and nonstandard steels are designated by the Unified Numbering System (UNS) developed by SAE and ASTM International. Details of this designation system can be found in Ref 2. Coverage in this book, however, is limited to steels of the first four classes—carbon, alloy, stainless, and tool steels—that are listed by SAE-AISI.

Why Steel Is So Important

It would be unjust to state that any one metal is more important than another without defining parameters of consideration. For example, without aluminum and titanium alloys, current airplanes and space vechicles could not have been developed.

Steel, however, is by far the most widely used alloy and for a very good reason. Among layman, the reason for steel's dominance is usually considered to be the abundance of iron ore (iron is the principal ingredient in all steels) and/or the ease by which it can be refined from ore. Neither of these is necessarily correct; iron is by no means the most abundant element, and it is not the easiest metal to produce from ore. Copper, for example, exists as nearly pure metal in certain parts of the world.

Steel is such an important material because of its tremendous flexibility in metal working and heat treating to produce a wide variety of mechanical, physical, and chemical properties.

Metallurgical Phenomena

The broad possibilities provided by the use of steel are attributed mainly to two all-important metallurgical phenomena: iron is an allotropic element; that is, it can exist in more than one crystalline form; and the carbon atom is only $\frac{1}{30}$ the size of the iron atom. These phenomena are thus the underlying principles that permit the achievements that are possible through heat treatment.

In entering the following discussion of constitution, however, it must be emphasized that a maximum of technical description is unavoidable. This portion of the subject is inherently technical. To avoid that would result in the discussion becoming uninformative and generally useless. The purpose of this chapter is, therefore, to reduce the prominent technical features toward their broadest generalizations and to present those generalizations and underlying principles in a manner that should instruct the reader interested in the metallurgical principles of steel. This is done at the risk of some oversimplification.

Constitution of Iron

It should first be made clear to the reader that any mention of molten metal is purely academic; this book deals exclusively with the heat treating range that is well below the melting temperature. The objective of this section is to begin with a generalized discussion of the constitution of commercially pure iron, subsequently leading to discussion of the iron-carbon alloy system that is the basis for all steels and their heat treatment.

All pure metals, as well as alloys, have individual constitutional or phase diagrams. As a rule, percentages of two principal elements are shown on the horizontal axis of a figure, while temperature variation is shown on the vertical axis. However, the constitutional diagram of a pure metal is a simple vertical line. The constitutional diagram for commercially pure iron is presented in Fig. 1. This specific diagram is a straight line as far as any changes are concerned, although time is indicated on the horizontal. As pure iron, in this case, cools, it changes from one phase to another at constant temperature. No attempt is made, however, to quantify time, but merely to indicate as a matter of interest that as temperature increases, reaction time decreases, which is true in almost any solid-solution reaction.

Pure iron solidifies from the liquid at 1538 °C (2800 °F) (top of Fig. 1). A crystalline structure, known as ferrite, or delta iron, is formed (point a, Fig. 1). This structure, in terms of atom arrangement, is known as a body-centered cubic lattice (bcc), shown in Fig. 2(a). This lattice has nine atoms—one at each corner and one in the center.

As cooling proceeds further and point b (Fig. 1) is reached (1395 °C, or 2540 °F), the atoms rearrange into a 14-atom lattice a shown in Fig. 2(b).

The lattice now has an atom at each corner and one at the center of each face. This is known as a face-centered cubic lattice (fcc), and this structure is called *gamma iron*.

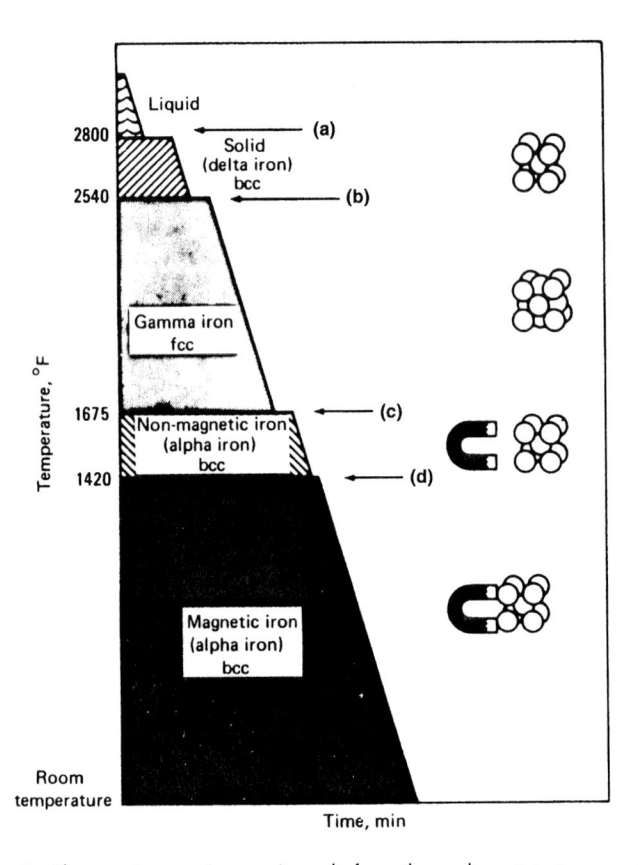

Fig. 1 Changes in pure iron as it cools from the molten state to room temperature. Source: Ref 3

As cooling further proceeds to 910 °C (1675 °F) (point c, Fig. 1), the structure reverts to the nine-atom lattice or alpha iron. The change at point d on Fig. 1 (770 °C, or 1420 °F) merely denotes a change from nonmagnetic to magnetic iron and does not represent a phase change. The entire field below 910 °C (1675 °F) is composed of alpha ferrite, which continues on down to room temperature and below. The ferrite forming above the temperature range of austenite is often referred to as *delta ferrite*; that forming below A_3 as *alpha ferrite,* though both are structurally similar. In this Greek-letter sequence, austenite is gamma iron, and the interchangeability of these terms should not confuse the fact that only two structurally distinct forms of iron exist.

Figures 1 and 2 thus illustrate the allotropy of iron. In the following sections of this chapter, the mechanism of allotropy as the all-important phenomenon relating to the heat treatment of iron-carbon alloys is discussed.

Alloying Mechanisms

Metal alloys are usually formed by mixing together two or more metals in their molten state. The two most common methods of alloying are by atom exchange and by the interstitial mechanism. The mechanism by which two metals alloy is greatly influenced by the relative atom size. The exchange mechanism simply involves trading of atoms from one lattice system to another. An example of alloying by exchange is the copper-nickel system wherein atoms are exchanged back and forth.

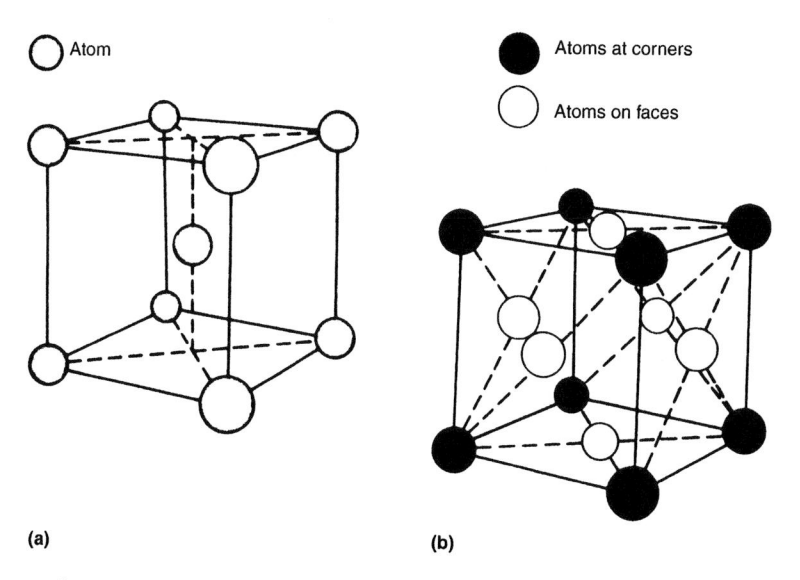

Fig. 2 Arrangement of atoms in the two crystalline structures of pure iron. (a) Body-centered cubic lattice. (b) Face-centered cubic lattice

Interstitial alloying requires that there be a large variation in atom sizes between the elements involved. Because the small carbon atom is 1/30 the size of the iron atom, interstitial alloying is easily facilitated. Under certain conditions, the tiny carbon atoms enter the lattice (the interstices) of the iron crystal (Fig. 2). A description of this basic mechanism follows.

Effect of Carbon on the Constitution of Iron

As an elemental metal, pure iron has only limited engineering usefulness despite its allotropy. Carbon is the main alloying addition that capitalizes on the allotropic phenomenon and lifts iron from mediocrity into the position of a unique structural material, broadly known as steel. Even in the highly alloyed stainless steels, it is the quite minor constituent carbon that virtually controls the engineering properties. Furthermore, due to the manufacturing processes, carbon in effective quantities persists in all irons and steels unless special methods are used to minimize it.

Carbon is almost insoluble in iron, which is in the alpha or ferritic phase (910 °C, or 1675 °F). However, it is quite soluble in gamma iron. Carbon actually dissolves; that is, the individual atoms of carbon lose themselves in the interstices among the iron atoms. Certain interstices within the fcc structure (austenite) are considerably more accommodating to carbon than are those of ferrite, the other allotrope. This preference exists not only on the mechanical basis of size of opening, however, for it is also a fundamental matter involving electron bonding and the balance of those attractive and repulsive forces that underlie the allotrope phenomenon.

The effects of carbon on certain characteristics of pure iron are shown in Fig. 3 (Ref 3). Figure 3(a) is a simplified version of Fig. 1; that is, a straight line constitutional diagram of commercially pure iron. In Fig. 3(b), the diagram is expanded horizontally to depict the initial effects of carbon on the principal thermal points of pure iron. Thus, each vertical dashed line, like the solid line in Fig. 3(a), is a constitutional diagram, but now for iron containing that particular percentage of carbon. Note that carbon lowers the freezing point of iron and that it broadens the temperature range of austenite by raising the temperature A_4 at which (delta) ferrite changes to austenite and by lowering the temperature A_3 at which the austenite reverts to (alpha) ferrite. Hence, carbon is said to be an austenitizing element. The spread of arrows at A_3 covers a two-phase region, which signifies that austenite is retained fully down to the temperatures of the heavy arrow, and only in part down through the zone of the lesser arrows.

In a practical approach, however, it should be emphasized that Fig. 1, as well as Fig. 3, represents changes that occur during very slow cooling, as would be possible during laboratory-controlled experiments, rather than under conditions in commercial practice. Furthermore, in slow heating of iron, these transformations take place in a reverse manner. Transformations occurring at such slow rates of cooling and heating are known as equilibrium transformations, due to the fact that temperatures indicated in Fig. 1.

Therefore, the process by which iron changes from one atomic arrangement to another when heated through 910 °C (1675 °F) is called a transformation. Transformations of this type occur not only in pure iron but also in many of its alloys; each alloy composition transforms at its own characteristic temperature. It is this transformation that makes possible the variety of properties that can be achieved to a high degree of reproducibility through use of carefully selected heat treatments.

Iron-Cementite Phase Diagram

When carbon atoms are present, two changes occur (see Fig. 3). First, transformation temperatures are lowered, and second, transformation takes place over a range of temperatures rather than at a single tempera-

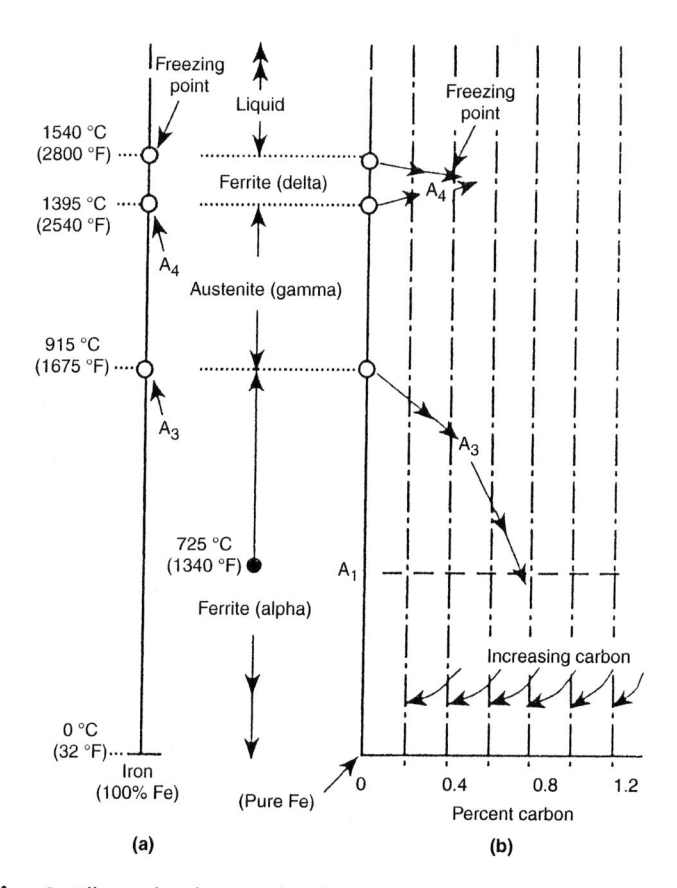

Fig. 3 Effects of carbon on the characteristics of commercially pure iron. (a) Constitutional diagram for pure iron. (b) Initial effects of carbon on the principal thermal points of pure iron. Source: Ref 3

ture. These data are shown in the well-known iron-cementite phase diagram (Fig. 4). However, a word of explanation is offered to clarify the distinction between phases and phase diagrams.

A phase is a portion of an alloy, physically, chemically, or crystallographically homogeneous throughout, which is separated from the rest of the alloy by distinct bounding surfaces. Phases that occur in iron-carbon alloys are molten alloy, austenite (gamma phase), ferrite (alpha phase), cementite, and graphite. These phases are also called constituents. Not all constituents (such as pearlite or bainite) are phases—these are microstructures.

A phase diagram is a graphical representation of the equilibrium temperature and composition limits of phase fields and phase reactions in an alloy system. In the iron-cementite system, temperature is plotted vertically, and composition is plotted horizontally. The iron-cementite diagram (Fig. 4), deals only with the constitution of the iron-iron carbide system, i.e., what phases are present at each temperature and the composition

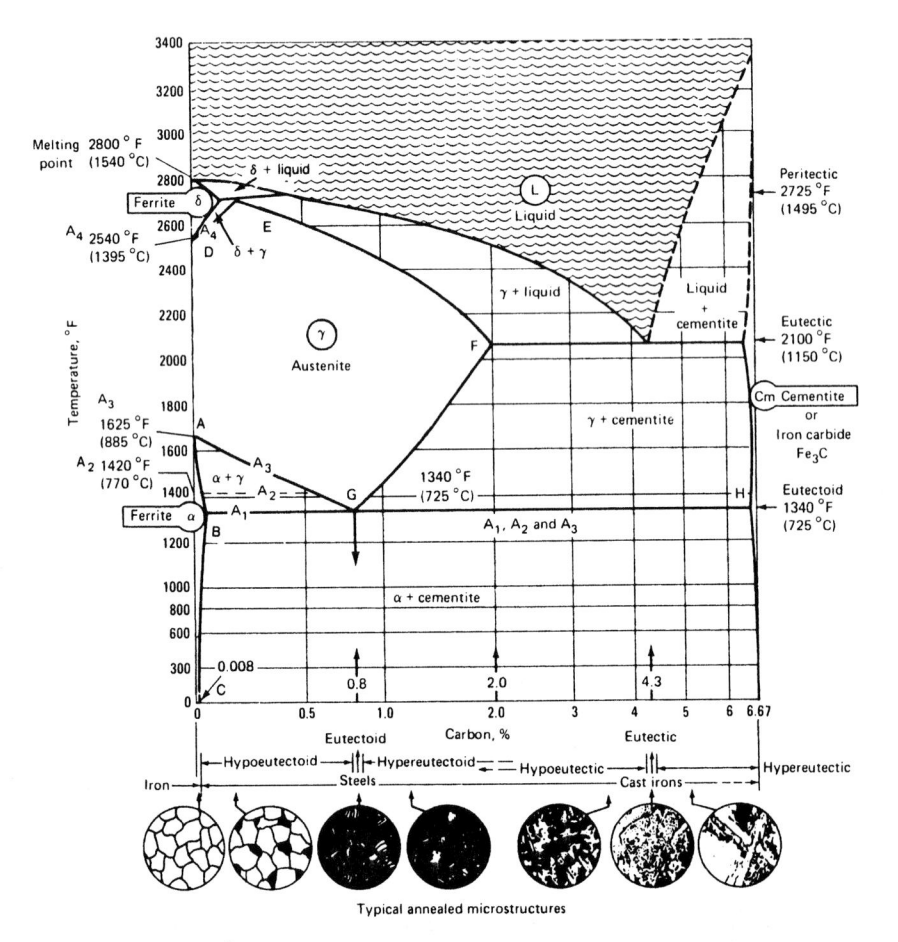

Fig. 4 Iron-cementite phase diagram. Source: Ref 3

limits of each phase. Any point on the diagram, therefore, represents a definite composition and temperature, each value being found by projecting to the proper reference axis.

Although this diagram extends from a temperature of 1870 °C (3400 °F) down to room temperature, note that part of the diagram lies below 1040 °C (1900 °F). Steel heat treating practice rarely involves the use of temperatures above 1040 °C (1900 °F). In metal systems, pressure is usually considered as constant.

Frequent reference is made to the iron-cementite diagram (Fig. 4) in this chapter and throughout this book. Consequently, understanding of this concept and diagram is essential to further discussion.

The iron-cementite diagram is frequently referred to incorrectly as the iron-carbon equilibrium diagram. Iron-"carbon" is incorrect because the phase at the extreme right is cementite, rather than carbon or graphite; the term equilibrium is not entirely appropriate because the cementite phase in the iron-graphite system is not really stable. In other words, given sufficient time (less is required at higher temperatures), iron carbide (cementite) decomposes to iron and graphite, i.e., the steel graphitizes. This is a perfectly natural reaction, and only the iron-graphite diagram (see Chapter 12, "Heat Treating of Cast Irons") is properly referred to as a true equilibrium diagram.

Solubility of Carbon in Iron

In Fig. 4, the area denoted as austenite is actually an area within which iron can retain much dissolved carbon. In fact, most heat treating operations (notably annealing, normalizing, and heating for hardening) begin with heating the alloy into the austenitic range to dissolve the carbide in the iron. At no time during such heating operations are the iron, carbon, or austenite in the molten state. A solid solution of carbon in iron can be visualized as a pyramidal stack of basketballs with golf balls between the spaces in the pile. In this analogy, the basketballs would be the iron atoms, while the golf balls interspersed between would be the smaller carbon atoms.

Austenite is the term applied to the solid solution of carbon in gamma iron, and, like other constituents in the diagram, austenite has a certain definite solubility for carbon, which depends on the temperature (shaded area in Fig. 4 bounded by AGFED). As indicated by the austenite area in Fig. 4, the carbon content of austenite can range from 0 to 2%. Under normal conditions, austenite cannot exist at room temperature in plain carbon steels; it can exist only at elevated temperatures bounded by the lines AGFED in Fig. 4. Although austenite does not ordinarily exist at room temperature in carbon steels, the rate at which steels are cooled from the austenitic range has a profound influence on the room temperature microstructure and properties of carbon steels. Thus, the phase known as austenite is fcc iron, capable of containing up to 2% dissolved carbon.

The solubility limit for carbon in the bcc structure of iron-carbon alloys is shown by the line ABC in Fig. 4. This area of the diagram is labeled alpha (α), and the phase is called ferrite. The maximum solubility of carbon in alpha iron (ferrite) is 0.025% and occurs at 725 °C (1340 °F). At room temperature, ferrite can dissolve only 0.008% C, as shown in Fig. 4. This is the narrow area at the extreme left of Fig. 4 below approximately 910 °C (1675 °F). For all practical purposes, this area has no effect on heat treatment and shall not be discussed further. Further discussion of Fig. 4 is necessary, although as previously stated, the area of interest for heat treatment extends vertically to only about 1040 °C (1900 °F) and horizontally to a carbon content of 2%. The large area extending vertically from zero to the line BGH (725 °C, or 1340 °F) and horizontally to 2% C is denoted as a two-phase area—α + Cm, or alpha (ferrite) plus cementite (carbide). The line BGH is known as the lower transformation temperature (A_1). The line AGH is the upper transformation temperature (A_3). The triangular area ABG is also a two-phase area, but the phases are alpha and gamma, or ferrite plus austenite. As carbon content increases, the A_3 temperature decreases until the eutectoid is reached—725 °C (1340 °F) and 0.80% C (point G). This is considered a saturation point; it indicates the amount of carbon that can be dissolved at 725 °C (1340 °F). A_1 and A_3 intersect and remain as one line to point H as indicated. The area above 725 °C (1340 °F) and to the right of the austenite region is another two-phase field—gamma plus cementite (austenite plus carbide).

Now as an example, when a 0.40% carbon steel is heated to 725 °C (1340 °F), its crystalline structure begins to transform to austenite; transformation is not complete however until a temperature of approximately 815 °C (1500 °F) is reached. In contrast, as shown in Fig. 4, a steel containing 0.80% C transforms completely to austenite when heated to 725 °C (1340 °F). Now assume that a steel containing 1.0% C is heated to 725 °C (1340 °F) or just above. At this temperature, austenite is formed, but because only 0.80% C can be completely dissolved in the austenite, 0.20% C remains as cementite, unless the temperature is increased. However, if the temperature of a 1.0% carbon steel is increased above about 790 °C (1450 °F), the line GF is intersected, and all of the carbon is thus dissolved. Increasing temperature gradually increases the amount of carbon that can be taken into solid solution. For instance, at 1040 °C (1900 °F), approximately 1.6% C can be dissolved (Fig.4).

Transformation of Austenite

Thus far the discussion has been confined to heating of the steel and the phases that result from various combinations of temperature and carbon content. Now what happens when the alloy is cooled? Referring to Fig. 4, assume that a steel containing 0.50% C is heated to 815 °C (1500 °F). All of the carbon will be dissolved (assuming, of course, that holding time is sufficient). Under these conditions, all of the carbon atoms will dissolve in the interstices of the fcc crystal (Fig. 2b). If the alloy is cooled slowly, transformation to the bcc (Fig. 2a) or alpha phase begins when the temperature drops below approximately 790 °C (1450 °F). As the temperature continues to decrease, the transformation is essentially complete at 725 °C (1340 °F). During this transformation, the carbon atoms escape from the lattice because they are essentially insoluble in the alpha crystal (bcc). Thus, in slow cooling, the alloy for all practical purposes, returns to the same state (in terms of phase) that it was before heating to form austenite. The same mechanism occurs with higher carbon steels, except that the austenite-to-ferrite transformation does not go through a two-phase zone (Fig. 4). In addition to the entry and exit of the carbon atoms through the interstices of the iron atoms, other changes occur that affect the practical aspects of heat treating.

First, a magnetic change occurs at 770 °C (1420 °F) as shown in Fig. 1. The heat of transformation effects may chemical changes, such as the heat that is evolved when water freezes into ice and the heat that is absorbed when ice melts. When an iron-carbon alloy is converted to austenite by heat, a large absorption of heat occurs at the transformation temperature. Likewise, when the alloy changes from gamma to alpha (austenite to ferrite), heat evolves.

What happens when the alloy is cooled rapidly? When the alloy is cooled suddenly, the carbon atoms cannot make an orderly escape from the iron lattice. This cause "atomic bedlam" and results in distortion of the lattice, which manifests itself in the form of hardness and/or strength. If cooling is fast enough, a new structure known as martensite is formed, although this new structure (an aggregate of iron and cementite) is in the alpha phase.

Classification of Steels by Carbon Content

It must be remembered that there are only three phases in steels, but there are many different structures. A precise definition of eutectoid carbon is unavoidable; it varies from 0.77% to slightly over 0.80%, depending on the reference used. However, for the objectives of this book, the precise amount of carbon denoted as eutectoid is of no particular significance.

Hypoeutectoid Steels. Carbon steels containing less than 0.80% C are known as hypoeutectoid steels. The area bounded by AGB on the iron-cementite diagram (Fig. 4) is of significance to the room temperature microstructures of these steels; within the area, ferrite and austenite each having different carbon contents, can exist simultaneously.

Assume that a 0.40% carbon steel has been slowly heated until its temperature throughout the piece is 870 °C (1600 °F), thereby ensuring a fully austenitic structure. Upon slow cooling, free ferrite begins to form from the austenite when the temperature drops across the line AG, into the area AGB, with increasing amounts of ferrite forming as the temperature continues to decline while in this area. Ideally, under very slow cooling conditions, all of the free ferrite separates from austenite by the time the temperature of the steel reaches A_1 (the line BG) at 725 °C (1340 °F). The austenite islands, which remain at about 725 °C (1340 °F), now have the same amount of carbon as the eutectoid steel, or about 0.80%. At or slightly below 725 °C (1340 °F) the remaining untransformed austenite transforms—it becomes pearlite, which is so named because of its resemblance to mother of pearl. Upon further cooling to room temperature, the microstructure remains unchanged, resulting in a final room temperature microstructure of an intimate mixture of free ferrite grains and pearlite islands.

A typical microstructure of a 0.40% carbon steel is shown in Fig. 5(a). The pure white areas are the islands of free ferrite grains described previously. Grains that are white but contain dark platelets are typical lamellar pearlite. These platelets are cementite or carbide interspersed through the ferrite, thus conforming to the typical two-phase structure indicated below the BH line in Fig. 4.

Eutectoid Steels. A carbon steel containing approximately 0.77% C becomes a solid solution at any temperature in the austenite temperature range, i.e., between 725 and 1370 °C (1340 and 2500 °F). All of the carbon is dissolved in the austenite. When this solid solution is slowly cooled, several changes occur at 725 °C (1340 °F). This temperature is a transformation temperature or critical temperature of the iron-cementite system. At this temperature, a 0.77% (0.80%) carbon steels transforms from a single homogeneous solid solution into two distinct new solid phases. This change occurs at constant temperature and with the evolution of heat. The new phases are ferrite an cementite, formed simultaneously; however, it is only at composition point G in Fig. 4 (0.77% carbon steel) that this phenomenon of the simultaneous formation of ferrite and cementite can occur.

The microstructure of a typical eutectoid steel is shown in Fig. 5(b). The white matrix is alpha ferrite and the dark platelets are cementite. All grains are pearlite—no free ferrite grains are present under these conditions.

Cooling conditions (rate and temperature) govern the final condition of the particles of cementite that precipitate from the austenite at 725 °C (1340 °F). Under specific cooling conditions, the particles become spheroidal instead of elongated platelets as shown in Fig. 5(b). Figure 5(c) shows a similar two-phase structure resulting from slowly cooling a eutectoid carbon steel just below A_1. This structure is commonly known as spheroidite but is still a despersion of cementite particles in alpha ferrite. There is no indication of grain boundaries in Fig. 5(c). The spheroidized structure is often preferred over the pearlitic structure because spheroidite has superior machinability and formability. Combination structures (that is, partly lamellar and partly spheroidal cementite in a ferrite matrix) are also common.

As noted previously, a eutectoid steel theoretically contains a precise amount of carbon. In practice, steels that contain carbon within the range of approximately 0.75 to 0.85% are commonly referred to as eutectoid carbon steels.

Hypereutectoid steels contain carbon contents of approximately 0.80 to 2.0%. Assume that a steel containing 1.0% C has been heated to 845 °C (1550 °F), thereby ensuring a 100% austenitic structure. When cooled, no

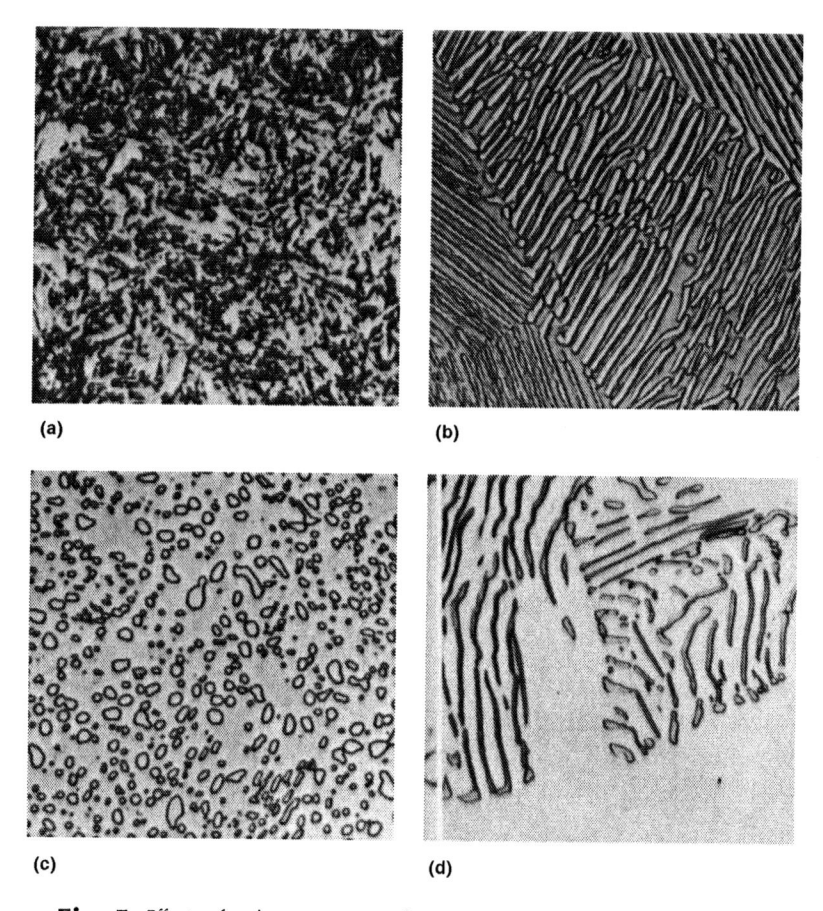

Fig. 5 Effects of carbon content on the microstructures of plain-carbon steels. (a) Ferrite grains (white) and pearlite (gray streaks) in a white matrix of a hypoeutectoid steel containing 0.4% C. 1000×. (b) Microstructure (all pearlite grains) of a eutectoid steel containing 0.77% C. 2000×. (c) Microstructure of a eutectoid steel containing 0.77% C with all cementite in the spheroidal form. 1000×. (d) Microstructure of a hypereutectoid steel containing ~1.0% C containing pearlite with excess cementite bounding the grains. 1000×. Source: Ref 4

change occurs until the line GF (Fig. 4), known as A_{cm} or cementite solubility line, is reached. At this point, cementite begins to separate out from the austenite, and increasing amounts of cementite separate out as the temperature of the 1% carbon steels descends below the A line. The composition of austenite changes from 1% C toward 0.77% C. At a temperature slightly below 725 °C (1340 °F), the remaining austenite changes to pearlite. No further changes occur as cooling proceeds toward room temperature, so that the room temperature microstructure consists of pearlite and free cementite. In this case, the free cementite exists as a network around the pearlite grains (Fig. 5d).

Upon heating hypereutectoid steels, reverse changes occur. At 725 °C (1340 °F), pearlite changes to austenite. As the temperature increases above 725 °C (1340 °F), free cementite dissolves in the austenite, so that when the temperature reaches the A_{cm} line, all the cementite dissolves to form 100% austenite.

Hysteresis in Heating and Cooling

The critical temperatures (A_1, A_6, and A_{cm}) are "arrests" in heating or cooling and have been symbolized with the letter A, from the French word *arret* meaning arrest or a delay point, in curves plotted to show heating or cooling of samples. Such changes occur at transformation temperatures in the iron-cementite diagram if sufficient time is given and cab be plotted for steels showing lags at transformation temperatures, as shown for iron in Fig. 4. However, because heating rates in commercial practice usually exceed those in controlled laboratory experiments, changes on heating usually occur at temperatures a few degrees above the transformation temperatures shown in Fig. 4 and are known as Ac temperatures, such as Ac_1 or Ac_3. The "c" is from the French word *chauffage,* meaning heating. Thus, Ac_1 is a few degrees above the ideal A_1 temperature.

Likewise, on slow cooling in commercial practice, transformation changes occur at temperatures a few degrees below those in Fig. 4. These are known as Ar, or Ar_3, the "r" originating from the French word *refroidissement,* meaning cooling.

This difference between the heating and cooling varies with the rate of heating or cooling. The faster the heating, the higher the Ac point; the faster the cooling, the lower the Ar point. Also, the faster the heating and cooling rate, the greater the gap between the Ac and Ar points of the reversible (equilibrium) point A.

Going one step further, in cooling a piece of steel, it is of utmost importance to note that the cooling rate may be so rapid (as in quenching steel in water) as to suppress the transformation for several hundreds of degrees. This is due to the decrease in reaction rate with decrease in temperature. As discussed subsequently, time is an important factor in transformation, especially in cooling.

Effect of Time on Transformation

The foregoing discussion has been confined principally to phases that are formed by various combinations of composition and temperature; little reference has been made to the effects of time. In order to convey to the reader the effects of time on transformation, the simplest approach is by means of a time-temperature-transformation (TTT) curve for some constant iron-carbon composition.

Such a curve is presented in Fig. 6 for a 0.77% (eutectoid) carbon steel. TTT curves are also known as "S" curves because the principal part of the curve is shaped like the letter "S." In Fig. 6, temperature is plotted on the vertical axis, and time is plotted on a logarithmic scale along the horizontal axis. The reason for plotting time on a logarithmic scale is merely to keep the width of the chart within a manageable dimension.

In analyzing Fig. 6, begin with line Ae_1 (725 °C or 1340 °F). Above this temperature, austenite exists only for a eutectoid steel (refer also to Fig. 4). When the steel is cooled and held at a temperature just below Ae_1 (705 °C or 1300 °F), transformation begins (follow line (P_s B_s), but very slowly at this temperature; 1 h of cooling is required before any significant amount of transformation occurs, although eventually complete transformation occurs isothermally (meaning at a constant temperature), and the transformation product is spheroidite (Fig. 5c). Now assume a lower temperature (650 °C or 1200 °F) on line P_s B_s (the line of beginning transformation); transformation begins in less than 1 min, and the transfor-

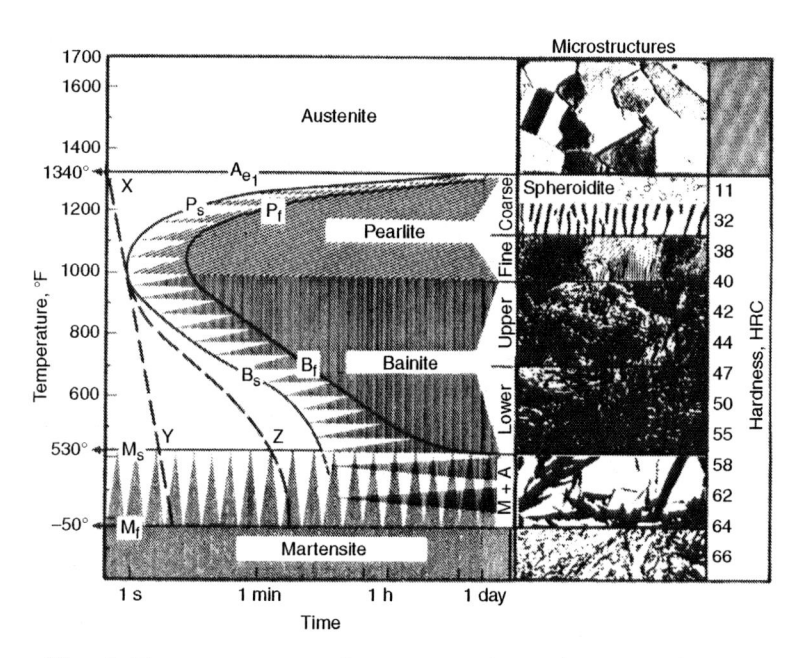

Fig. 6 Time-temperature-transformation (TTT) diagram for a eutectoid (0.77%) carbon steel. Source: Ref 3

mation product is coarse pearlite (near the right side of Fig. 6). Next, assume a temperature of 540 °C (1000 °F); transformation begins in approximately 1 s and is completely transformed to fine pearlite in a matter of a few minutes. The line $P_f B_f$ represents the completion of transformation and is generally parallel with $P_s B_s$. However, if the steel is cooled very rapidly (such as by immersing in water) so that there is not sufficient time for transformation to begin in the 540 °C (1000 °F) temperature vicinity, then the beginning of transformation time is substantially extended. For example, if the steel is cooled to and held at 315 °C (600 °F), transformation does not begin for well over 1 min. It must be remembered that all of the white area to the left of line $P_s B_s$ represents the austentic phase, although it is highly unstable. When transformation takes place isothermally within the temperature range of approximately 290 to 425 °C (550 to 800 °F), the transformation product is a microstructure called bainite (upper or lower as indicated toward the right of Fig. 6). A bainitic microstructure is shown in Fig. 7(b). In another example, steel is cooled so rapidly that no transformation takes place in the 540 °C (1000 °F) region and rapid cooling is continued (note line XY in Fig. 6) to and below 275 °C (530 °F) or M_s. Under these conditions, martensite is formed. Point M_s is the temperature at which martensite begins to form, and M_f indicates the complete finish of transformation. It must be remembered that martensite is not a phase but is a specific microstructure in the ferritic (alpha) phase. Martensite is formed from the carbon atoms jamming the lattice of the austenitic atomic arrangement. Thus, martensite can be considered as an aggregate of iron and cementite (Fig. 7a).

In Fig. 6, the microstructure of austenite (as it apparently appears at elevated temperature) is shown on the right. It is also evident that the lower the temperature at which transformation takes place, the higher the hardness (see Chapter 3, "Hardness and Hardenability").

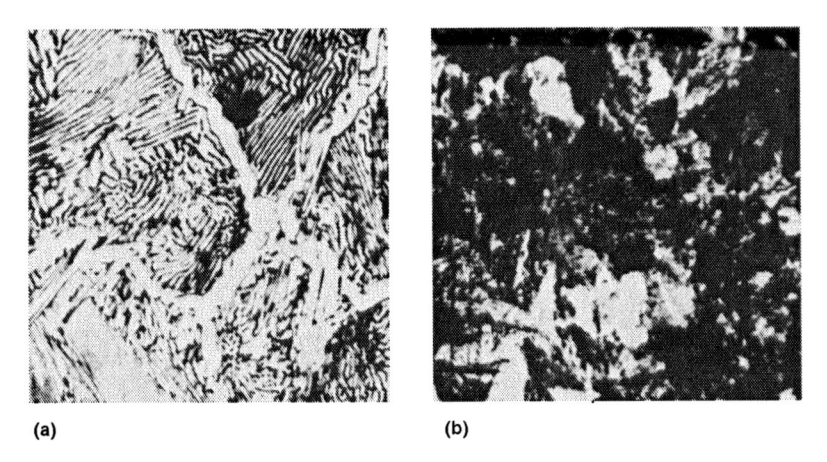

Fig. 7 (a) Microstructure of quenched 0.95% carbon steel. 1000×. Structure is martensitic. (b) Bainitic structure in a quenched 0.95% carbon steel. 550×. Source: Ref 4

It is also evident that all structures from the top to the region where martensite forms (Ae$_1$) are time-dependant, but the formation of martensite is not time-dependent.

Each different steel composition has its own TTT curve; Fig. 6 is presented only as an example. However, patterns are much the same for all steels as far as shape of the curves is concerned. The most outstanding difference in the curves among different steels is the distance between the vertical axis and the nose of the S curve. This occurs at about 540 °C (1000 °F) for the steel in Fig. 6. This distance in terms of time is about 1 s for a eutectoid carbon steel, but could be an hour or more for certain high-alloy steels, which are extremely sluggish in transformation.

The distance between the vertical axis and the nose of the S curve is often called the "gap" and has a profound effect on how rapidly the steel must be cooled to form the hardened structure—martensite. Width of this gap for any steel is directly related to the critical cooling rate for that specific steel. Critical cooling rate is defined as the rate at which a given steel must be cooled from the austenite to prevent the formation of nonmartensitic products.

In Fig. 6, it is irrelevant whether the cooling rate follows the lines X to Y or X to Z because they are both at the left of the beginning transformation line P$_s$ B$_s$. Practical heat treating procedures are based on the fact that once the steel has been cooled below approximately 425 °C (800 °F), the rate of cooling may be decreased. These conditions are all closely related to hardenability, which is dealt with in Chapter 3.

REFERENCES

1. H.E. Boyer, Chapter 1, *Practical Heat Treating,* 1st ed., American Society for Metals, 1984, p 1–16
2. *Metals & Alloys in the Unified Numbering System,* 10th ed, SAE International and ASTM International, 2004
3. H.N. Oppenheimer, Heat Treatment of Carbon Steels, Course 42, Lesson 1, *Practical Heat Treating,* Materials Engineering Institute, ASM International, 1995
4. H.E. Boyer, Chapter 2, *Practical Heat Treating,* 1st ed., American Society for Metals, 1984, p 17–33

SELECTED REFERENCES

- C.R. Brooks, *Principles of the Heat Treatment of Plain Carbon and Low Alloy Steels,* ASM International, 1996
- *Atlas of Time-Temperature Diagrams for Irons and Steels,* G.F. Vander Voort, Ed., ASM International, 1991
- G. Krauss, *Steels: Processing, Structure, and Performance,* ASM International, 2005

CHAPTER **3**

Hardness and Hardenability

THE PREVIOUS CHAPTER dealt exclusively with the fundamentals involved in the heat treatment of steel, which included the basic mechanism of hardening. However, when one begins to apply this fundamental knowledge to the hardening of steel, several oddities immediately become evident. First, it may be observed that one steel becomes a great deal harder than another, in spite of the fact that both have been hardened according to the correct principles. Second, it may be observed that, when two steels are hardened, one may be as hard in the center of a bar as it is on the surface, but another may be a great deal softer in the center than it is at the surface. This occurs in spite of the fact that both bars are identical in diameter and have been correctly processed. Naturally, one wants to know why such differences exist. In order to explain this, it is necessary to obtain and understand an accurate and reliable measuring device.

There are several methods for measuring or evaluating the results attained from heat treating—tension testing, impact testing, bend testing, and the various types of shear testing, to name a few. However, as a rule, the types of tests mentioned above are restricted to special applications. The hardness test is, by far, the most universally used device for measuring the results of heat treating for several reasons—the hardness test is simple to perform; it does not usually impair the usefulness of the workpiece being tested; and accurately made and properly interpreted hardness tests can be used to evaluate (or at least estimate) other mechanical properties. For example, hardness varies directly with strength and inversely with ductility.

What Is Hardness?

Since childhood, we have been acquainted with hardness, in one way or another, and we assume that we are well informed on so simple a matter.

If, however, we should ask our neighbors for a definition of hardness, the results would be quite surprising. To some, a hard substance is one that resists wear; to others, something that does not bend; to someone else, an object that resists penetration; and to still others, an object that breaks or scratches other substances. From the variety of answers given, it is apparent that there are many different opinions about what actually constitutes hardness. Thus, if an attempt is made to evaluate the answers and arrive at some common definition, it becomes apparent that hardness is an elusive property, far more complex than most people would believe.

The definition provided in most dictionaries is "the relative capacity of a substance for scratching another or for being scratched or indented by another." While this definition is somewhat vague, it is probably the best sense of the term that can be conveyed in a few words. For example, if steel and glass are considered with respect to the various aforementioned definitions, the difficulty of evaluating hardness is better appreciated. For example, both glass and steel have been used as paving materials in sidewalks, and their resistance to wear has been observed on many occasions; from the viewpoint of pure wear, it is known that both endure equally well. Therefore, it might be said that one is as hard as the other. It is true, moreover, that glass actually can scratch steel, which would appear to place glass in a harder category. Steel, however, can crush or break glass, and this would appear to make it harder than glass. The hopelessness of trying to evaluate hardness on the basis of numerous properties, without a concise definition of what property we want to measure as hardness, is readily apparent.

The difficulty of defining and measuring hardness was recognized early in recorded history and continues to exist today. Any attempt to make an all-inclusive definition leads only to confusion. For this reason, in both ferrous and nonferrous metallurgy, hardness is defined and measured on the basis of: a static test in which a standard penetrator is pressed into the specimen, using a standard load, and the resistance of the metal to penetration is the measure of the hardness; and a dynamic test in which a small hammer is dropped against the surface of the material from a fixed height and the hardness expressed in terms of height to which the hammer rebounds.

History of Hardness Testing

According to the records, the first method of evaluating hardness was developed in 1772, when Reaumur devised a procedure of pressing the edges of two prisms together, as shown in Fig. 1(a). The pressure applied to each prism was the same and thus allowed a direct comparison of hardness between the two metals.

In 1897, Foeppl altered the original work by Reaumur by pressing together two semicylindrical bars instead of two prisms, as shown in Fig.

1(b). By measuring the area of contact of the flattened surfaces and dividing this area into the load, he obtained a hardness evaluation (Ref 1).

A few years later, Brinell introduced his well-known method, consisting of pressing a hard steel ball into the surface to be tested, determining the surface area of the impression made, and dividing this value into the load, thus arriving at a hardness value. This test still is extensively used and represents the first hardness test that yields reproducible results. Numerous modifications of hardness tests have been developed that utilize a wide variety of penetrators and loads imposed on the penetrators (indenters). In most instances, hardness was evaluated by using a constant load, then determining either the depth or projected area of the indentation. The reverse has also been used; that is, forcing the penetrator to a preestablished depth, then measuring the required load. Each of the variations has fields of usefulness that are quite valuable and illustrate that, even though it is possible to define hardness (at least as applied to metals), a single universal testing procedure is not feasible.

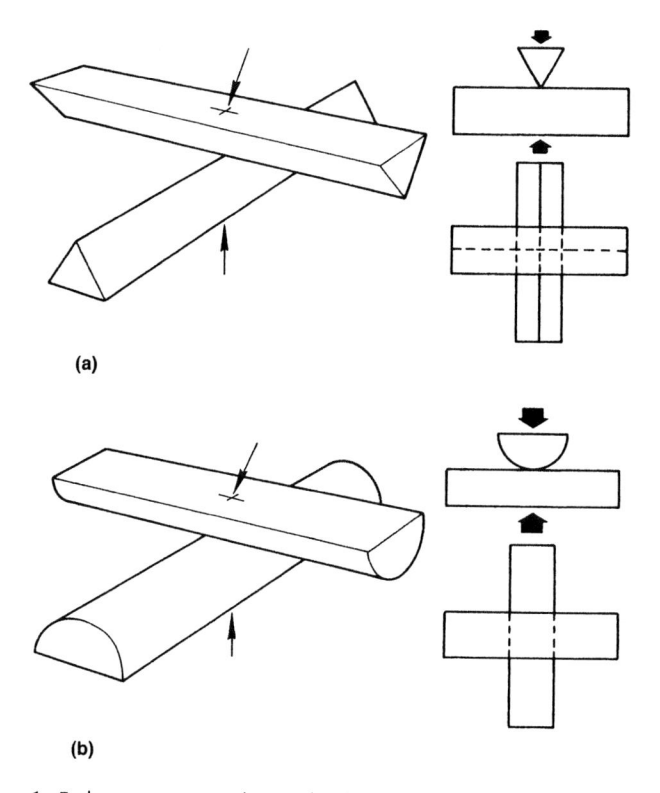

Fig. 1 Early attempts at evaluating hardness of metals. (a) Reaumur's method of determining comparative hardness. Edges of two metal prism specimens are pressed together. (b) Foeppl's method of hardness testing. Two semicylindrical bars are pressed together and the flattened area of contact is measured. This technique is also used in hot hardness testing. Source: Ref 1

Hardness Testing Systems

Many different systems have been devised for testing the hardness of metals, but only a few have achieved commercial importance. They are the Brinell, Rockwell, Vickers, Scleroscope, and various microhardness testers that employ Vickers or Knoop indenters. All of these systems, except the Scleroscope, depend strictly on the principle of indentation for measurement.

For the reader to have a clear understanding of heat treating, it is essential that he understand the commonly used hardness testing systems. Therefore, each of the aforementioned systems is discussed subsequently. For detailed descriptions of these systems and their application, see Ref 2 and 3.

Brinell Testing. The Brinell test is simple and consists of applying a constant load, usually 500 to 3000 kg (1100 to 6600 lb), or a hardened steel ball type of indenter, 10 mm (0.4 in.) in diameter, to the flat surface of a workpiece (Fig. 2). The 500 kg (1100 lb) load usually is used for testing nonferrous metals such as copper and aluminum alloys, whereas the 3000 kg (6600 lb) load is most often used for testing harder metals such as steels and cast irons. The load is held for a specified time (10 to 15 s for iron and steel and about 30 s for softer metals), after which the diameter of the recovered indentation is measured in millimeters. This time period is required to ensure that plastic flow of the work metals has stopped.

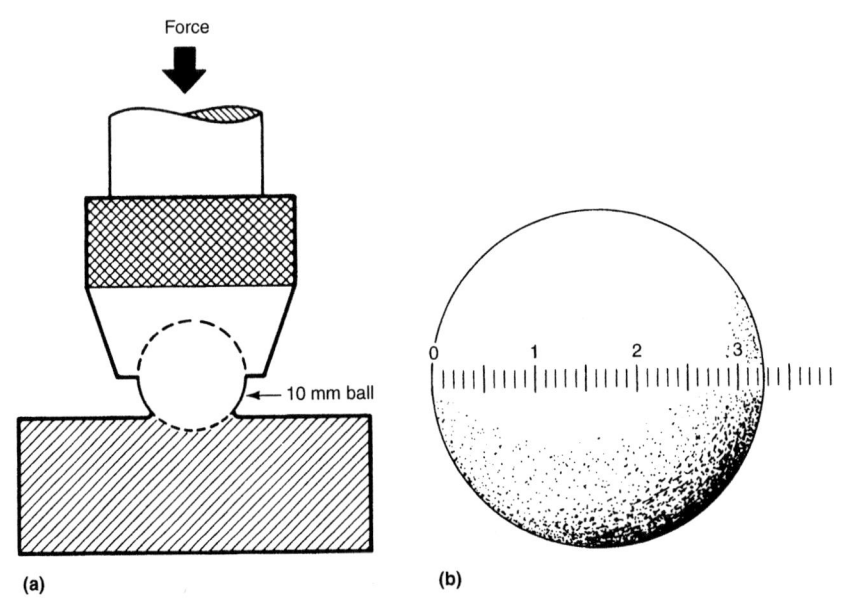

Fig. 2 Brinell indentation process. (a) Schematic of the principle of the Brinell indenting process. (b) Brinell indentation with measuring scale in millimeters. Source: Ref 3

Hardness is evaluated by taking the mean diameter of the indentation (two readings at right angles to each other) and calculating the Brinell hardness number (HB) by dividing the applied load to the surface area of the indentation according to the following formula:

$$HB = \frac{L}{\frac{\pi D}{2}(D - \sqrt{D^2 - d^2})}$$

where L is the load, in kilograms; D is the diameter of the ball, in millimeters; and d is the diameter of the indentation, in millimeters. It is not necessary, however, to make the calculation for each test. Tables, such as Table 1 (for loads ranging from 500 to 3000 kg), are provided that give the hardness in Brinell numbers for each diameter of indentation. Highly hardened steels cannot be tested by the Brinell method because the steel ball indenter will deform. The practical upper limit for Brinell testing with hardened steel balls is an indentation of about 2.90 mm (0.11 in.) diameter of 444 HB. Tungsten carbide indenters are usable up to 627 HB (2.40 mm, or 0.10 in., diam indentation). The Brinell indenter can be loaded by any one of a number of different machines. There are several machines designed for the purpose, although Brinell testing can be performed in a conventional hydraulic tension-compression testing machine. Disadvantages of the Brinell test (in addition to the upper limit of hardness) are:

- A great deal of time is required—at least 1 min for applying the load then reading the impression diameter by means of a specially designed microscope
- Because the impression made by the indenter is relatively large, small workpieces cannot be tested
- The impression left by such a large indenter may be damaging to the workpiece.

To reduce the amount of time required for testing, however, high-speed, automatic, digital readout machines now are available (Fig. 3).

Rockwell Testing. The Rockwell test is the most versatile of all hardness testers. By means of various applied loads and a number of different indenters, the Rockwell test can be used for testing all metals from soft solder to highly hardened steels and carbides.

Rockwell hardness testing differs from Brinell hardness testing in that hardness is determined by the depth of indentation made by a constant load impressed upon an indenter. Although a number of different indenters are used for Rockwell hardness testing, the most common type is a diamond, ground to a 120° cone with a spherical apex having a 0.2 mm (0.008 in.) radius, which is known as a Brale indenter.

As shown in Fig. 4(a), the Rockwell hardness test consists of measuring the additional depth to which an indenter is forced by a heavy (major) load (Fig. 4b) beyond the depth of a previously applied light (minor) load (Fig. 4a). Application of the minor load eliminates backlash in the load train and causes the indenter to break through slight surface roughness and to crush particles of foreign matter, thus contributing to much greater accuracy in the test. The basic principle involving minor and major loads illustrated in Fig. 4 applies to steel ball indenters as well as to diamond indenters.

The minor load is applied first, and a reference or "set" position is established on the dial gage of the Rockwell hardness tester. Then the

Table 1 Brinell hardness numbers for a 10 mm diameter ball

Diam of ball impression mm	\multicolumn{5}{c}{Brinell hardness number for a load of kg}	Diam of ball impression mm	\multicolumn{5}{c}{Brinell hardness number for a load of kg}
	500	1000	1500	2000	2500	3000		500	1000	1500	2000	2500	3000
2.00	158	316	473	632	788	945	4.25	33.6	67.2	101	134	167	201
2.05	150	300	450	600	750	899	4.30	32.8	65.6	98.3	131	164	197
2.10	143	286	428	572	714	856	4.35	32.0	64.0	95.9	128	160	192
2.15	136	272	408	544	681	817	4.40	31.2	62.4	93.6	125	156	187
2.20	130	260	390	520	650	780	4.45	30.5	61.0	91.4	122	153	183
2.25	124	248	372	496	621	745	4.50	29.8	59.6	89.3	119	149	179
2.30	119	238	356	476	593	712	4.55	29.1	58.2	87.2	116	145	174
2.35	114	228	341	456	568	682	4.60	28.4	56.8	85.2	114	142	170
2.40	109	218	327	436	545	653	4.65	27.8	55.6	83.3	111	139	167
2.45	104	208	313	416	522	627	4.70	27.1	54.2	81.4	108	136	163
2.50	100	200	301	400	500	601	4.75	26.5	53.0	79.6	106	133	159
2.55	96.3	193	289	385	482	578	4.80	25.9	51.8	77.8	104	130	156
2.60	92.6	185	278	370	462	555	4.85	25.4	50.8	76.1	102	127	152
2.65	89.0	178	267	356	445	534	4.90	24.8	49.6	74.4	99.2	124	149
2.70	85.7	171	257	343	429	514	4.95	24.3	48.6	72.8	97.2	122	146
2.75	82.6	165	248	330	413	495	5.00	23.8	47.6	71.3	95.2	119	143
2.80	79.6	159	239	318	398	477	5.05	23.3	46.6	69.8	93.2	117	140
2.85	76.8	154	230	307	384	461	5.10	22.8	45.6	68.3	91.2	114	137
2.90	74.1	148	222	296	371	444	5.15	22.3	44.6	66.9	89.2	112	134
2.95	71.5	143	215	286	358	429	5.20	21.8	43.6	65.5	87.2	109	131
3.00	69.1	138	207	276	346	415	5.25	21.4	42.8	64.1	85.6	107	128
3.05	66.8	134	200	267	334	401	5.30	20.9	41.8	62.8	83.6	105	126
3.10	64.6	129	194	258	324	388	5.35	20.5	41.0	61.5	82.0	103	123
3.15	62.5	125	188	250	313	375	5.40	20.1	40.2	60.3	80.4	101	121
3.20	60.5	121	182	242	303	363	5.45	19.7	39.4	59.1	78.8	98.5	118
3.25	58.6	117	176	234	293	352	5.50	19.3	38.6	57.9	77.2	96.5	116
3.30	56.8	114	170	227	284	341	5.55	18.9	37.8	56.8	75.6	95.0	114
3.35	55.1	110	165	220	276	331	5.60	18.6	37.2	55.7	74.4	92.5	111
3.40	53.4	107	160	214	267	321	5.65	18.2	36.4	54.6	72.8	90.8	109
3.45	51.8	104	156	207	259	311	5.70	17.8	35.6	53.5	71.2	89.2	107
3.50	50.3	101	151	201	252	302	5.75	17.5	35.0	52.5	70.0	87.5	105
3.55	48.9	97.8	147	196	244	293	5.80	17.2	34.4	51.5	68.8	85.8	103
3.60	47.5	95.0	142	190	238	285	5.85	16.8	33.6	50.5	67.2	84.2	101
3.65	46.1	92.2	138	184	231	277	5.90	16.5	33.0	49.6	66.0	82.5	99.2
3.70	44.9	89.8	135	180	225	269	5.95	16.2	32.4	48.7	64.8	81.2	97.3
3.75	43.6	87.2	131	174	218	262	6.00	15.9	31.8	47.7	63.6	79.5	95.5
3.80	42.4	84.8	127	170	212	255	6.05	15.6	31.2	46.8	62.4	78.0	93.7
3.85	41.3	82.6	124	165	207	248	6.10	15.3	30.6	46.0	61.2	76.7	92.0
3.90	40.2	80.4	121	161	201	241	6.15	15.1	30.2	45.2	60.4	75.3	90.3
3.95	39.1	78.2	117	156	196	235	6.20	14.8	29.6	44.3	59.2	73.8	88.7
4.00	38.1	76.2	114	152	191	229	6.25	14.5	29.0	43.5	58.0	72.6	87.1
4.05	37.1	74.2	111	148	186	223	6.30	14.2	28.4	42.7	56.8	71.3	85.5
4.10	36.2	72.4	109	145	181	217	6.35	14.0	28.0	42.0	56.0	70.0	84.0
4.15	35.3	70.6	106	141	177	212	6.40	13.7	27.4	41.2	54.8	68.8	82.5
4.20	34.4	68.8	103	138	172	207	6.45	13.5	27.0	40.5	54.0	67.5	81.0

Source: Ref 3

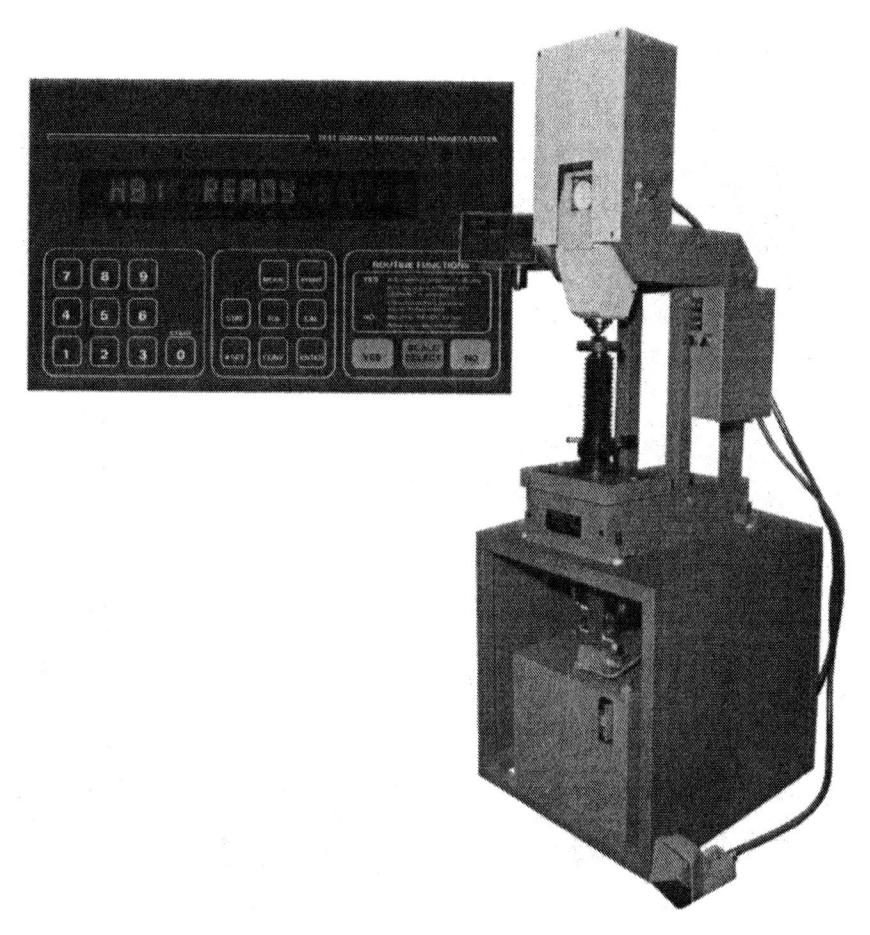

Fig. 3 Automatic Brinell hardness tester system with digital readout. Source: Ref 4

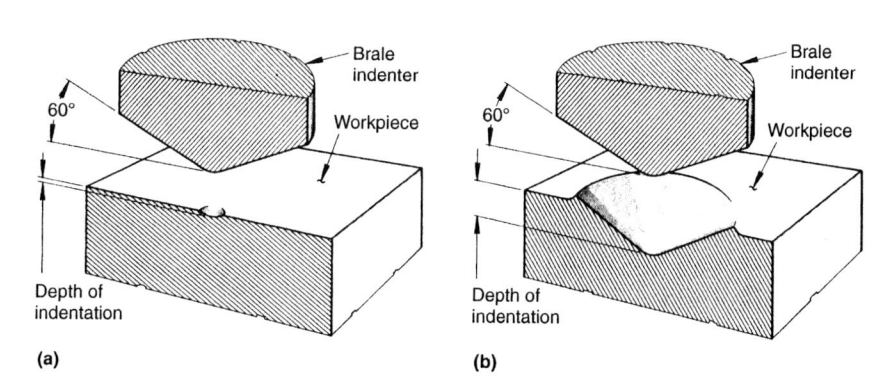

Fig. 4 Indentation in a workpiece made by application of (a) the minor load and (b) the major load, on a diamond Brale indenter in Rockwell hardness testing. The hardness value is based on the difference in depths of indentation produced by the minor and major loads. Source: Ref 1

major load is applied. Without moving the workpiece being tested, the major load is removed, and the Rockwell hardness number is automatically indicated on the dial gage. The entire operation takes 5 to 10 s.

Diamond indenters are used mainly for testing material such as hardened steels and cemented carbides. Steel ball indenters, available with diameters of 1/16, 1/8, 1/4, and 1/2 in. (~1.6, 3.2, 6, and 12.5 mm) are used when testing materials such as soft steel, copper alloys, aluminum alloys, and bearing metals. Each load and indenter combination has a specified letter designation as indicated in Tables 2 and 3. Regardless of the scale, hardness readings are written with the number first (dial indication); the letters HR (hardness Rockwell) follow, and finally, the scale designation. For example, 60 Rockwell "C" (diamond indenter with a 150 kg, or 330 lb, load) is written as "60 HRC." Likewise, a reading of 90 Rockwell "B" would be written as "90 HRB." The Rockwell "C" scale was developed first and is still the most widely used regardless of whether the test is actually conducted with a diamond indenter and a 150 kg (330 lb) load, or whether the actual test is made with another scale and converted to the "C" scale (conversion tables are readily available; see Ref 5). When hard-

Table 2 Rockwell standard hardness scales

Scale symbol	Indenter	Major load, kgf	Typical applications
A	Diamond (two scales—carbide and steel)	60	Cemented carbides, thin steel, and shallow case-hardened steel
B	1.588 mm (1/16 in.) ball	100	Copper alloys, soft steels, aluminum alloys, malleable iron
C	Diamond	150	Steel, hard cast irons, pearlitic malleable iron, titanium, deep case-hardened steel, and other materials harder than 100 HRB
D	Diamond	100	Thin steel and medium case-hardened steel and pearlitic malleable iron
E	3.175 mm (1/8 in.) ball	100	Cast iron, aluminum, and magnesium alloys, bearing metals
F	1.588 mm (1/16 in.) ball	60	Annealed copper alloys, thin soft sheet metals
G	1.588 mm (1/16 in.) ball	150	Phosphor bronze, beryllium copper, malleable irons. Upper limit 92 HRG to avoid possible flattening of ball
H	3.175 mm (1/8 in.) ball	60	Aluminum, zinc, lead
K	3.175 mm (1/8 in.) ball	150	Bearing metals and other very soft or thin materials. Use smallest ball and heaviest load that do not produce anvil effect
L	6.350 mm (1/4 in.) ball	60	Bearing metals and other very soft or thin materials. Use smallest ball and heaviest load that do not produce anvil effect
M	6.350 mm (1/4 in.) ball	100	Bearing metals and other very soft or thin materials. Use smallest ball and heaviest load that do not produce anvil effect
P	6.350 mm (1/4 in.) ball	150	Bearing metals and other very soft or thin materials. Use smallest ball and heaviest load that do not produce anvil effect
R	12.70 mm (1/2 in.) ball	60	Bearing metals and other very soft or thin materials. Use smallest ball and heaviest load that do not produce anvil effect
S	12.70 mm (1/2 in.) ball	100	Bearing metals and other very soft or thin materials. Use smallest ball and heaviest load that do not produce anvil effect
V	12.70 mm (1/2 in.) ball	150	Bearing metals and other very soft or thin materials. Use smallest ball and heaviest load that do not produce anvil effect

Source: Ref 4

ness readings are reported, it is essential that it be indicated where conversions have been made to obtain the listed values.

Rockwell testers vary widely in design. Figure 5 shows the components of a common type of Rockwell tester. Regardless of tester design, the

Table 3 Rockwell superficial hardness scales

Scale symbol	Indenter	Major load, kgf
15N	Diamond	15
30N	Diamond	30
45N	Diamond	45
15T	1.588 mm (1/16 in.) ball	15
30T	1.588 mm (1/16 in.) ball	30
45T	1.588 mm (1/16 in.) ball	45
15W	3.175 mm (1/8 in.) ball	15
30W	3.175 mm (1/8 in.) ball	30
45W	3.175 mm (1/8 in.) ball	45
15X	6.350 mm (1/4 in.) ball	15
30X	6.350 mm (1/4 in.) ball	30
45X	6.350 mm (1/4 in.) ball	45
15Y	12.70 mm (1/2 in.) ball	15
30Y	12.70 mm (1/2 in.) ball	30
45Y	12.70 mm (1/2 in.) ball	45

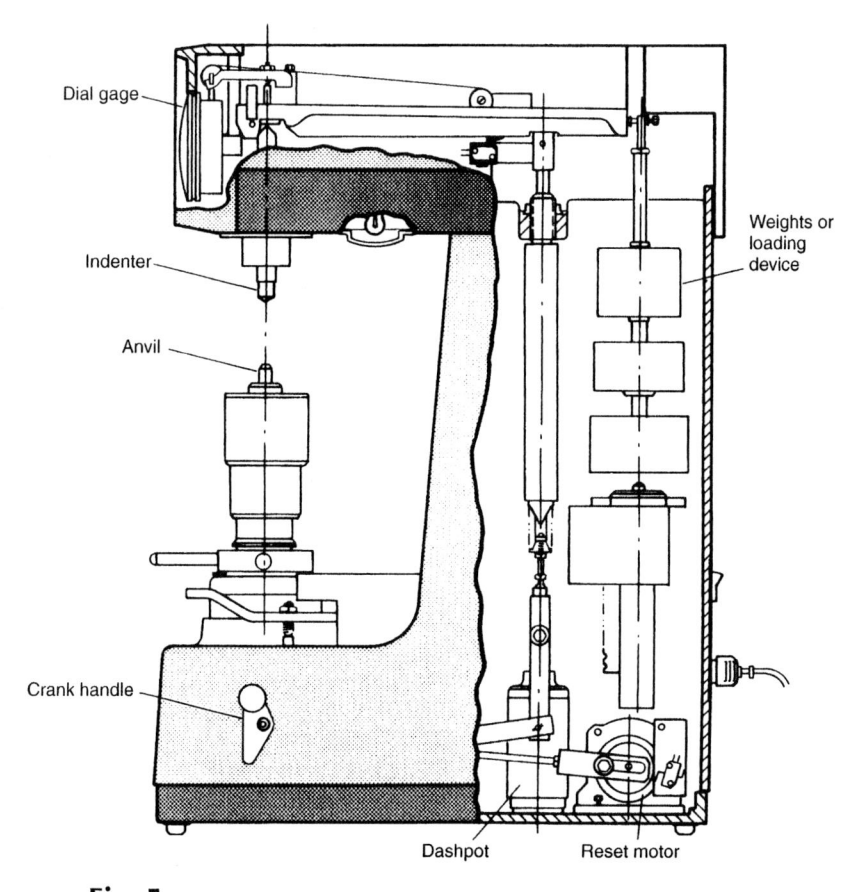

Source: Ref 4

Fig. 5 Principal components of a Rockwell testing machine. Source: Ref 4

principles are the same. Consequently, testers are now available in an almost infinite number of designs to accommodate a large variety of testing requirements. Several models are available with digital readout and/or automation with computer analysis of the hardness data. A fully automatic unit that can test multiple specimens is shown in Fig. 6.

Vickers Testing. The Vickers hardness test is similar to the Brinell principle in that an indenter of definite shape is pressed into the material to be tested, the load removed, the diagonals of the resulting indentation are measured, and the hardness number is calculated based on the area of indentation. The Vickers indenter is square-based pyramid that has an angle of 136° between faces, as shown in Fig. 7.

With the Vickers indenter, the depth of indentation is about one-seventh of the diagonal length of the indentation. The Vickers hardness number (HV) is the ratio of the load applied to the indenter to the surface area of the indentation. However, as is the case for Brinell testing, tables are posted on testers that provide the Vickers hardness number based on load and the diagonal readings made with a microscope—an integral part of the Vickers tester.

The range of hardness that can be accommodated by the Vickers test method is quite broad because of the various loads that can be used. The Vickers test is, however, less versatile than the Rockwell test and more time-consuming. In a Vickers tester, the load range is 1 to 120 kg (2 to 265 lb); however, for most hardness testing, 50 kg (110 lb) is maximum. The Vickers test also can be used for microhardness testing as described later.

Scleroscope Testing. The Scleroscope hardness test is essentially a dynamic indentation test wherein a diamond-tipped hammer is dropped from a fixed height onto the surface of the material being tested. The height of rebound of the hammer is a measure of the hardness of the material. The Scleroscope scale consists of units that are determined by dividing the average rebound of the hammer from a quenched (to maximum hardness) and untempered water-hardening tool steel into 100 units. The scale is continued above 100 to permit testing of materials having hardness greater than that of fully hardened tool steel. Scleroscope hardness testing can be conducted rapidly, and some testing instruments are portable so that they can be used for testing large workpieces that would be difficult to bring to the tester.

Other Hardness Tests. Because the test accuracy of the portable Scleroscope is often very sensitive to testing technique, in recent years, microprocessor-based, digital, portable hardness testers such as the Leeb tester and ultrasonic testers have gained widespread acceptance among heat treaters. Leeb testers, also known as Equotip testers, operate on a dynamic rebound principal similar to the Scleroscope. Ultrasonic microhardness testing offers an alternative to the more conventional methods based on visual (microscopic) evaluation of an indentation. Both Leeb scale (Equotip) and ultrasonic hardness testing methods are described in Ref 3 and 6.

Microhardness Testing

Microhardness testing is usually done with loads that do not exceed 1 kg (2 lb) and may be as little as 1g, although the most common load range

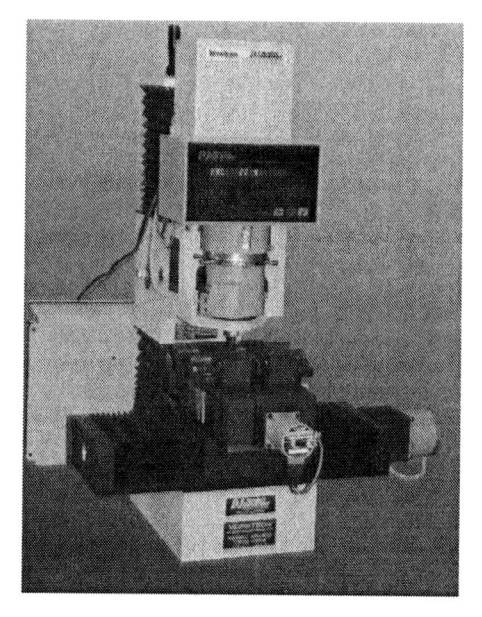

Fig. 6 Automatic Rockwell testing system for high production work (Jominy specimens are being tested). Source: Ref 3

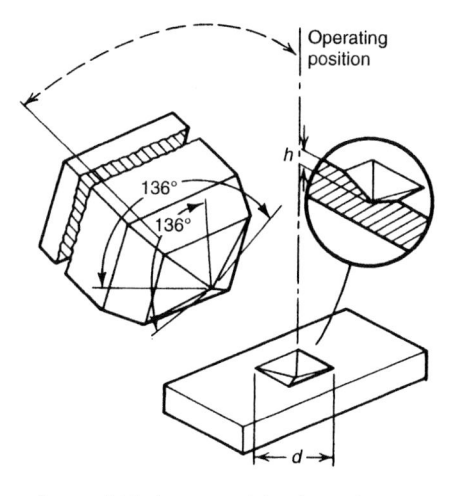

Fig. 7 Diamond pyramid indenter used for the Vickers test and resulting indentation in the workpiece. d, mean diagonal of the indentation in millimeters. Source: Ref 4

is 100 to 500 g. In general, however, the term *microhardness* relates to the size of the indentation rather than to the applied load.

Microhardness testing is capable of providing information on hardness characteristics that cannot be provided by the more conventional methods. It is not, however, strictly a research tool and is used extensively for control of many production operations. Specific fields of application include:

- Measuring hardness of precision workpieces that are too small to be measured by other methods
- Measuring hardness of product forms such as foils that are too thin to be measured by other methods
- Monitoring of carburizing or nitriding operations, which usually is accomplished by hardness surveys taken on cross sections of test pieces that accompanied the workpieces through production operations
- Measuring hardness of individual microconstituents
- Measuring hardness close to edges, detecting undesirable surface conditions such as grinding burn and decarburization
- Measuring hardness of surface layers such as plating or bonded layers

Microhardness testing can be performed with either the Vickers indenter (Fig. 7) or the Knoop indenter shown in Fig. 8.

Knoop indentation testing is performed with a diamond, ground to pyramidal form, that produces a diamond-shaped indentation having an approximate ratio between long and short diagonals of 7 to 1. The pyramid

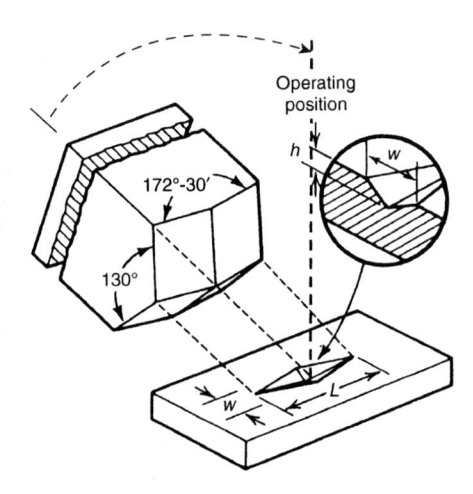

Fig. 8 Schematic of the pyramidal-shaped diamond indenter used for the Knoop test and an example of the indentation it produces. Source: Ref 7

shape employed has an included longitudinal angle of 172° 30′ and an included transverse angle of 130° (see Fig. 8). The depth of indentation is about ¹⁄₃₀ its length. Because of the shape of the indenter, indentations of accurately measurable length are obtained with light loads.

The Knoop hardness number (HK) is the ratio of the load applied to the indenter to the unrecovered projected area of indentation. By formula:

$$HK = \frac{P}{A} = \frac{P}{CL^2}$$

where P is the applied load, in kilograms; A is the unrecovered projected area of indentation, in square millimeters; and C is 0.07028, a constant of the indenter relating projected area of the indentation to the square of the length of the long diagonal.

This formula is, however, mainly for academic interest, as tables are provided that provide the Knoop hardness for the size of any indentation (measured in microns) and the applied load. Superiority of the Vickers versus the Knoop indenter is often controversial. One advantage of the Knoop indenter is that only the length of the indentation is measured (Fig. 9), whereas for greatest accuracy, both diagonals of the Vickers indentation should be taken and then averaged.

Figure 9 presents a comparison of the indentations made by Knoop and Vickers indenters. Each has some advantages over the other. For example, the Vickers indenter penetrates about twice as far into the workpiece as does the Knoop indentation. Therefore, the Vickers indenter is less sensitive to minute differences in surface condition than the Knoop indenter. However, the Vickers indentation, because of the shorter diagonal, is more sensitive to errors in measuring than the Knoop indentation.

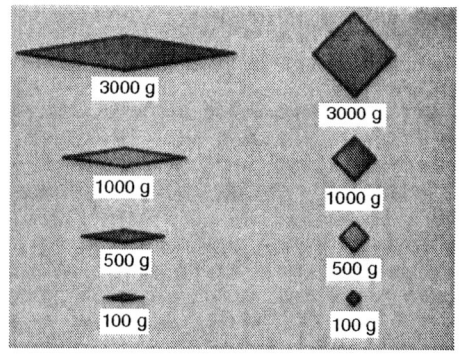

Fig. 9 Comparison of indentations made by Knoop and Vickers indenters in the same work metal and at the same loads. Source: Ref 1

For Knoop loads of 100 g or less, the HK may be a function of load as shown in Fig. 10. The solution to the problem of elastic recovery of the material is to calibrate the tester for the actual load being used.

Several types of microhardness testers are available. The most accurate operate through the direct application of load by dead weight, or by weights and levers. Principal components of a typical microhardness tester are shown in Fig. 11, which demonstrates how the indentation is made on the specimen, as well as the repositioning of the specimen for measurement by a special type of microscope. Programmable, digital, and direct-reading microhardness testing equipment is available for both Vickers and Knoop hardness testing. These testers can also provide hardness values versus depth profiles and computerized data storage.

Effect of Carbon Content on Annealed Steels

In Chapter 2, "Fundamentals of the Heat Treating of Steel," two metallurgical phenomena—the allotropy of iron and the small size of the carbon atom—are described as key factors in the heat treatment of steel. Chapter 2 also discusses how carbon atoms exit from the face-centered cubic (fcc) iron crystal in a slow and orderly manner, resulting in an annealed (soft) condition. Rapid cooling entraps the carbon atoms in the lattice; consequently, the structure changes from an fcc to a body-centered cubic (bcc) lattice structure, as high strength and hardness result. Also, referring to the time-temperature-transformation (TTT) diagram (Fig. 6)

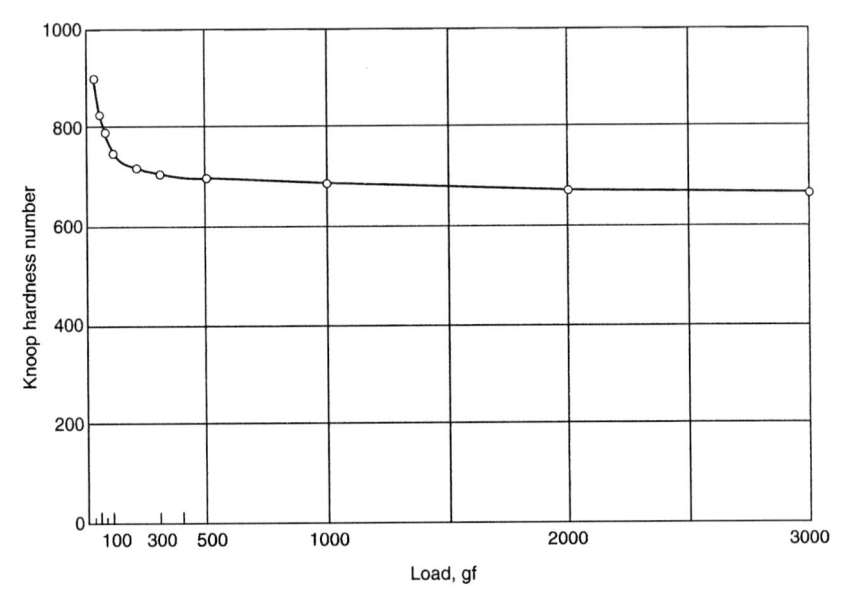

Fig. 10 Relationship of Knoop hardness number and load. Note the increase in Knoop hardness for hardened steel with decreasing load. Source: Ref 8

in Chapter 2, note that the hardness of martensite (the fully hardened structure) is 65 HRC.

Logically, the reader might wonder at this point what affect increasing carbon content has on annealed steels. To be sure, there is an effect. Referring to the iron-cementite phase diagram (Fig. 4) in Chapter 2, near the extreme left—at approximately 0.20% C, about one-fourth of the grains are pearlite, while the remainder is nearly pure ferrite (see microstructures associated with the iron-iron carbide diagram, Fig. 5). The ferrite grains are very soft (below the HRC scale), while the pearlite grains are considerably harder because they contain a dispersion of cementite in the ferrite. Moving to the right of the iron-cementite diagram, the number of pearlite grains increases in proportion to the number of ferrite grains until the eutectoid (0.77% C) is reached. At this point, 100% pearlite grains exist. Thus, hardness, as measured by indentation methods, increases as the ratio of pearlite-to-ferrite grains increases, because the cementite particles act as reinforcement to the ferrite and resist being "pushed" by an indenter. This condition generally exists even when the cementite particles are spheroidal (see Fig. 5 in Chapter 2). However, the spheroidal particles offer less resistance to force from the indenter compared with a steel of the same carbon content with a structure of lamellar pearlite. Therefore, the spheroidal structure registers as the softer of the two materials.

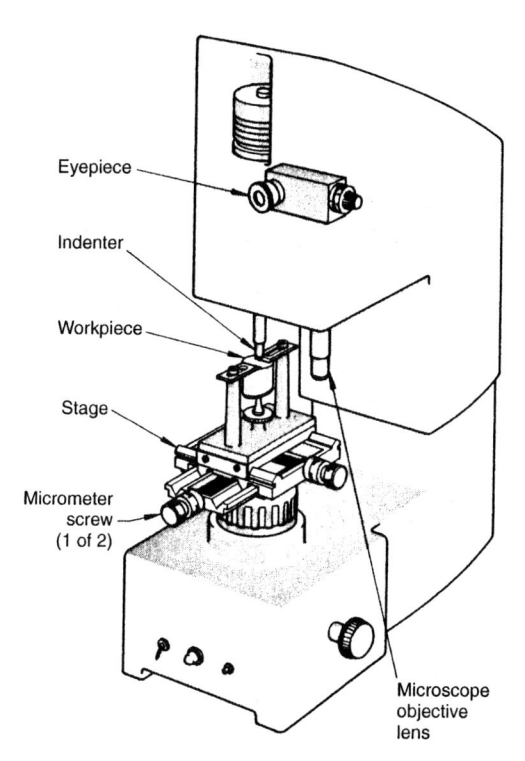

Fig. 11 Principal components of a typical microhardness tester. Source: Ref 1

Now, let us go a step further and assume that a steel containing 1.0% C or higher (a hypereutectoid steel) is cooled slowly so that lamellar pearlite is formed. With the carbon content higher than a eutectoid alloy, there is an excess of cementite that often gathers at the grain boundaries as shown in Fig. 5(d) in Chapter 2. This cementite is relatively hard, thus offering increased resistance to indentation. Therefore, hardness does increase as carbon content increases, but not dramatically for annealed steels; that is, not as measured by a conventional hardness tester. Likewise, yield strength and tensile strength increase in annealed carbon steels as carbon content increases.

It should be noted at this point that there are two types of hardness—apparent and true. Apparent hardness is the actual reading taken from an instrument such as a Brinell or Rockwell tester. When a relatively large indenter is forced into the metal specimen, the indenter can register total resistance of the mass into which it is being forced; when two or more microconstituents are involved, the result is an average of these values.

Conversely, true hardness involves testing with a microhardness tester so as to obtain the true hardness values of the various microconstituents. For example, cementite is far harder than ferrite. This accounts for the fact that hypereutectoid steels have far greater resistance to abrasive wear than hypoeutectoid or eutectoid steels, even though all steels are in the annealed condition. The difference in apparent hardness is not great. Actually, the tiny particles of cementite embedded in a matrix act like sand embedded in a plastic surface.

Role of Carbon in Hardened Steels

As emphasized in Chapter 2, carbon is the key to the hardening of steels by the heating and quick cooling (quenching) mechanism. The carbon content of a steel determines the maximum hardness attainable. The effect of carbon on attainable hardness is demonstrated in Fig. 12. The maximum

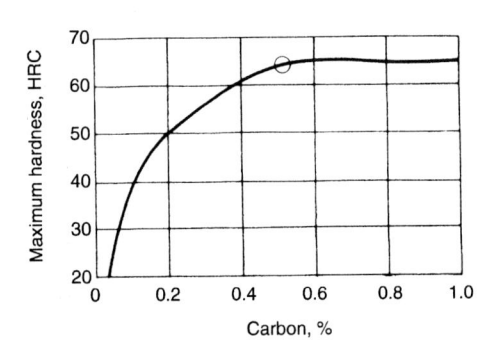

Fig. 12 Relationship between carbon content and maximum hardness. Full hardness can be obtained with as little as 0.60% C as noted by data point. Source: Ref 9

attainable hardness requires only 0.60% C, which probably seems odd to the reader. From the iron-cementite diagram (Fig. 4 in Chapter 2), it would seem logical that hardness would not become a straight line until about 0.77% C is reached. However, there is essentially no change in attainable hardness above about 0.60% C according to the data shown in Fig. 12. This can be accounted for through some unavoidable deficiencies concerned with indentation hardness testing. Despite this apparent oddity concerning Fig. 12, these data are accurate and absolutely reproducible for extremely thin sections of carbon steel.

In fact, the data shown in Fig. 12 are precise to the extent that they sometimes are used in reverse; that is, they can be used as a quick means of determining carbon content of an unknown steel. For example, when the carbon content of a steel is unknown and analytical means are not available, cutting wafer-thin sections of the unknown steel, heating them above the transformation temperature, and quenching them in water helps aid in identification of the material by observing and measuring the physical and mechanical properties that result. The specimens then are measured for hardness; if specimens show hardness values of approximately 50 HRC, the carbon content is about 0.20%; for readings for around 60 HRC, the carbon content is slightly over 0.40%. Obviously, as shown in Fig. 12, this method of carbon determination would not be valid much above 60 HRC because the hardness line starts to become a straight line.

In conventional heat treating practice, it should be remembered that the data shown in Fig. 12 are based on heat treating of wafer-thin sections that are cooled from their austenitizing temperature to room temperature within a matter of seconds, thus developing 100% martensite throughout their sections. Therefore, the ideal condition shown in Fig. 12 is seldom attained in practice. Figure 13 shows a better example of hardness versus carbon content, because it is a more accurate condition, one expected in commercial practice.

The most important factor influencing the maximum hardness that can be attained is mass of the metal being quenched. In a small section, the

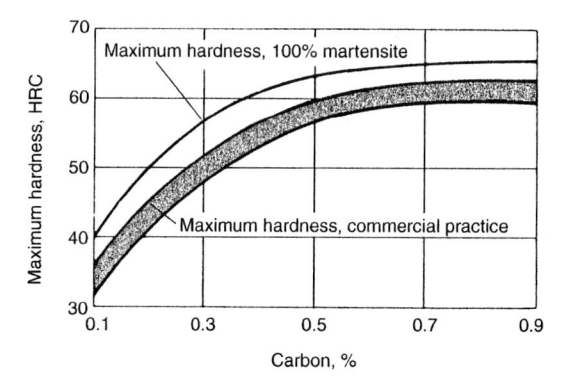

Fig. 13 Relationship between carbon content and maximum hardness usually attained in commercial hardening. Source: Ref 9

heat is extracted quickly, thus exceeding the critical cooling rate of the specific steel. The critical cooling rate is the rate of cooling that must be exceeded to prevent formation of nonmartensitic products. As section size increases, it becomes increasingly difficult to extract the heat fast enough to exceed the critical cooling rate and thus avoid formation of nonmartensitic products. A typical condition is shown in Fig. 14, which illustrates the effect of section size on surface hardness and is a good example of the mass effect. For small sections up to 13 mm (0.5 in.), full hardness of about 63 or 65 HRC is attainable. As the diameter of the quenched piece is increased, cooling rates and hardness decrease because the critical cooling rate for this specific steel is not exceeded.

Hardenability

The term *hardenability* is used freely throughout the remainder of this book; thus, it is important that the reader understand its precise meaning. One might think that hardenability means the ability to be hardened. Although true to a degree, the precise meaning of hardenability is much broader. In student classes on heat treating, two questions are often asked: (a) a specific steel, after heating above A_3 and quenching in water, shows a hardness of 65 HRC (near the limit for most steels as measured by Rockwell testing); does this establish it as a high-hardenability steel? and (b) after heating and quenching, a specific steel registers only 35 HRC; is it a low-hardenability steel? Invariably a student will answer yes to the first question and no to the second. The correct answer for both is "not necessarily." Information on section thickness in each instance is required before either question can be correctly answered. Therefore, hardenability does not necessarily mean the ability to be hardened to a certain Rockwell or Brinell value.

For example, just because a given steel is capable of being hardened to 65 HRC does not necessarily mean that it has high hardenability. Also, a

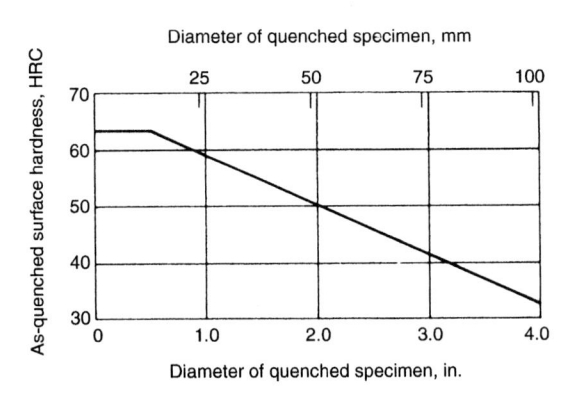

Fig. 14 Effect of section size on surface hardness of a 0.54% carbon steel quenched in water from 830 °C (1525 °F). Source: Ref 9

steel that can be hardened to only 40 HRC may have very high hardenability. Hardenability refers to capacity of hardening (depth) rather than to maximum attainable hardness values.

As stated previously, carbon controls the maximum attainable hardness for any conventional steel, but carbon has only a minor effect on hardenability. Thus, Fig. 14 also serves as an excellent example of a low-hardenability steel. Plain carbon steels are characterized by their low hardenability, with critical cooling rates lasting only for brief periods. Hardenability of all steels is directly related to critical cooling rates. The longer the time of critical cooling rate, the higher is the hardenability for a given steel, almost regardless of carbon content (see TTT curve, Fig. 6, in Chapter 2).

Quantifying Hardenability

To understand how one can calculate and quantify hardenability into a single number for a specific steel analysis without heat treating and testing a series of various diameter specimens, a brief and basic discussion of the concept of the ideal critical diameter (D_I) and austenitic grain size of steels is necessary.

The ideal critical diameter (D_I) is defined as the diameter of bar (in inches), which, if heated to above the upper critical temperature (A_{c3}) and quenched at an infinite cooling rate ($H = \infty$), the center of the bar would quench to 50% martensite.

When a steel is made, the dissolved oxygen must be removed from the molten metal before it solidifies. Some of the additions that aid in this deoxidation process are elements that easily combine with oxygen such as silicon or aluminum. Depending on which element or elements are added during this process, an austenitic grain size is established for the particular batch of steel being treated.

The austenitic grain size of steels can be determined using a McQuaid-Ehn grain size procedure. Coarse grain steels have standard grain sizes of 0 to 5 as rated on the standard SAE or ASTM grain size charts. Fine grain steels have typical grain sizes of 6 to 8 and some tool steels can have grain sizes as fine as 15.

The relationship between grain size and carbon level on D_I is shown in Fig. 15. Coarser grain sizes (lower numbers) and higher carbon levels result in greater base D_I values and thus greater hardenability. This base D_I value must then be multiplied by the factor for each of the alloying elements present in order to obtain the overall ideal critical diameter for the specific chemical analysis of steel in question.

Effect of Alloying Elements on Hardenability

The principal reason for using alloying elements in the standard grades of steel is to increase hardenability. The alloying elements used in conventional steels are confined to manganese, silicon, chromium, nickel, molybdenum, and vanadium. Because of the small amounts used, boron is not usually called an alloy. Steels that contain boron, whether they are carbon or alloy grades, are more often termed *boron-treated* steels. The use of cobalt, tungsten, zirconium, and titanium is generally confined to tool or other specialty steels. Tool steels and other highly alloyed steels are covered in other chapters of this book.

Manganese, silicon, chromium, nickel, molybdenum, and vanadium all have separate and unequal effects on hardenability. However, the individual effects of these alloying elements may be completely altered when two or more of these are used together. In periods of alloy shortages, extensive investigations were conducted, and it has been established that more hardenability can be attained with less total alloy content when two or more alloys are used together. This practice is clearly reflected in the standard alloy steel compositions (see Chapter 7, "Heat Treating of Alloy Steels"). This approach not only saves alloys that are often in scarce supply, but also results in more hardenability at lower cost.

Therefore, all of the alloy steels listed in Chapter 7 have significantly greater hardenability than the carbon steels. It must be further emphasized

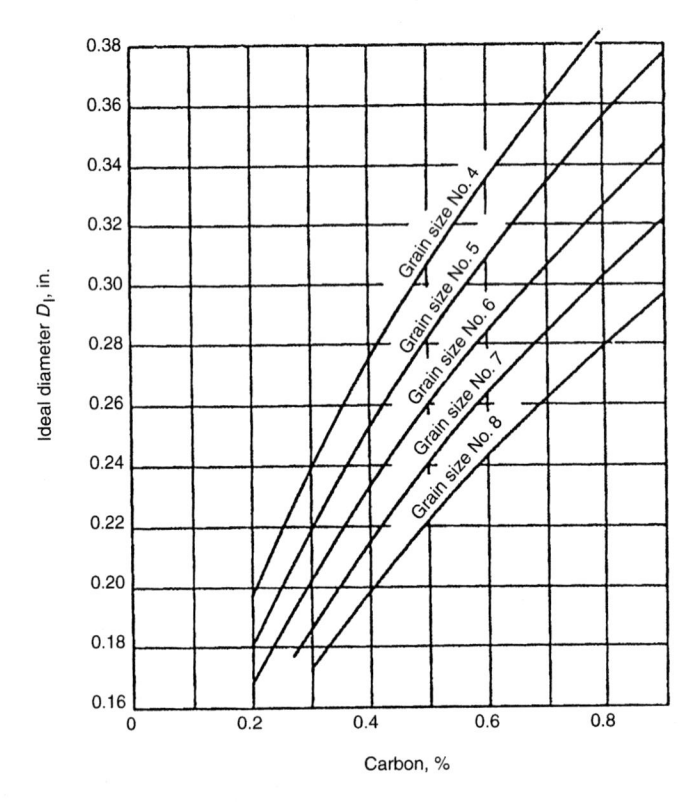

Fig. 15 Hardenability, expressed as ideal critical size, as a function of austenite grain size and carbon content of iron-carbon alloys. Source: Ref 10

that the hardenability varies widely among the alloy grades, which is a principal reason for the existence of so many grades.

The quantitative effect of alloying elements has been determined for each of the elements commonly used. The relative effects of the various elements are expressed as multiplying factors based on the percentage of the element, as shown in Fig. 16. These values are then each used as a multiplier to the base D_I value—as determined from Fig. 15—to arrive at the overall ideal critical diameter for the specific grain size and steel analysis. Table 4 shows the range of D_I values for various carbon and alloy grades of steel. Such calculations are somewhat complex, but if the reader is interested, analog calculators or computer programs such as SteCal from ASM International are available. For further information on precise calculation of hardenability, see Ref 10.

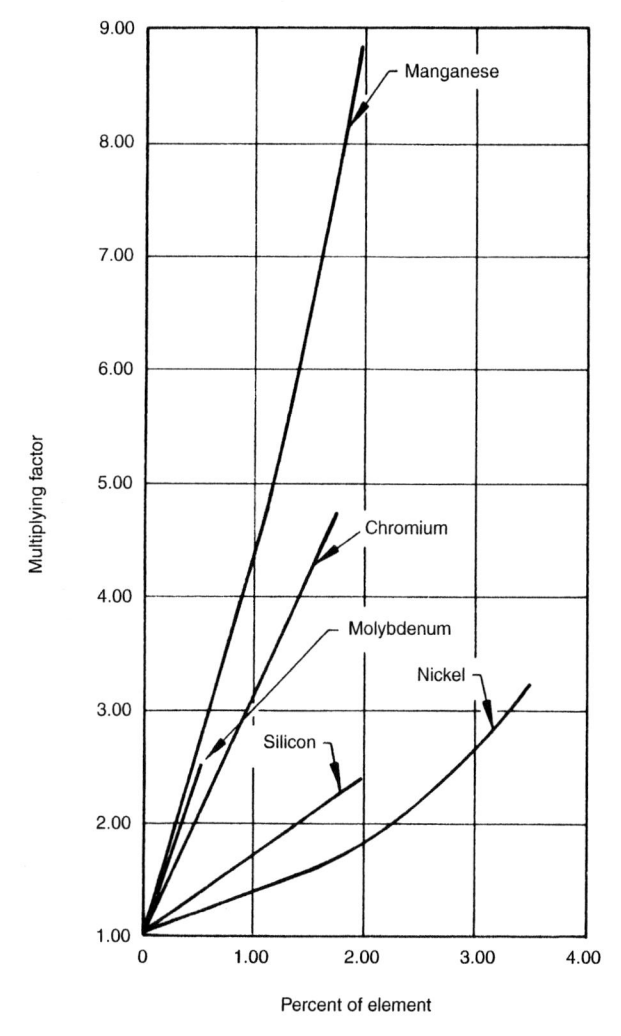

Fig. 16 Multiplying factors for five common alloying elements. Source: Ref 1

Methods of Evaluating Hardenability

Many test methods have been devised for evaluating hardenability, although a majority of these test procedures are applicable for only a few (or sometimes only one) grades of steel. Other tests are arbitrary in nature and are not discussed in this book.

One hardenability test that has proved to be useful and reproducible is the hardness penetration diagram test. This is accomplished by plotting the cross-sectional hardness of a bar, or other section, on suitable graph paper. In this procedure, the specimen is hardened and sectioned. Hardness is then determined at various depths along the radius and plotted against distance from the surface, as in Fig. 17(a). Several hardness traverses are made along a number of radii, and the various values at each point are averaged before being plotted. After plotting, the mirror image is used to complete the symmetrical curve shown in Fig. 17(b). The pertinent point of this procedure is that a perfectly symmetrical curve results, based on an average of several readings. The hardness penetration diagram is adaptable to shallow, medium, or deep hardening steels.

A refinement of the hardness penetration diagram is designated as the S-A-C test. In this procedure, a 1 in. (25 mm) diameter bar is quenched under standardized conditions, and the resultant hardness distribution is plotted as a symmetrical U-shaped curve. From this curve, the surface hardness is reported as S, the area under the curve as A (the units of area are "Rockwell inches"), and the hardness at the center of the piece as C. The test is commonly known as the Rockwell inch test. Because this test

Table 4 Hardenabilities (stated as a range of D_I values) for various steels

Steel	D_I	Steel	D_I	Steel	D_I
1045	0.9–1.3	4135H	2.5–3.3	8625H	1.6–2.4
1090	1.2–1.6	4140H	3.1–4.7	8627H	1.7–2.7
1320H	1.4–2.5	4317H	1.7–2.4	8630H	2.1–2.8
1330H	1.9–2.7	4320H	1.8–2.6	8632H	2.2–2.9
1335H	2.0–2.8	4340H	4.6–6.0	8635H	2.4–3.4
1340H	2.3–3.2	X4620H	1.4–2.2	8637H	2.6–3.6
2330H	2.3–3.2	4620H	1.5–2.2	8640H	2.7–3.7
2345	2.5–3.2	4621H	1.9–2.6	8641H	2.7–3.7
2512H	1.5–2.5	4640H	2.6–3.4	8642H	2.8–3.9
2515H	1.8–2.9	4812H	1.7–2.7	8645H	3.1–4.1
2517H	2.0–3.0	4815H	1.8–2.8	8647H	3.0–4.1
3120H	1.5–2.3	4817H	2.2–2.9	8650H	3.3–4.5
3130H	2.0–2.8	4820H	2.2–3.2	8720H	1.8–2.4
3135H	2.2–3.1	5120H	1.2–1.9	8735H	2.7–3.6
3140H	2.6–3.4	5130H	2.1–2.9	8740H	2.7–3.7
3340	8.0–10.0	5132H	2.2–2.9	8742H	3.0–4.0
4032H	1.6–2.2	5135H	2.2–2.9	8745H	3.2–4.3
4037H	1.7–2.4	5140H	2.2–3.1	8747H	3.5–4.6
4042H	1.7–2.4	5145H	2.3–3.5	8750H	3.8–4.9
4047H	1.8–2.7	5150H	2.5–3.7	9260H	2.0–3.3
4047H	1.7–2.4	5152H	3.3–4.7	9261H	2.6–3.7
4053H	2.1–2.9	5160H	2.8–4.0	9262H	2.8–4.2
4063H	2.2–3.5	6150H	2.8–3.9	9437H	2.4–3.7
4068H	2.3–3.6	8617H	1.3–2.3	9440H	2.4–3.8
4130H	1.8–2.6	8620H	1.6–2.3	9442H	2.8–4.2
4132H	1.8–2.5	8622H	1.6–2.3	9445H	2.8–4.4

Source: Ref 10

employs 1 in. (25 mm) diam bars, its usefulness is limited to steels that do not harden throughout such a section. If deeper hardening steels are to be tested by this procedure, use of a larger round is necessary.

End-Quench Testing. While the simple test just described and several other hardenability tests are commonly used, the end-quench test (Jominy) is by far the most generally accepted and widely used method for evaluating the hardenability of carbon and alloy steels. The test is relatively simple to perform and can produce much useful information for the designer, as well as the fabricator.

Although variations are sometimes made to accommodate specific requirements, the test bars for the end-quench test are normally 1 in. (25 mm) in diameter by 4 in. (100 mm) long. A 1⅛ in. (28.5 mm) diameter collar is left on one end to hold it in a quenching jig, as illustrated in Fig. 18.

In this test, water flow is controlled by a suitable valve so that the amount striking the end of the specimen (Fig. 18) is constant in volume

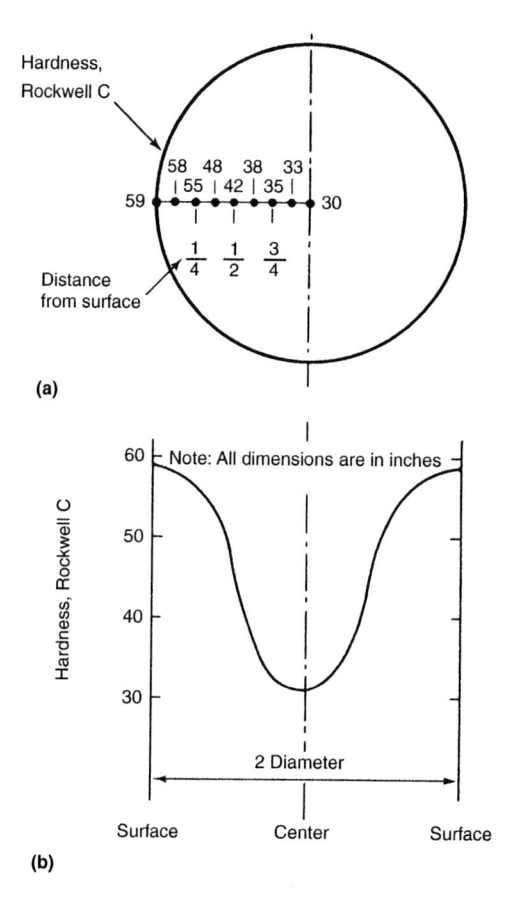

Fig. 17 Hardness penetration diagram. (a) Method of taking hardness traverse. (b) Plot of results of averaging several hardness traverses and mirror image. Source: Ref 1

and velocity. The water impinges on the end of the specimen only, then drains away. By this means, cooling rates vary from a very rapid rate on the quenched end to a very slow rate, essentially equal to cooling in still air, on the opposite end. This results in a wide range of hardnesses along the length of the bar.

After the test bar has been heated and quenched, two opposite and flat parallel surfaces are ground along the length of the bar to a depth of 0.015 in. (0.380 mm). Rockwell C hardness determinations are then made every $\frac{1}{16}$ in. A specimen-holding indexing fixture is helpful for this operation for convenience as well as accuracy. Such fixtures are available as accessory attachments for conventional Rockwell testers. The next step is to record the readings and plot them on graph paper to develop a curve, as illustrated in Fig. 19. By comparing the curves resulting from end-quench tests of different grades of steel, their relative hardenability may be established. The steels having higher hardenability will be harder at a given distance from the quenched end of the specimen than steels having lower hardenability. Thus, the flatter the curve, the greater is the hardenability. On the end-quench curves, hardness is not usually measured beyond approximately 2 in. (50 mm) because hardness measurements beyond this distance are seldom of any significance. At about this 2 in. (50 mm) distance from the quenched end, the effect of water on the quenched end deteriorates, and the effect of cooling from the surrounding air becomes significant. An absolutely flat curve demonstrates conditions of very high hardenability, which characterizes an air-hardening steel such as some of the highly alloyed tool steels.

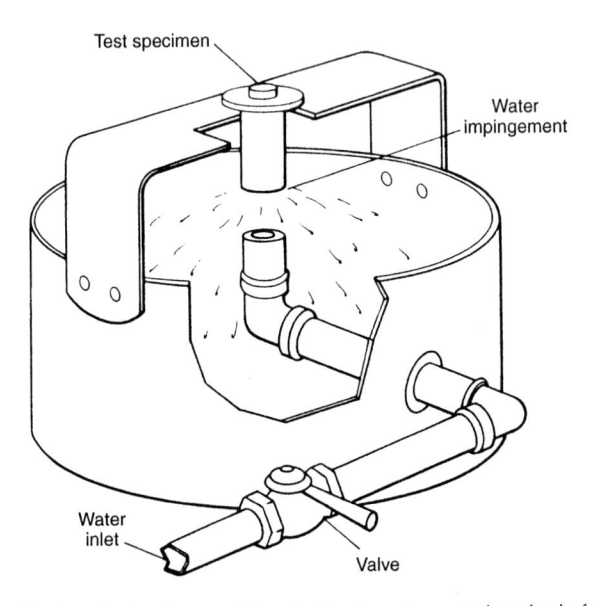

Fig. 18 Standard end-quench (Jominy) test specimen and method of quenching in quenching jig. Source: Ref 9

Variations in Hardenability. Because hardenability is a principal factor in steel selection and because hardenability varies over a broad range for the standard carbon and alloy steels, many grades are available. As a rule, hardenability of the standard carbon grades is very low, although there is still a great deal of variation in hardenability among the different grades. This variation depends to a great extent on the manganese content and sometimes to a smaller extent on the residual alloys that are sometimes present. A hardenability curve for a high-manganese grade of carbon steel, 1541, is shown in Fig. 20. This curve represents near maximum hardenability that can be obtained from any standard carbon grade.

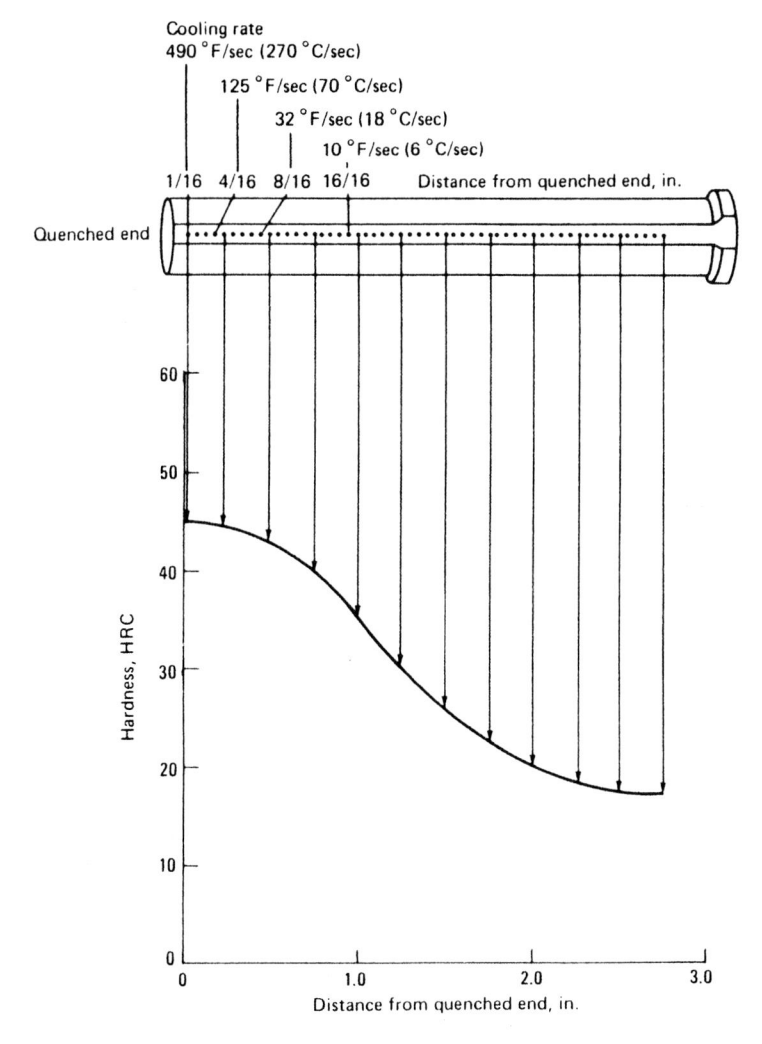

Fig. 19 Method of developing end-quench curve by plotting hardness versus distance from quenched end. Hardness plotted every ¼ in. for clarity, although Rockwell C readings were taken in increments of ¹⁄₁₆ in., as shown at top of illustration. Source: Ref 9

In contrast to the curve shown in Fig. 20, typical hardenability curves for four different 0.50% carbon alloy steels are presented in Fig. 21. These data emphasize the fact that maximum attainable hardness is provided by the carbon content, while the differences in alloy content markedly affect hardenability.

H-Steels

Because of the normal variations within prescribed limits of composition, it would be unrealistic to expect that the hardenability of a given grade would always follow a precise curve, such as is shown in Fig. 20 and 21. Instead, the hardenability of any grade varies considerably, which results in a hardenability band such as the 4150H band in Fig. 22. This

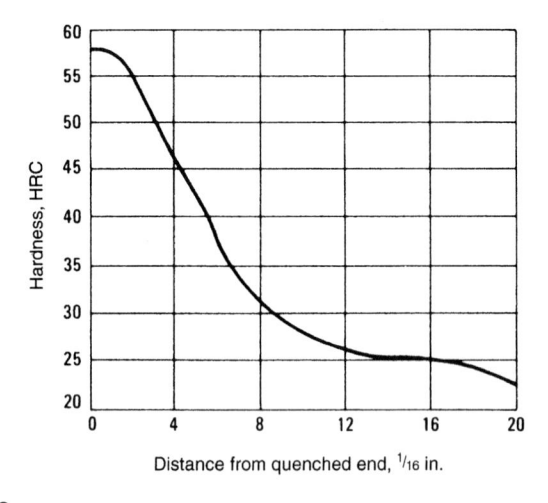

Fig. 20 End-quench hardenability curve for 1541 carbon steel. Source: Ref 9

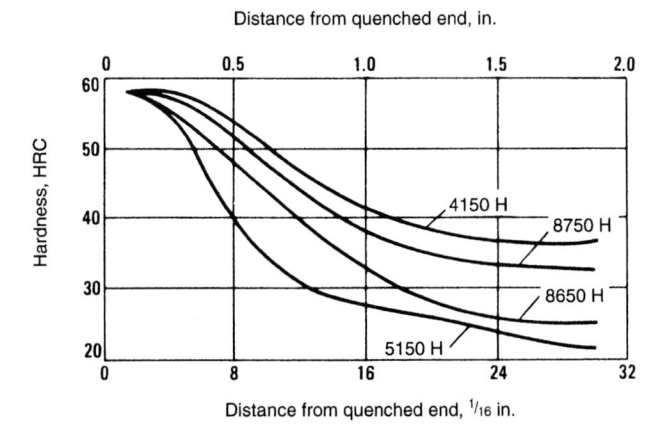

Fig. 21 Hardenability curves for several alloy steels. Source: Ref 9

steel was normalized at 870 °C (1600 °F), then annealed at 845 °C (1550 °F) before end quenching. The upper and lower curves that represent the boundaries of the hardenability band not only show the possible variation in hardness at the quenched end, caused by the allowable carbon range of 0.47 to 0.54%, but also the difference in hardenability as a result of the alloying elements being on the high or low side of the prescribed limits.

The need for hardenability data for steel users has long been recognized. Cooperative work by SAE and AISI has been responsible for devising hardenability bands for a large number of carbon and alloy steels, principally the latter. Steels that are sold with guaranteed hardenability bands are known as the H-steels. The numerical parts of the designation are the same as for the other standard grades, but the suffix letter H, such as 4140H, identifies it as a steel that meets prescribed hardenability limits. Not all of the steels listed by SAE-AISI are available as H-grades.

In order to give steel producers the latitude necessary in manufacturing for common hardenability limits, the chemical compositions of normal grades have been modified to form the H-steels. These modifications permit adjustments in manufacturing ranges of chemical composition. These adjustments correct melting practice for individual plants that might otherwise influence the hardenability bands. However, the modifications are not great enough to influence the general characteristics of the original compositions of the steels.

Steels are often available either made to a "chemistry only" specification (e.g., 8620) or made to the hardenability band requirement (8620H). Figure 23 shows that the hardenability band for the H steel is 25 to 33% narrower in the J-4 to J-6 positions than that of the "chemistry only" steel. This more uniform hardenability means a more uniform response to a given heat treatment. For this reason, one should always purchase the H-grade steel if the option is available.

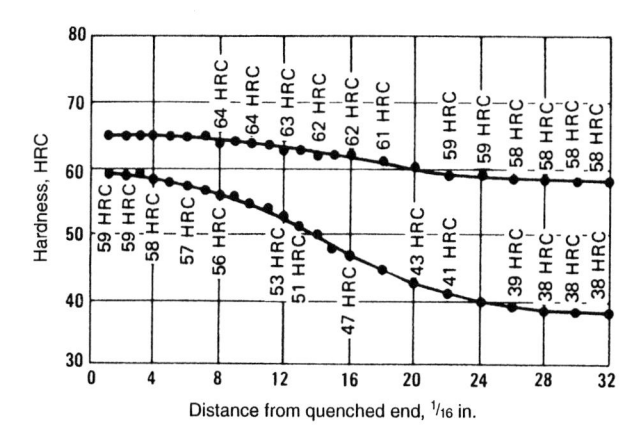

Fig. 22 Hardenability band for an alloy steel. 4150H: 0.47 to 0.54% C, 0.65 to 1.10% Mn. Normalized at 870 °C (1600 °F). Annealed at 845 °C (1550 °F). Source: Ref 9

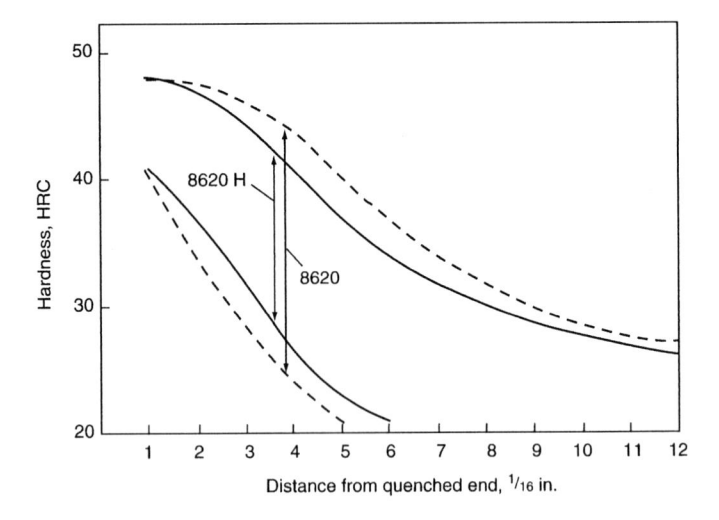

Fig. 23 Comparison of Jominy hardenability curves for 8620H and 8620 steels. Chemistry at maximum and minimum of the chemistry range. Source: Ref 11

REFERENCES

1. H.E. Boyer, Chapter 3, *Practical Heat Treating,* 1st ed., American Society for Metals, 1984
2. G. Revankar, et al., Hardness Testing, *Mechanical Testing and Evaluation,* Vol 8, *ASM Handbook,* ASM International, 2000, p 195–287
3. H. Chandler, Ed., *Hardness Testing,* 2nd ed., ASM International, 1999
4. E.L. Tobolski, Macroindentation Hardness Testing, *Mechanical Testing and Evaluation,* Vol 8, *ASM Handbook,* ASM International, 2000, p 203–220
5. Hardness Conversions for Steels, *Mechanical Testing and Evaluation,* Vol 8, *ASM Handbook,* ASM International, 2000, p 282–287
6. E.L. Tobolski, Miscellaneous Hardness Tests, *Mechanical Testing and Evaluation,* Vol 8, *ASM Handbook,* ASM International, 2000, p 252–259
7. G.F. Vander Voort, Microindentation Hardness Testing, *Mechanical Testing and Evaluation,* Vol 8, *ASM Handbook,* ASM International, 2000, p 221–231
8. V.E. Lysaght, *Indentation Hardness Testing,* Wilson Instrument Division, 1968
9. H. Chandler, Ed., Carbon & Alloy Steels Introduction, *Heat Treater's Guide: Practices and Procedures for Irons and Steels,* 2nd ed., ASM International, 1995, p 111–120
10. M.A. Grossmann and E.C. Bain, *Principles of Heat Treatment,* 5th ed., American Society for Metals, 1964
11. J.L. Dossett, *Distortion Prediction and Control Presentation,* ASM International Satellite Seminar, 1996

CHAPTER 4

Furnaces and Related Equipment for Heat Treating

HEAT TREATING FURNACES and related equipment include the heating devices (furnaces), fixtures and/or holding devices, quenching systems, and atmosphere and temperature control systems—all of which are required for the majority of heat treating operations. Specialized equipment for induction and flame heat treating are covered in Chapter 9, "Flame and Induction Hardening."

Types of Heat Treating Furnaces

Sizes and designs of heat treating furnaces vary over such a wide range that any precise classification is virtually impossible. In size, furnaces vary from a small model that sits on a bench and has a work space capacity for only a few ounces (often used for heat treating instrument parts) to a large car-bottom furnace that is capable of handling hundreds of tons in a single heat (Fig. 1).

Regardless of size, furnaces may be directly fired with fuel, where the work is exposed to combustion gases, or indirectly fired where the work is separated from combustion gases. Furnaces may also be heated by electrical resistance.

Ovens and Furnaces

The Industrial Heating Equipment Association (IHEA) classifies heating devices as ovens and furnaces. This separation is made on the basis of operating temperature—up to about 540 °C (1000 °F) is an oven, and any unit that operates at temperatures exceeding 540 °C (1000 °F) is called a furnace. This separation, based on operating temperature, is related directly to heating mode.

Modes of Heat Transmission

The three basic modes of heat transmission are conduction, convection, and radiation. In industrial heat treating, these modes may be used singly, or in combination.

Conduction of heat in a solid such as a metal workpiece is the transfer of heat from one part of the solid to another, under the influence of a temperature gradient and without appreciable displacement of the particles. For instance, if the temperature of the surface of a part is elevated, the heat flow to the center is by a molecular mechanism. Conduction involves the transfer of kinetic energy from one molecule to another in a chain reaction. Heat flow continues until equilibrium occurs. The time involved depends on the conductivity of the given metal, but in general, the speed of conduction within metals is relatively fast.

In most heat treating processes, while conduction plays only a minor role in the total heat transfer from the source to the workpiece, it is the sole mode of transferring heat from the surface to the center of a workpiece. An exception to the minor role conduction normally plays is the immersed electrode salt bath or a fluidized bed. While all three modes of heating are utilized in a molten salt, molten metal bath, or a fluidized bed,

Fig. 1 Car-bottom batch furnace for homogenizing large cylindrical parts. Source: Ref 1

conduction plays an important role because the hot medium is in direct contact with the work-metal surfaces.

Convection involves the transfer of heat by mixing one parcel of fluid (fluid refers to either liquid or gas) with another. The motion of the fluid may be entirely the result of density differences resulting from temperature difference, as in natural convection, or it may be produced by mechanical means, as in power convection. Fans commonly are used to increase the overall heat transfer coefficient of the system. A hot-air home heating system is an excellent example of heating by convection.

Tempering of steel is a common application of convection heating in heat treating. A typical installation is shown in Fig. 2. This furnace is heated by recirculating, forced air convection. The electric heating elements are located apart from the work chamber. The air in the furnace is forced through the heating chamber at high velocity and then through the work chamber. With this system, accurate control of temperature is easily achieved. Also, for any heat treating application (tempering or other), this method of heating is highly efficient up to about 480 °C (900 °F). It is used for processing at somewhat higher temperatures, although efficiency of convection heating decreases as the temperature is increased beyond approximately 480 °C (900 °F). Most ovens, as defined by IHEA, are heated by the convection mode.

Radiation. A body emits radiant energy in all directions by means of electromagnetic waves, the wavelength ranging from 4 to 7 µm. When this energy strikes another body, some of the energy is absorbed, raising the level of molecular activity and producing heat. Some of the energy is reflected. The amount absorbed depends on the emissivity of the surface of the receiver. The sender gives up heat or energy. On this basis, if two

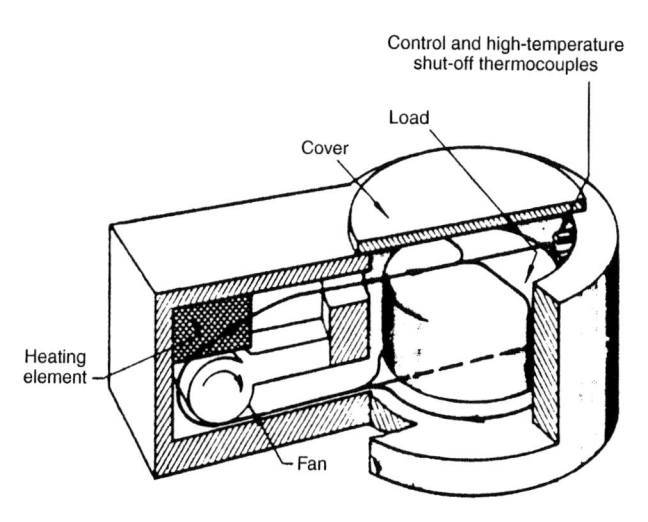

Fig. 2 Low-temperature heat processing furnace heated by convection. Source: Ref 3

pieces of metal, one hot and one cold, are placed in a completely insulated enclosure, the hot piece cools and the cold one is heated. The exchange of energy takes place until both objects come to equilibrium or to the same temperature level. Even after equilibrium of temperature is established, the process continues with each piece radiating and absorbing energy from each other.

Transfer of heat by radiation, therefore, relates directly to emissivity, which is the ratio of loss of heat per unit area of a surface at a given temperature to the rate of heat loss per unit area of a black body at the same temperature in the same surroundings. The practical meaning is that when a workpiece is placed in a furnace and exposed to radiant heat, its rate of heating depends on its surface. A highly reflective object (polished stainless steel, for example) absorbs heat at a lower rate compared with a dark workpiece.

As a rule, in practical heat treating, no attempt is made to alter the workpiece surfaces to make them more receptive to radiant energy, although workpieces have been intentionally blackened to improve heating efficiency.

Most heat treating furnaces that operate at temperatures higher than approximately 595 °C (1100 °F) are heated primarily by radiation. This is generally true regardless of their size, or whether they are heated by electrical resistance elements, directly by means of radiation from burners and the furnace walls, or indirectly by tube-contained burners (radiant tubes). An older model of a radiation-heated heat treating furnace is the simple box-type batch furnace shown in Fig. 3. Electrical resistance elements can be seen on the side walls of the furnace. Circulating fans may be added to larger furnaces to provide a convection heating assist and/or to provide for atmosphere circulation

Classification of Furnaces by Heat Transfer Medium

One means of classifying heat treating furnaces is by the type of heat transfer medium employed; this classification method is valid regardless of size and most common variables of furnace components. The types of heat-transfer media are gaseous (air or vacuum), liquid (molten metal or molten salt bath), or solid (as with fluidized bed furnaces). Those classifications, for the most part, are used in the remainder of this chapter.

Many types and designs of heat treating furnaces, regardless of the mode of heat transfer or the medium employed for heat transfer, are available as standard models. When an existing predesigned and/or prebuilt model is suitable for specific customer requirements, the cost is naturally lower, because the engineering has been completed.

In many instances, standard models may require minor factory modifications to meet specific customer requirements. This increases cost but seldom equals the cost of designing and building a custom furnace. However, hundreds of heat treating furnaces exist that have been designed and constructed for a specific application, notably, the larger furnaces (Fig. 1).

Batch-Type versus Continuous-Type Furnaces

Batch-Type Furnaces. In some plants, heat treating furnaces are classified as batch or continuous types. A batch-type furnace refers to one that is loaded with a charge and then closed for the preestablished heating cycle. After completion of the heating cycle, the workload may be cooled in the furnace at a planned cooling rate (such as for annealing), removed, and cooled in still air (as in normalizing), or quickly cooled (quenching) as by immersion in oil or water. Figures 1, 2, 3, and 4 illustrate specific types of batch-type furnaces.

Car-bottom furnaces incorporate a railcar for the hearth of the furnace. This hearth must be well insulated to keep the heat from reaching the

Fig. 3 Small batch-type furnace. Note steel frame, insulating brick, electrical resistors, and rolltop table in front of furnace for handling work. Source: Ref 3

wheel roller surfaces. The furnace usually is built at floor level and frequently is equipped with a lift door. The work to be heated is placed on the hearth, the car moved into the furnace on rails, and the door closed. Troughs filled with sand usually provide a seal between the lower edges of the furnace and the car bottom. Another variation of this design is the elevator car-bottom furnace, where the furnace shell may be lifted while the car bottom is being positioned. The furnace shell is then lowered and rests on the car hearth during the heating period. The problem of proper sealing is greatly simplified with this construction. The primary disadvantage of the car-bottom furnace is that the furnace cannot be heated with the car out of position because the latter forms the insulated floor of the furnace. Such furnaces are most commonly used for very large workpieces that require stress relieving, annealing, or normalizing.

Another batch-type furnace is shown in Fig. 2. This unit is heated by convection and frequently is used for tempering, stress relieving, or process (subcritical) annealing. As indicated in Fig. 2, this type of furnace is a top loader.

Regardless of the type of energy employed for heating, the heating mode is principally radiation. Such furnaces are generally available in a broad range of sizes and can be heated with gas, oil, or electricity. The furnace shown in Fig. 3 is heated by electrical heating elements (metallic), which allows efficient operation up to 980 °C (1800 °F) or slightly higher.

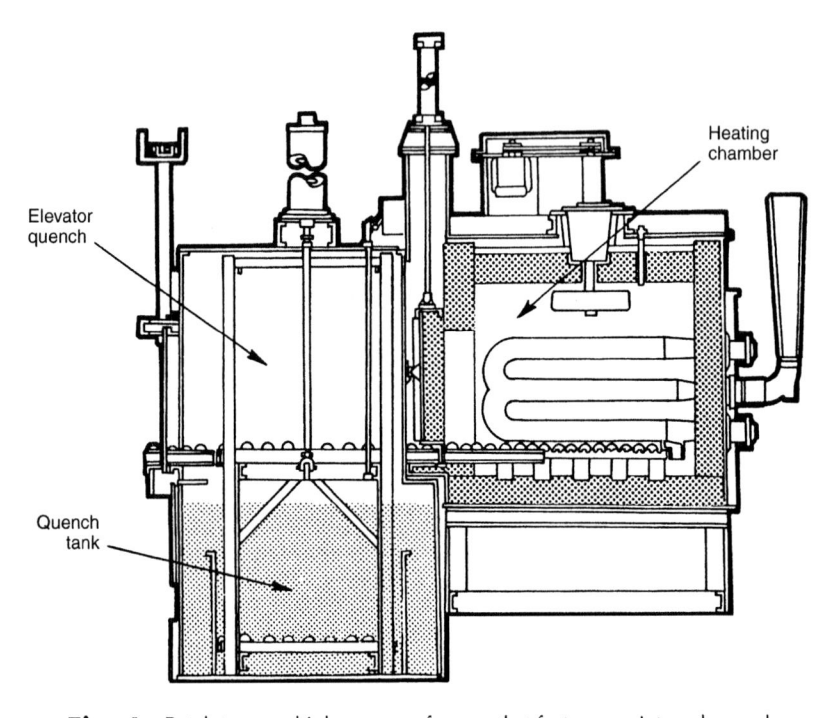

Fig. 4. Batch-type multiple-purpose furnace that features an integral quenching system. Source: Ref 4

However, by use of silicon carbide heating elements, the operating temperature can be extended to 1260 °C (2300 °F) or slightly higher, thus permitting use for heating highly alloyed tool steels (see Chapter 11, "Heat Treating of Tool Steels"). Therefore, such a furnace is quite versatile and can be used for almost any heat treating operation. However, it generally is not recommended for heating at temperatures lower than approximately 650 °C (1200 °F), unless equipped with one or more circulating fans. This type of furnace can be operated with a "natural" atmosphere or with any one of several prepared atmospheres as required.

Another example of a popular, standard-design batch type furnace is shown in Fig. 4. This batch furnace can be adapted to almost any heat treating operation. The furnace is a front loader with an integral (part of the furnace) quenching system and a top fan for a convection heating assist and atmosphere circulation. The parts are generally heated by gas-fired radiant tubes. Furnaces of this type are amenable to use with a variety of prepared atmospheres.

Bell-type furnaces (not illustrated) are also widely used as batch-type units. In bell-type furnaces, the round hearth or base that supports the load is stationary at floor level, while the furnace can be lifted off and transferred from one base to another by an overhead crane. After the work is placed on the base, the furnace is lowered and properly positioned on the base by guideposts. Sealing is effected by sand or oil in a circular trough around the outside of the base. Often an inner metal retort or muffle with a skirt at the bottom to seal in a protective atmosphere is first used to cover the work before the furnace is lowered into place. This type of furnace is widely used with an inner muffle and protective atmosphere for annealing material in coil form, such as steel sheet or wire, and nonferrous products.

Continuous furnaces may be any of many designs including their conveying mechanisms, but basically they portray an "in-one-end" and "out-the-other-end" type of unit. A continuous furnace is generally intended for continuous high production of similar parts or for parts requiring similar process cycles. The capabilities that can be designed into a continuous furnace are virtually limitless in terms of varied heat treating cycles.

Roller-Hearth Continuous Furnaces. The charging end of a large continuous furnace is illustrated in Fig. 5. As shown in Fig. 5, the work is conveyed through the furnace by means of rollers. The ends of the rolls project through the walls of the furnace to external air- or water-cooled bearings. Usually, the rolls are power driven by a common source through a chain and sprocket mechanism. Frequently, the driven rolls extend to some distance beyond the furnace at both the loading and discharging end.

The work is placed on the revolving rollers at the charge end and is carried through and out of the furnace by the friction between the work surface and the revolving roll surfaces. This obviously works only if the charge material is sufficiently long relative to the roll spacings. Work of smaller size can be stacked on trays, loaded into baskets, or hung on special fixtures carried by the rollers. Figure 5 shows the entrance end of a roller hearth furnace fired with radiant-tube burners. The entrance vestibule is equipped with a labyrinth of vertical ceramic fiber curtains to help confine the internal protective atmosphere. A similar vestibule is provided at the discharge end when work is cooled inside the furnace.

Belt-Type Continuous Furnaces. In this construction, a continuous conveyor belt acts as a moving hearth to carry the work through the furnace. Several belt designs are used, depending on the size and weight of the work to be handled, how the hot belt is supported inside the furnace, and the operating temperature. Belts may be constructed of woven wire mesh, flat cast alloy links, or a more open slat design. They may have a relatively smooth, even surface, or may have surfaces that are recessed, grooved, or channeled to control spacing between individual parts of the load. The conveyor usually is driven by large diameter drums at each end of the furnace. The belt may return either outside or inside the furnace, depending on the function the furnace is performing. This type of furnace is suited to continuous annealing, carburizing, carbonitriding, tempering, sintering, and hardening operations. If desired, the heated workpieces may be dumped directly into an enclosed quenching tank without losing atmosphere protection.

Pusher-type continuous furnaces are usually designed to carry higher unit loads than belt-type furnaces. The work is placed on the hearth of the

Fig. 5 Roller hearth furnace showing charging end. Rolls leading into the furnace for transporting work are similar to those throughout the length of the furnace. Source: Ref 3

furnace and pushed ahead periodically by a mechanical ram operating at the charging end. Work is generally loaded onto sturdy cast alloy trays, baskets, or other fixtures that ride on skid rails, tracks, or rollers built into the floor of the furnace. The pusher can be actuated by an air or hydraulic cylinder, or a rack and pinion mechanism driven by an electric motor. The action of the ram and opening and closing of interior and/or exterior furnace doors must be properly coordinated by an interlocking safety system.

Rotary-hearth continuous furnaces are exception to the "straight through" type of continuous furnace, as illustrated in Fig. 6. In this type of furnace, the hearth is a flat ring, similar to the revolving floor of a merry-go-round. The furnace in Fig. 6 is heated with gas-fired radiant tubes, but such a furnace can be heated electrically. As also indicated in Fig. 6, the work is charged into and removed through a single opening. The length of the heat cycle is governed by establishing the speed of hearth rotation. While there are no fixed limitations on applications of rotary hearth furnaces, they usually are used for heating workpieces weighing over 2.2 kg (5 lb).

Liquid Bath Furnaces

Heating by immersing the workpieces in a liquid represents an entirely different concept compared with gaseous atmosphere furnaces. Heating parts in molten metal (usually molten lead) is an age-old practice. A pot-type furnace containing molten lead provides an effective means of heating steel parts, but there are certain disadvantages:

- Molten lead is heavy, consequently, steel parts float if not anchored.
- Lead adheres to steel parts, which impairs the quenching action and poses a cleaning problem.

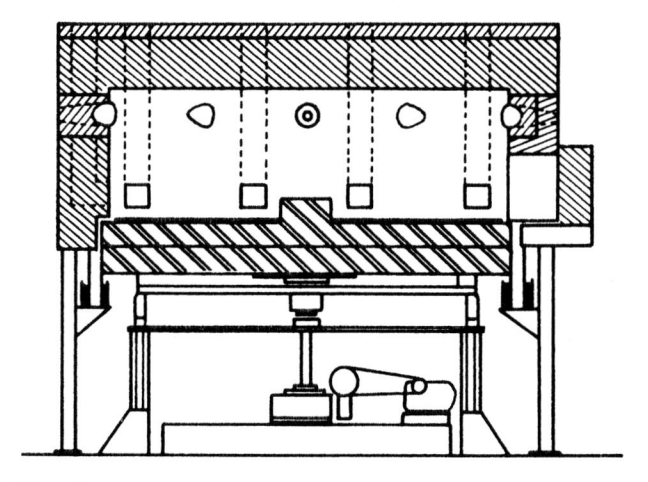

Fig. 6 Rotary hearth forging or heat treating furnace. Source: Ref 3

Very few lead-pot furnaces are still being used, and for the most part, they have been replaced by molten salt bath furnaces.

Molten salt baths offer several distinct advantages: (a) salts are available for operation in the temperature range of 175 to 1260 °C (350 to 2300 °F); (b) parts do not scale or otherwise result in deteriorated surfaces because they are fully protected while they are in the molten bath; (c) a thin film of salt remains on the work during transfer from heat to quench so that clean hardening is facilitated; and (d) a wide variety of salts are available, including salts that change the surface chemistry of the steel (see Chapter 8, "Case Hardening of Steel").

The principal disadvantage of heating parts in molten salt is the necessary cleaning after heat treating, which can be difficult, especially for parts of complex design.

Types of Salt Bath Equipment. The simplest form of molten salt bath involves heating a metal pot filled with a low-melting-temperature salt with electric immersion heaters. This type of salt bath, however, is applicable only to temperatures from approximately 175 to 345 °C (350 to 650 °F).

As temperature requirements increase, more sophisticated equipment is required. A common type of fuel-fired pot furnace that can be used for either molten metal (usually lead) or molten salt is shown in Fig. 7(a). This type of furnace can be used for heating metals up to about 900 °C (1650 °F). Higher temperatures can be achieved, but deterioration of the pot and other components becomes excessive.

The furnace shown in Fig. 7(b) is similar to Fig. 7(a), except it is heated by electrical resistance. Because the pot in this type of furnace is constantly surrounded with a strongly oxidizing atmosphere, pot life is shortened when used in the higher temperature ranges.

Both of the aforementioned types of furnaces are extremely versatile, but they are best suited to limited production of a variety of small parts. To attain acceptable pot life, the pots must be made from an expensive nickel-chromium alloy material.

Most liquid bath heat treating requiring the temperature range of about 760 to 1260 °C (1400 to 2300 °F) is done in furnaces such as those illustrated in Fig. 7(c) and (d), immersed and submerged electrode types, respectively. In both types, heat is generated by resistance to current flow through the molten salt from one electrode to the other. This creates a stirring action at the electrodes, thus providing uniform temperatures within the bath.

The two types (immersed and submerged electrode) generally are competitive with each other, each having certain advantages and disadvantages. Either type lends itself to batch or continuous operation.

A principal disadvantage of the immersed electrode type (over-the-top electrodes shown in Fig. 7c) is that the electrodes deteriorate just above the salt line where the heat is intense. The submerged electrode type overcomes this disadvantage; however, the submerged type furnace requires a molded-in ceramic type of pot, and ceramic pots are not compatible with all salt compositions.

Salt bath furnaces heat by radiation and conduction. In the immersed and submerged electrode types, convection heating is also added because of the stirring action. Therefore, the heating rate in any salt bath is much greater than in the gaseous atmosphere furnaces discussed earlier in this chapter.

Molten salt baths do, therefore, offer an efficient means of heating metals—principally steels—although some salt compositions are compatible with nonferrous metals and alloys. The degree of economy in heating is particularly realized with the immersed and submerged electrode types, if they can be properly applied. Both of these furnace types are best adapted to continuous production because of the difficulty in restarting them when salts are allowed to solidify. Thus, such furnaces are not well suited to intermittent production. For weekends or other required downtime, they are idled at a temperature just above the solidification temperature of the salt.

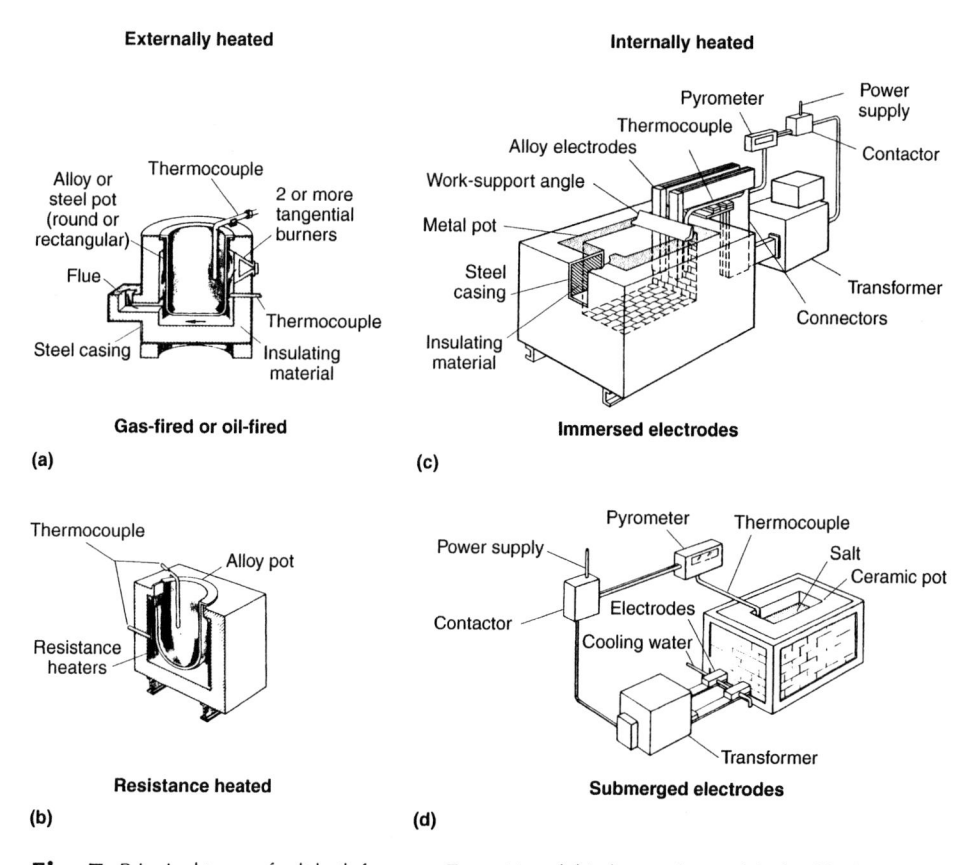

Fig. 7 Principal types of salt bath furnaces. Types (a) and (b) also can be used for lead baths. Source: Ref 5

Fluidized Bed Furnaces (Ref 6)

Fluidized bed heating is carried out in a bed of mobile inert particles, usually aluminum oxide. These particles are suspended by the combustion of a fuel/air mixture flowing upward through the bed. Components are immersed into the fluidized bed as if it were a liquid and are heated by the hot fluid bed. Heat transfer rates in a fluidized bed are up to ten times that which can be achieved in conventional direct-fired furnaces. The combination of high heat transfer, excellent heat capacity, and uniformity of behavior over a wide temperature range provides a constant temperature bath for many applications. The advent of fuel-fired fluidized beds using a gas/air mixture as both the heating and fluidization medium has resulted in fluidized bed furnaces being available to perform most standard heat treating operations.

A fluidized bed is held in a metallic or refractory container and is started by initially lighting the combustion mixture at the top. The flame-front gradually moves down the depth of the bed until it stabilizes above the ceramic plate. The combustion occurs spontaneously within about 25 mm (1 in.) of the plate surface. The ceramic distribution plate ensures uniform properties within the bed.

There are two types of fluidized beds available: internally fired for high-temperature applications (760 to 1215 °C (1400 to 2220 °F), and externally fired for low-temperature applications (760 °C, or 1400 °F, and below).

In the internally fired bed (Fig. 8), the fuel and air are mixed in near stoichiometric proportions and passed through a porous ceramic plate

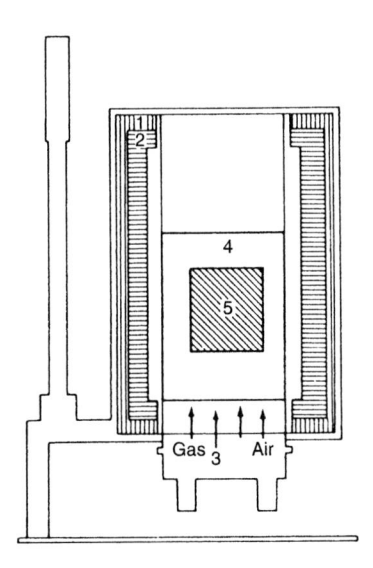

Fig. 8 Gas-fired fluidized-bed furnace with internal combustion. 1, insulating lagging; 2, refractory material; 3, air and gas distribution box; 4, fluidized bed; 5, parts to be treated. Source: Ref 6

above which the particles are fluidized in the gas stream. The gas stream imparts its thermal energy to the bed particles, which in turn impart thermal energy to the object being processed.

For processing temperatures below 760 °C (1400 °F), the externally fired bed is used. Such a bed is illustrated in Fig. 9. In this system, an excess air burner fires into a plenum chamber, above which the fluidized bed is supported by a porous metallic plate. The bed is fluidized by the products of combustion from the plenum chamber.

The temperature of the bed is automatically controlled by a proportioning controller linked to a motorized valve that meters the appropriate amount of gas/air mixture to the distribution tile. The control is arranged so that its range is well above the minimum fluidization velocity of the particles. The fluidization gas is also the furnace atmosphere in which the metal is treated. By regulating the air/gas ratio to the distributor tile, the atmosphere may be inert, oxidizing, or reducing.

Applications for fluidized beds are numerous. The most obvious in heat treating are neutral hardening, where they may be used in place of neutral salt baths because the rate of heat transfer is even faster than in any salt bath. Additionally, particles from the fluidized bed do not adhere to workpieces so that there is no cleaning problem. There is no dragout of particles from a fluidized bed, compared with the constant dragout (and need for replenishment) of salt from a molten salt bath furnace. Therefore, this can be a significant cost factor in favor of the fluidized bed. Actually, fluidized beds can be adapted to virtually all heat treating operations, which include hardening of steel (even high-speed tool steels), normalizing, annealing, stress relieving, carburizing, carbonitriding, and tempering. Fluidized beds are also well suited to heat treating of nonferrous metals.

Vacuum Furnaces

Heating of metal parts in a vacuum furnace consists of carrying out various thermal operations in a heated chamber evacuated to a vacuum pressure suitable to the particular material and process desired. Although originally developed for the processing of electron-tube and space-age materials, it has been found to be extremely useful in many less-exotic metallurgical areas as vacuum technology has progressed. Vacuum heat treating can be used to:

- Prevent reactions at the surface of the work, such as oxidation or decarburization, thus retaining a clean surface intact
- Remove surface contaminants such as oxide films and residual traces of lubricants resulting from fabricating operations. The latter often are severe contaminants to the furnace

- Add a substance to the surface layers of the work, such as by carburization
- Remove dissolved contaminating substances from metals, using the degassing effect of a vacuum, such as hydrogen or oxygen from titanium
- Join metals by brazing or diffusion bonding

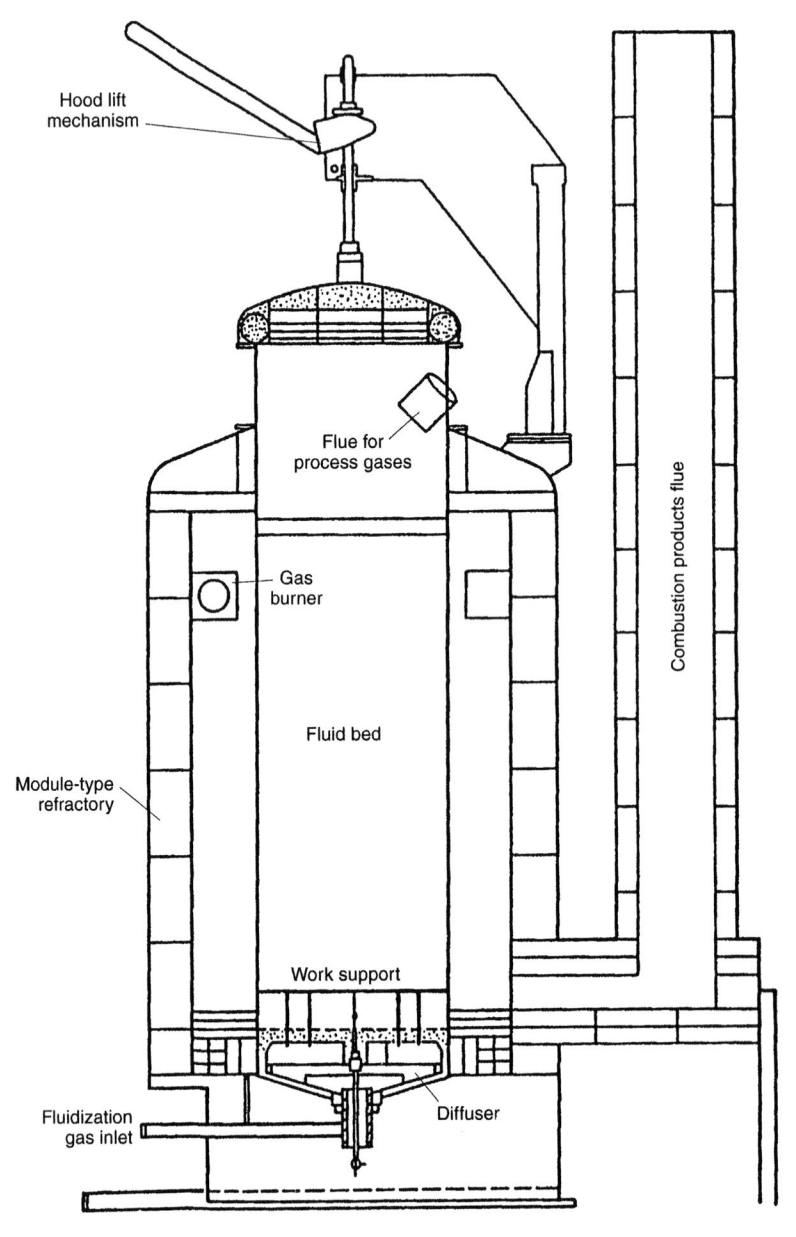

Fig. 9 Externally gas-fired fluidized-bed furnace. Source: Ref 6

Operations performed in vacuum furnaces include hardening, tempering, carburizing, annealing, stress relieving, and sintering. In vacuum furnaces, a complete vacuum (absolutely none of the original air) is virtually impossible to attain. One standard atmosphere at sea level equals 760 mm (30 in.) of mercury. The degree of vacuum used for most heat treating operations is about $1/760$ of an atmosphere. Under these conditions, the amount of original air remaining in the work chamber is approximately 0.1%. This degree of vacuum can normally be obtained by "pumping down" with a mechanical pump. There are, however, some heat treating operations involving highly alloyed materials that require a "harder" vacuum; that is, less than 0.1% of the original air. Under these conditions, mechanical pumping is followed by use of the highly sophisticated oil-diffusion pump.

Sectional views of a cold-wall vacuum furnace are shown in Fig. 10, which is a three-chamber furnace that includes a loading vestibule (left side of view), heating chamber (center of side view), and an elevator quenching arrangement at the extreme right side of the view. Vacuum furnaces are heated by electric resistors, frequently the graphite type shown in Fig. 10.

Another configuration of a vacuum furnace utilizing a pressure-quench is shown in cross section in Fig. 11. Depending on the gas used for quenching, this furnace could be used for hardening some smaller section sizes of AISI 4140 or 4340 steel.

The technology has also been developed for carburizing (see Chapter 8) in a partial vacuum. The cross section of a typical vacuum furnace used for carburizing is shown in Fig. 12.

Vacuum furnaces offer a number of advantages, which may include surface cleanliness and minimal part distortion. The primary disadvan-

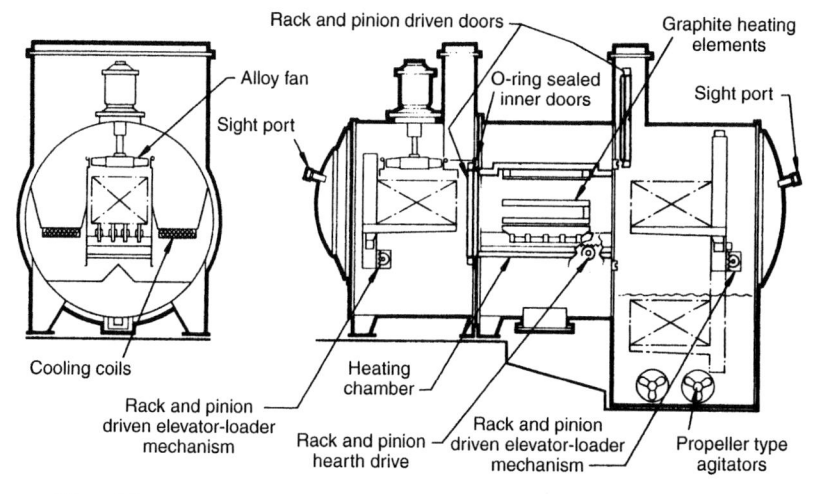

Fig. 10 Three-chamber cold-wall vacuum oil-quench furnace. Source: Ref 7

tages are high initial capital investment, the requirement for greater operator skill, and possibly higher maintenance costs.

Furnace Parts and Fixtures (Ref 10)

While generally the interior components of heat treating furnaces are, insofar as possible, constructed of refractory materials, there are many components, especially in highly sophisticated systems, that are necessarily made from metal.

The heat-resistant materials suitable for use as furnace parts or fixtures are listed in Tables 1 and 2. The selection of a cast or fabricated component for furnace parts and fixtures depends primarily on the operating conditions associated with heat treating equipment in the specific processes, and secondarily on the stresses that may be involved. The factors of temperature, loading conditions, work volume, rate of heating, and furnace cooling or quenching need to be examined for the operating and economic trade-offs. Other factors that enter into the selection include furnace and fixture design, type of furnace atmosphere, length of service life, and pattern availability or justification.

Some of the factors affecting the service life of alloy furnace parts, not necessarily in order of importance, are alloy selection, design, maintenance procedures, furnace and temperature control, atmosphere, contamination of atmosphere or workload, number of shifts operated, thermal cycle, and overloading. High-alloy parts may last from a few months to many years, depending on operating conditions. In the selection of a heat-resistant alloy for a given application, all properties should be considered in relation to the operating requirements to obtain the most economical life.

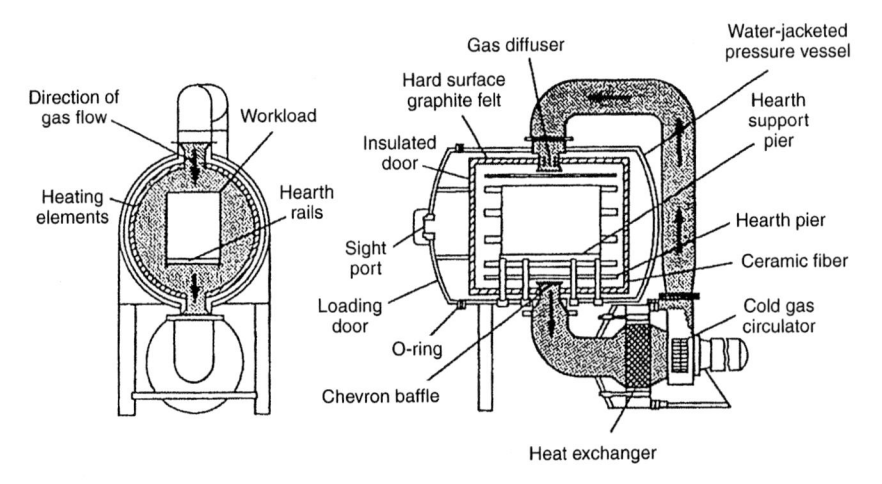

Fig. 11 Single-chamber batch-type pressure-quench vacuum furnace. Source: Ref 8

If either cast or wrought alloy fabrications can be used practically, both should be considered. Similar alloy compositions in cast or wrought form may have varying mechanical properties, different initial costs, and inherent advantages and disadvantages. Castings are more adaptable to complicated shapes, and fabrications to similar parts, but a careful comparison should be made to determine the overall costs of cast and fabricated parts. Initial costs, including pattern or tooling costs; maintenance expenses; and estimated life are among the factors to be included in such a comparison.

Trays and Grids. Many parts to be heat treated are irregular in shape and as such must be conveyed through the continuous-heat-treating furnaces or loaded and unloaded from the batch furnaces on grids or trays (Fig. 13). These trays or grids must withstand exposure to the same furnace conditions as the product: They are subjected to repeated heating and cooling, as well as repeated compression and tensile loading. Heat-resistant alloys are used extensively for these parts, although there are instances in which dispensable carbon or low-alloy-steel fabricated trays are employed.

Baskets and Fixtures. In many situations, parts being heat treated are of a size that does not permit them to be loaded directly on a furnace hearth, tray, or grid. They require some type of container, such as a basket.

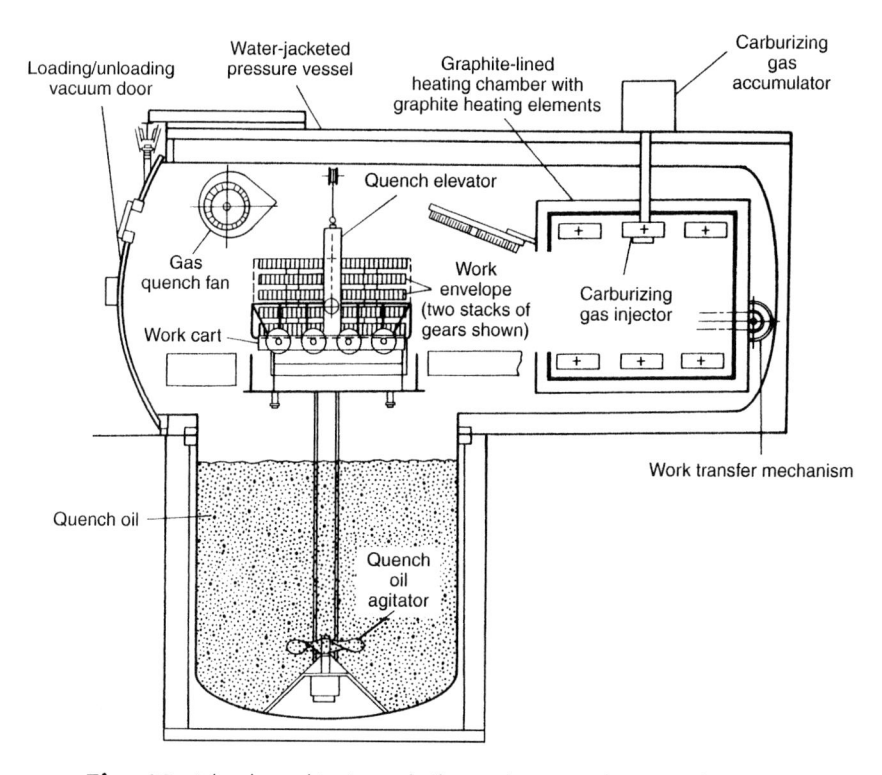

Fig. 12 A batch-graphite integral oil-quench vacuum furnace with vacuum-carburizing capability. Source: Ref 9

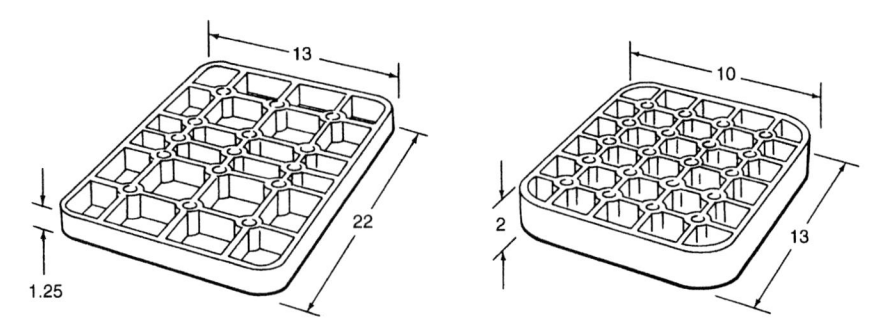

Fig. 13 Typical heat treat alloy carburizing furnace trays. Dimensions given in inches. Source: Ref 10

Table 1 Compositions of selected heat-resistant alloys

Grade	UNS No.	C	Cr	Ni	Si (max)
HF	J92603	0.20–0.40	19–23	9–2	2.00
HH	J93503	0.20–0.50	24–28	11–14	2.00
HK	J94224	0.20–0.60	24–38	18–22	2.00
HN	J94213	0.20–0.50	19–23	23–27	2.00
HT	N08605	0.35–0.75	13–17	33–37	2.50
HU	N08005	0.35–0.75	17–21	37–41	2.50
HX	N06050	0.35–0.75	15–19	64–68	2.50

(a) Balance Fe in all compositions

Table 2 Compositions of selected heat-resistant alloys

Grade	UNS No.	Fe	C	Cr	Ni	Co	Mo	W	Other
309S	S30908	Bal	0.08 max	22–24	12–15	...	...	...	...
310S	S31008	Bal	0.08 max	24–26	19–22	...	...	...	...
RA 85H	S30615	Bal	0.2	28.5	14.5	...	...	...	3.6 Si, 1.0 Al
RA 330	N08330	Bal	0.08 max	17–20	34–37	...	...	...	...
RA 330 HC	...	Bal	0.4 max	17–22	34–37	...	...	...	...
RA 333	N06333	Bal	0.08 max	24–27	44–47	3	3	3	...
253 MA	S30815	Bal	0.08 max	21	11	...	...	...	1.7 Si, 0.17 N, 0.04 Ce
Fecralloy A	...	Bal	0.03	15.8	...	...	...	...	4.8 Al, 0.3 Y
HR-120	N08120	Bal	0.05	25	37	...	...	...	0.7 Nb, 0.2 N
HR-160	N12160	2	0.05	28	Bal	29	...	...	2.75 Si
556	R30556	Bal	0.1	22	20	18	3	2.5	0.6 Ta, 0.2 N, 0.02 La
214	N07214	3	0.05	16	Bal	...	...	...	4.5 Al, 0.002–0.40 Y
230	N06230	...	0.1	22	Bal	...	2	14	0.005 B, 0.02 La
Incoloy 800	N08800	Bal	0.1	19–23	30–35	...	...	...	0.15–0.60 Al, 0.15–0.60 Ti
Incoloy 802	N08802	Bal	0.2–0.5	19–23	30–35	...	...	...	...
Incoloy 600	N06600	6–10	0.15 max	14–17	72 min	...	...	...	...
Incoloy 601	N06601	Bal	0.10 max	21–25	58–63	...	...	...	1.0–1.7 Al
Incoloy 617	N06617	1.5	0.07	22	Bal	12.5	9	...	1.2 Al
Incoloy MA 956	S67956	Bal	...	20	...	...	...	...	0.5 Y$_2$O$_3$, 4.5 Al, 0.5 Ti

The design of these baskets varies because each product is developed for a specific application and loading and must function with a specific type of furnace equipment.

Baskets and fixtures can be produced from cast or wrought alloys. Fabricated parts are used in light-to-medium loading applications, intricate designs, complex shapes, and generally with lighter metal sections.

Baskets (Fig. 14) and fixtures (Fig. 15) are used successfully for the reduction of part distortion during heat treatment. For some smaller parts, random loading in the baskets may be acceptable, but other parts may require stacking with a preferred orientation in the basket. In either case, the loaded baskets are then stacked together to make up the furnace load. In the case of parts with shafts, they may be oriented vertically by hanging or positioning the parts so they remain in the same position for the complete heat treat process. Proper use of baskets and fixtures can have a

Fig. 14 Bar frame type basket. Source: Ref 3

Fig. 15 Tray/fixture for carburizing ring gears. Courtesy of Castalloy Corporation

significant impact in reducing the required subsequent operations (operational costs) after heat treatment.

Suppliers of alloy baskets and fixtures—cast, wrought, or welded structures—have an array of standard designs that can be used and also are the best source for assistance in materials application and design for the parts being processed.

Quenching Mediums and Systems (Ref 11)

Quenching is the rapid cooling of a metal or an alloy from a suitable elevated temperature. This usually is accomplished by immersion in water, oil, or a polymer, although forced or still air is sometimes used and is considered quenching.

There are two major applications of quenching. One is cooling of quench-hardenable steels to develop an acceptable as-quenched microstructure and mechanical properties that meet minimum specifications after the parts are tempered. The second is the rapid cooling of certain iron-base alloys and nonferrous metals from elevated temperatures to retain a uniform solid solution in the metal. Such treatments are used to obtain a fully annealed condition to permit forming or to permit precipitation hardening by a subsequent aging process. The major portion of this discussion covers quench hardening of steels.

Metallurgical Aspects. Most steels are quenched to control the transformation of austenite and to subsequently form the desired microconstituents. Microstructures that may be obtained are indicated on the combination time-temperature-transformation (TTT) diagram.

Martensite is the as-quenched microstructure that is usually desired, as indicated by curve A in Fig. 16. Along cooling curves B, C, and D, some

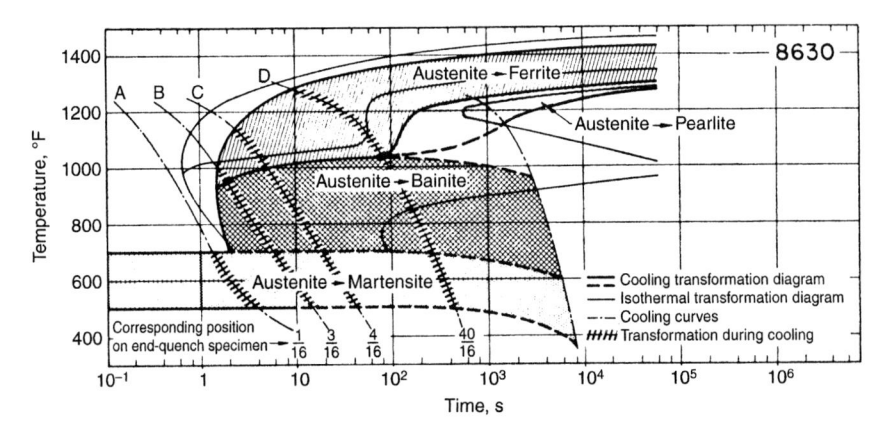

Fig. 16 Transformation diagrams and cooling curves for AISI 8630 steel, indicating the transformation of austenite to other constituents as a function of cooling rate. Source: Ref 3

transformation to bainite and ferrite occurs, with a corresponding decrease in the amount of martensite formed and the hardness developed. Formation of these mixed structures commonly is referred to as "slack quenching." See Chapter 2, "Fundamentals of the Heat Treating of Steel," for an in-depth discussion of TTT curves.

Cooling Rates. When carbon steel is quenched from the austenitizing temperature, a cooling rate equal to or greater than about 55 °C/s (100 °F/s) (measured at 705 °C, or 1300 °F) is necessary for avoiding the nose of the TTT curve. The actual cooling rate required depends on the hardenability of the steel. The entire cross section of the workpiece must cool at this rate to obtain the maximum amount of martensite. Under ideal conditions, water provides a cooling rate of about 280 °C/s (500 °F/s) at the surface of steel cylinders 13 mm (0.5 in.) in diameter by 100 mm (4 in.) long. This rate decreases rapidly below the surface. Thus for carbon steel, only thin sections with a high ratio of surface area to volume can be fully hardened throughout the cross section.

Quenching Mediums

Many different mediums have been used for quenching. Most of them are included in the list that follows, and some of these are used only to a very limited extent:

- Water
- Brine solutions (aqueous)
- Caustic solutions
- Polymer solutions (synthetics)
- Oils
- Molten salts
- Molten metals
- Gases, including still or moving
- Fog quenching
- Dry dies, commonly water cooled

In cooling power, nonagitated water is arbitrarily rated as (1.0); other mediums are rated by comparison with this value. The relative cooling powers of the most common quenching mediums (water, oil, and brine or caustic solutions) are presented in Table 3, which also shows the effect of agitation on cooling power. With agitation, the cooling power of any medium is greatly increased. Cooling powers of synthetic quenching mediums (polymers) are not shown in Table 3 because they can vary from the relatively low power of oil to the high quenching power of brine. Synthetic materials are available as proprietary materials for mixing with water. Synthetic materials are somewhat difficult to use because quench bath composition is easily degenerated through dragout (the materials adhere to the quenched workpieces).

Table 3 shows that brine or caustic (usually 5 to 10% in water) is an extremely severe quenching medium. When strongly agitated, it has a quenching power of 5 and may, therefore, be preferred over water. A disadvantage of brine is its corrosivity to the equipment. Caustic is not usually corrosive, but can endanger personnel if not handled properly.

Water is inexpensive and is a very effective quenching medium. However, it is susceptible to formation of steam pockets that may result in soft spots on the quenched workpieces. This is particularly likely if the temperature of the quenching system is not closely controlled. In a water quenching system, the water should never be allowed to exceed 32 °C (90 °F).

Oil generally has a quenching power (depending on which quench oil) of 25 to 40% of that of plain water, assuming similar agitation. Oil is less temperature sensitive than water and, depending on the oil, the temperature range for effective quenching is 32 to 70 °C (90 to 160 °F). Often the ideal temperature is 60 to 65 °C (140 to 150 °F) due to poorer mobility at the lower temperatures.

Other quenching mediums listed generally are restricted to use for specialized applications. Air cooling (quenching) often is used for very high-hardenability steels.

Quenching Systems

Equipment requirements for quenching may vary over a wide range. For instance, in a small shop where only a few small parts made from carbon steel are heat treated each day, the equipment might be as simple as a steel barrel containing water or brine. No heating or cooling facilities are necessary, and the agitation is accomplished by the operator moving the workpiece about in the medium during cooling. Such a simple system may be adequate.

At the other extreme, for continous operation, the quenching system may contain all or most of the following components regardless of the quenching medium used:

Table 3 Effect of agitation on the effectiveness of quenching

Circulation or agitation	H-value or quenching power		
	Oil	Water	Caustic soda or brine
None	0.25–0.30	0.9–1.0	2
Mild	0.30–0.35	1.0–1.1	2–2.2
Moderate	0.35–0.40	1.2–1.3	...
Good	0.4–0.5	1.4–1.5	...
Strong	0.5–0.8	1.6–2.0	...
Violent	0.8–1.1	4	5

Source: Ref 3

- Tanks or quenching machines
- Agitation equipment
- Fixtures for quenching
- Cooling systems
- Heaters
- Pumps

A simple, but effective, tank system for water quenching is shown in Fig. 17. In this system, a supply of fresh water is continuously fed into the bath and allowed to overflow to the drain in order to maintain the water temperature. The system is intended for use where water is plentiful and inexpensive. Where water is scarce, however, this system may be connected to a central or closed water-cooling system that would cool the water for reuse.

The ultimate in a sophisticated integral, oil-quenching system is that shown in Fig. 18. The oil is heated by a submerged, gas-fired, radiant tube or electrically with immersion heaters. An air/oil heat exchanger would provide the necessary oil cooling. The use of "water type" heat exchangers is not recommended due to the fire or explosion hazard should the oil become contaminated with water.

All of the components listed previously are included in the system design. Such installations are common where production rates are high and demands for quenching are essentially continuous.

Furnace Atmospheres

In many heat treating operations, austenitizing must be performed under conditions that provide some form of surface protection for the workpieces. If this is not done, workpieces may become severely oxidized (have severe scale buildup) and/or become decarburized. As steel is heated, the surfaces become more active chemically as the temperature

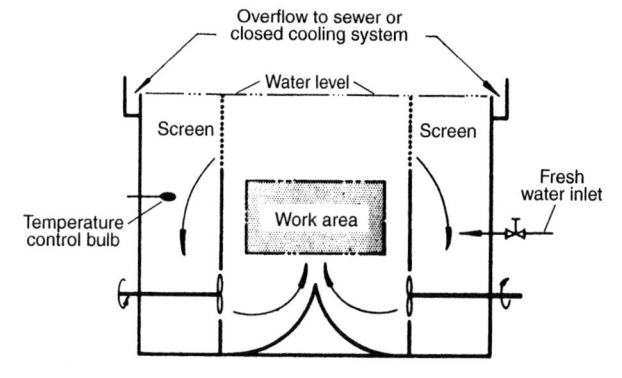

Fig. 17 Temperature-controlled overflow tank for water quenching. Source: Ref 3

increases. Severe oxidation of carbon steels commonly begins at about 425 °C (800 °F). At temperatures exceeding 650 °C (1200 °F), the rate of oxidation becomes exponential with an increase in temperature. This higher temperature range is where parts can become decarburized as well as oxidized.

Decarburization, by definition, is the loss of base carbon from the surface of the steel. This condition is the worst at the surface where the carbon can, depending on the atmosphere, time, and temperature conditions, be reduced to 0 02% carbon (ferrite). Surface oxidation (scaling) can be removed after heat treatment by cleaning using an abrasive media, and although there is a slight loss in part size, this generally is not a problem for most forgings, castings, and weldments. Decarburization, however, is a problem caused by carbon diffusion, which, after it is created must be removed by machining or the lost carbon can be replaced by a carbon restoration process. Decarburization is a severe problem for the heat treater because it can either cause or be a contributing factor to "cracking" during the steel-hardening process

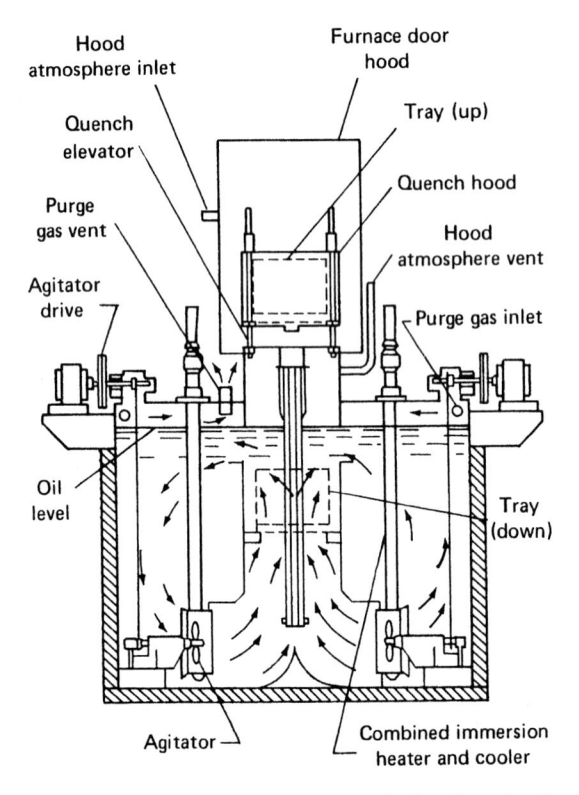

Fig. 18 Schematic of a typical installation for high-volume batch quenching of carburized or hardened parts on trays. Directional vanes in the oil stream distribute the oil flow uniformly. Unit contains combined heating and cooling elements and provision for blanketing the surface of the oil with an inert gas atmosphere. Radiant tube is used for heating. Source: Ref 12

Generally, furnace atmospheres serve two requirements: they provide heat treated workpieces whose surfaces are clean and essentially unchanged from their original conditions, and they serve to attain a controlled condition of surface change, as in certain case hardening operations. When workpieces are heated in salt baths or fluidized beds, atmospheres are automatically provided because workpieces may also be accomplished in molten salts or fluidized beds.

Principal types of gaseous atmospheres that may be used in atmosphere furnaces are listed here in the general order of increasing cost:

- Natural (conventional air)
- Atmosphere derived from products of combustion in a direct fuel-fired furnace
- Exothermic (generated)
- Endothermic (generated)
- Nitrogen base
- Vacuum
- Dissociated ammonia
- Dry hydrogen (dried bottled gas)
- Argon (from bottles)

Discussion of furnace atmospheres can be extensive and complex. Therefore, at the risk of some oversimplification, an attempt is made subsequently to briefly describe the types of atmospheres available, emphasizing applications and limitations. For more detailed information on furnace atmospheres, see Ref 13 and 14.

Natural Atmospheres (Air). A natural atmosphere is the air we breathe, which is essentially composed of 79% nitrogen and 20% oxygen. Such atmospheres exist in any heat treating furnace where the work chamber does not contain products of combustion, or where specially prepared atmospheres are not admitted (see Fig. 3). In many heat treating applications, this type of atmosphere is acceptable, and it may be satisfactory when workpieces are to be machined after heat treating because natural atmospheres are strongly oxidizing. The heavily oxidized surfaces created by use of this type of atmosphere inhibit cooling rate during quenching.

Products of combustion in direct-fired furnaces automatically provide some atmosphere protection compared with exposure to air. When fuels are mixed with air and burned at the ideal ratio, a condition results wherein minimal reaction with the steel surfaces occurs. This ideal ratio varies for different fuels; for example, in burning natural gas (methane), this ratio is approximately 10 parts air to 1 part gas; whereas for propane, the ideal ratio is 23 parts air to 1 part gas. When an excess of air exists in the mixture, loose scale forms. When the mixture contains an excess of fuel, a tight adherent oxide is formed on the workpiece surfaces. In all instances, a certain amount of water vapor develops in the mixture, which causes decarburization in higher-carbon steels. Although the atmosphere protection derived by control of combustion is by no means a "perfect atmosphere," it is inexpensive and is sufficient for many applications.

Exothermic atmospheres are widely used prepared atmospheres due to their low cost. There are several modifications of this general type of atmosphere, but rich exothermic gas is produced by combustion of a hydrocarbon fuel such as natural gas or propane with the air/fuel ratio closely controlled. This air/gas mixture is burned in a confined combustion space to maintain a reaction temperature of at least 980 °C (1800 °F) for sufficient time to permit the combustion reaction to reach equilibrium. Heat is obtained directly from combustion, hence the term exothermic. The resultant gas is then cooled to remove part of the water vapor formed by burning and to permit convenient transportation and metering. In this process, the simplified theoretical reaction of methane with air is:

$$CH_4 + 1.25O_2 + 4.75N_2 \rightarrow 0.375CO_2 + 0.625CO + 0.88H_2 + 4.75N_2 + 1.12H_2O + heat$$

where 1 volume of fuel and 6 volumes of air yield 6.63 volumes of product gas mixture, with water vapor removed. In practice, exothermic gas generators are seldom operated with an air/gas ratio lower than about 6.6 to 1 to prevent sooting.

Exothermic atmospheres serve a variety of applications, including clean (even bright) annealing and clean hardening of steel, as well as some nonferrous metals. They are not well suited to processing high-carbon steels, unless the workpieces are subjected to stock removal in finishing, because exothermic atmospheres decarburize high-carbon steels.

Endothermic atmospheres are more costly than are exothermic, but they are more flexible than the exothermic types. Endothermic atmospheres are produced in generators that use air and a hydrocarbon gas as the input gas. These two gasses are mixed in a controlled ratio, slightly compressed, and then passed through a nickel-bearing ceramic catalyst. The reaction chamber is heated to approximately 1040 °C (1900 °F). The gases react in the chamber to form a gas consisting of approximately 40% nitrogen, 40% hydrogen, and 20% carbon monoxide, with trace amounts of carbon dioxide and water vapor. The gas must then be immediately cooled to room temperature to prevent the formation of "soot" (carbon) in the gas. Figure 19 is a schematic diagram of an endothermic gas generator.

Endothermic atmospheres can be used in virtually all furnace processes that operate above 760 °C (1400 °F). The most common use is as carrier gases in gas carburizing and carbonitriding applications. Because of the wide range of possible carbon equivalencies, however, endothermic atmospheres also are used for bright hardening of steel, for carbon restoration in forgings and bar stock, and for sintering of powder compacts that require reducing atmospheres.

Commercial nitrogen-base atmospheres are used for many heat treating applications. Commercial nitrogen-base atmosphere systems employed by the metalworking and heat treating industry use gases and equipment that are common among all applications. In most instances, the major atmosphere component is industrial liquid nitrogen, which is supplied to the furnace from a system consisting of a storage tank, gas vaporizer, and a station controlling pressure and flow rate. The nitrogen serves as a pure, dry, inert gas that provides for efficient purging and blanketing function within the heat treating furnace. The nitrogen stream is often enriched with a reactive component, and the resulting composition and flow rate are determined by the specific furnace design, temperature, and material being heat treated. Although there is similarity in the components of commercial nitrogen-base atmosphere systems, the flexibility of controlling atmosphere composition and flow rate independently over a wide range provides very different end-use characteristics. Therefore, the classification of commercial nitrogen-base atmosphere systems is appropriately made according to three major categories of atmosphere function—protection, reactivity, and carbon control—rather than by gas or equipment components.

The basic components of industrial gas atmosphere systems are illustrated in Fig. 20. Figure 20(a) shows a nitrogen-hydrogen protective atmosphere system used for annealing, brazing, and sintering. Figure 20(b) is a nitrogen-methanol system that may be used to create an endothermic-type atmosphere typical of those used for carburizing, carbonitriding, or neutral hardening. In both instances, there are three basic parts to the commercial-nitrogen system: the storage vessels containing the elemental atmosphere components, the blend panel used to control the flow rate of each constituent gas, and the piping and wiring required for safe operation compatible with the furnace design.

Nitrogen is the main component of most commercial nitrogen-base systems. The nitrogen can be from either "off-site" generated liquid nitrogen,

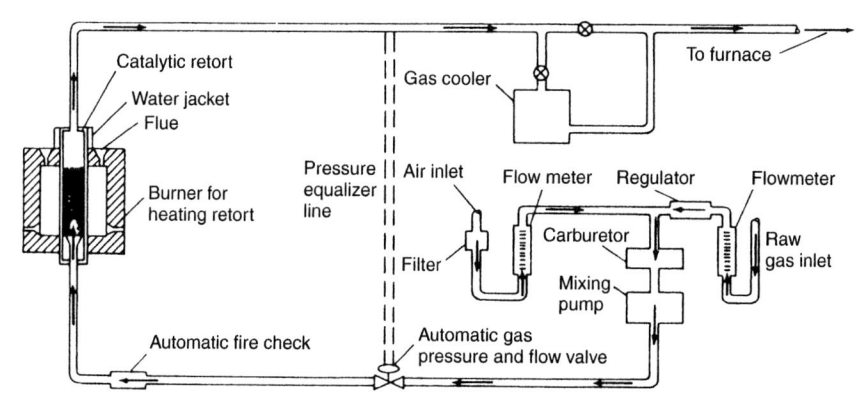

Fig. 19 Schematic flow diagram of an endothermic gas generator. Source: Ref 13

which has higher purity, or from "on-site" generated gaseous nitrogen that is slightly less pure.

Because nitrogen is noncorrosive, special materials of construction are not required, except that they must be suitable for the temperatures of liquid nitrogen. Tanks may be spherical or cylindrical in shape.

Hydrogen is used as a reactive reducing gas for many heat treating atmosphere applications. In commercial nitrogen-base systems, hydrogen is normally blended as a gas with nitrogen to form an atmosphere composition of 90% nitrogen and 10% hydrogen.

Vacuum atmospheres are flexible, in terms of applications, and easily generated. Equipment cost is high, however.

Dissociated ammonia provides a relatively high-cost prepared furnace atmosphere providing a dry, carbon-free source of reducing gas. Typical composition is 75% hydrogen, 25% nitrogen, less than 300 ppm residual ammonia, and less than $-50\ °C$ ($-60\ °F$) dew point.

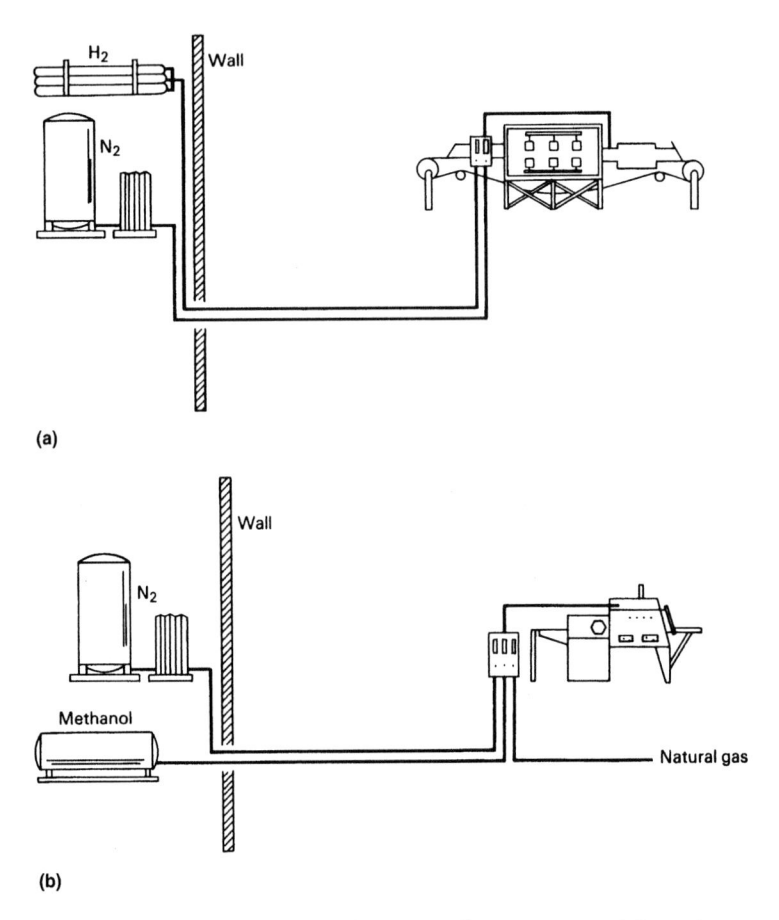

Fig. 20 Industrial-gas nitrogen-base atmosphere processes. (a) Protective atmosphere using H_2 and N_2. (b) Carbon-controlled atmosphere using methanol and natural gas. Source: Ref 13

Principal uses of dissociated-ammonia furnace atmosphere include bright copper and silver brazing; bright heat treating of selected nickel alloys, copper alloys, stainless steels, and carbon steels; bright annealing of electrical components; and as a carrier mixed gas for certain nitriding processes. Dissociated ammonia ($N_2 + 3H_2$) is produced from commercially supplied anhydrous ammonia (NH_3) with an ammonia dissociator. This equipment raises the temperature of ammonia vapor in a catalyst-filled retort to approximately 900 to 980 °C (1650 to 1800 °F). The gas is then cooled for metering and transport as a prepared atmosphere. At these reaction temperatures in the presence of catalyst, ammonia vapor dissociates into separate constituents of hydrogen and nitrogen.

Dry Hydrogen Atmospheres. Commercially available hydrogen is 98 to 99.9% pure. All cylinder hydrogen contains traces of water vapor and oxygen. Methane, nitrogen, carbon monoxide, and carbon dioxide may be present as impurities in very small amounts, depending on the method of manufacture. Hydrogen is produced commercially by a variety of methods, including the electrolysis of water, the catalytic conversion of hydrocarbons, the decomposition of ammonia, and the water gas reaction.

Dry hydrogen is used in the annealing of stainless and low-carbon steels, electrical steels, and several nonferrous metals. It is used also in the sintering of refractory materials such as tungsten carbide and tantalum carbide, in the nickel brazing of stainless steel and heat-resistant alloys, and in copper brazing, direct reduction of metal ores, annealing of metal powders, and the sintering of powder metallurgy compacts.

Argon, as received in tanks (such as used for gas shielded-arc welding), provides an excellent neutral atmosphere. However, due to high cost, its use is confined mainly to heat treating of certain exotic alloys in small furnaces.

Steam Atmospheres. Steam may be used as an atmosphere for scale-free tempering and stress relieving of ferrous metals in the temperature range of 345 to 650 °C (650 to 1200 °F). The steam causes a thin, hard, and tenacious blue-black oxide to form on the metal surface. This oxide film, which is about 0.00127 to 0.008 mm (0.00005 to 0.0003 in.) thick, improves certain properties of various metal parts.

Before parts are processed in steam atmospheres, their surfaces must be clean and oxide-free to permit the formation of a uniform coating. To prevent condensation and rusting, steam should not be admitted until workpiece surfaces are above 100 °C (212 °F). Air must be purged from the furnace before the temperature exceeds 425 °C (800 °F) to prevent the formation of a brown coating instead of the desired blue-black coating.

REFERENCES

1. J.W. Smith, Types of Heat-Treating Furnaces, *Heat Treating,* Vol 4, *ASM Handbook,* ASM International, 1991, p 465–472

2. *Energy Conversion in Heat Treating and Forging Operations,* Course 20, Lesson 4, Metals Engineering Institute, American Society for Metals, 1980
3. H.E. Boyer, Chapter 4, *Practical Heat Treating,* 1st ed., American Society for Metals, 1984, p 59–90
4. H.N. Oppenheimer, Types of Furnaces and Associated Equipment, Course 42, Lesson 8, *Practical Heat Treating,* Materials Engineering Institute, ASM International, 1995
5. W.J. Laird, Jr., Salt Bath Equipment, *Heat Treating,* Vol 4, *ASM Handbook,* ASM International, 1991, p 475–483
6. R.F. Sagon-King, Fluidized-Bed Equipment, *Heat Treating,* Vol 4 *ASM Handbook,* ASM International, 1991, p 484–491
7. R.C. Anderson et al., Heat Treating in Vacuum Furnaces and Auxiliary Equipment, *Heat Treating,* Vol 4, *ASM Handbook,* ASM International, 1991, p 492–509
8. F.G. Lambating, Vacuum Heat Treating Processes, Course 42, Lesson 15, *Practical Heat Treating,* Materials Engineering Institute, ASM International, 1995
9. J. St. Pierre, Vacuum Carburizing, *Heat Treating,* Vol 4, *ASM Handbook,* ASM International, 1991, p 348–351
10. G.Y. Lai, Heat-Resistant Materials for Furnace Parts, Trays, and Fixtures, *Heat Treating,* Vol 4, *ASM Handbook,* ASM International, 1991, p 510–518
11. *Heat Processing Technology,* Course 6, Lesson 15, Metals Engineering Institute, American Society for Metals, 1978
12. C.E. Bates, G.E. Totten, and R.L. Brennan, Quenching of Steel, *Heat Treating,* Vol 4, *ASM Handbook,* ASM International, 1991, p 67–120
13. P. Johnson, Furnace Atmospheres, *Heat Treating,* Vol 4, *ASM Handbook,* ASM International, 1991, p 542–567
14. F.G. Lambating, Furnace Atmospheres, Course 42, Lesson 9, *Practical Heat Treating,* Materials Engineering Institute, ASM International, 1995

CHAPTER 5

Instrumentation and Control of Heat Treating Processes

HEAT TREATING OPERATIONS generally consist of three separate functions: material movement, the application of energy, and the supervision of process conditions. In a typical heat treating operation, work is moved into a furnace, heated according to a time-temperature program while exposed to defined surrounding conditions; cooled or quenched under certain conditions; and, finally, moved out of the quench vessel or furnace. The temperature, and frequently the atmosphere condition, must be controlled precisely in order to achieve the desired metallurgical results. In order to ensure the repeatability of the operation, the heat treating system must have the necessary sensors, timers, and variable (temperature, atmosphere, etc.) controllers to hold the process within prescribed or specified limits.

Temperature-Control Systems

Temperature control in heat treating is of paramount importance for maintenance of quality. Process temperatures generally should be controlled to within about ±2.5 °C (±5 °F). Although this close range is possible at temperatures below 675 °C (1250 °F), a more practical control range is nearer ±8 °C (±15 °F) for higher temperatures.

In any temperature-control system, three steps must be executed. Before control can be established, the variable must first be "sensed" by some device that responds to changes in the quality or value of the variable. This quantity, or its change, must then be indicated or recorded prior to being controlled. The last step in the sequence is the transmission of the controller output to the "final element," which is a component of the process itself. Final elements relay the output of the controller and cause corrective changes in the process.

Temperature Sensors

As is often the case, one variable is measured then translated, or converted, to another. For example, ambient temperatures actually are measured by expansion or contraction of a column of fluid or of a metal (e.g., mercury). By means of calibration, these variables are converted to numerical temperature readings. These simple devices, however, are not suitable for the higher temperatures involved in most heat treating operations. The temperature sensors used in heat treating can be divided into contact and the noncontact types.

Resistance temperature detectors (RTDs) are contact-type sensors. Their electrical resistance is proportional to temperature. Typical detector materials are platinum, copper, and nickel. They are more stable, accurate, and interchangeable than thermocouples, but even the platinum detectors have an upper temperature limit of approximately 750 °C (1380 °F), which limits their use in the metals industry. Resistance temperature detectors are normally larger in size and slower in response than thermocouples.

Thermocouples are the most widely used contact sensors for measuring temperatures of heat treating processes. In this instance, however, the same approach is used; that is, one variable is measured then converted to another.

Thermocouples consist of two dissimilar metal wires that are metallurgically homogeneous. They are joined at one end—the measuring, or hot, junction. The other end, which is connected to the copper wire of the measuring instrument circuitry, is called the reference, or cold, junction. The electrical signal output in millivolts (mV) is proportional to the difference in temperature between the measuring junction (hot) and the reference junction (cold) (Fig. 1). The voltage is then transmitted to the indicating instrument.

There are a number of standard metal combinations that may be used for temperature measurement. The most common combinations, along with their letter type designations and applicable temperature ranges, are listed in Table 1. Type K is, by far, the most widely used.

To obtain accurate temperature sensing, thermocouples must be placed near the work. In many instances, more than one thermocouple may be used within a given furnace.

Thermocouple elements with ceramic insulation inside a protection tube are available (Fig. 1). Heavier protective tubes may be used to provide mechanical support and further protect the thermocouple from the process environment. Protection tubes are made from metallic or ceramic materials and are used in vacuum, air, or for other gaseous atmospheres. See Ref 1 for the recommended protection tube materials.

The response of the thermocouple to changes in process temperature should be as fast or faster than that of the work in the furnace. If the workpieces are small and the thermocouple is contained in a heavy gage protection tube, the thermocouple temperature will lag behind the actual work temperature, and workpieces will be overheated when they are initially brought up to control temperature.

Temperature measurement represents the second step in a temperature-control system. Measurement instruments measure the output signal (millivolts) of the temperature sensor and convert it to a temperature indication. Measurement instruments may also have the capability of controlling and/or recording the temperature. The indicating phase of the system must, however, be calibrated and set for the two dissimilar metals used in the thermocouple.

The accuracy of the measurement depends primarily on the accuracy of the temperature sensor and the connecting lead wire (Table 2). The accuracy of the digital measurement instrument generally is $\pm 0.01\%$ of the full scale reading.

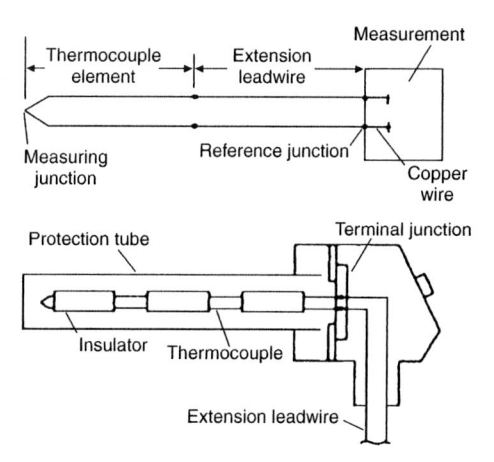

Fig. 1 Simple thermocouple (upper view) and cutaway of a thermocouple assembly (lower view). Source: Ref 1

Table 1 Thermocouple types, nominal temperature ranges, and material combinations

Type	Nominal temperature range °C	°F	Typical thermocouple material(a)
B	50–1818	120–3300	PLATINUM, 30% RHODIUM-platinum, 6% rhodium
E	0–870	32–1600	CHROMEL-constantan
J	−185 to 760	−300 to 1400	IRON-constantan
K	0–1260	32–2300	NICKEL, CHROMIUM-nickel, aluminum, CHROMEL-alumel
R	0–1480	32–2700	PLATINUM, 13% RHODIUM-platinum
S	0–1480	32–2700	PLATINUM, 10% RHODIUM-platinum
T	−185 to 370	−300 to 700	COPPER-constantan
W5	−20 to 2205	0–4000	TUNGSTEN, 5% RHENIUM-tungsten, 26% rhenium

(a) Uppercase letters indicate the positive lead. Source: Ref 2

All of the newer digital measurement and control instruments have digital displays and are recording or nonrecording types. Older analog displays included meters and motor-driven pointers. Analog strip chart or round chart recorders included analog temperature indication. Digital displays are available with a digital interface or communications ports.

Temperature measurement instruments incorporate reference junction compensation in their circuitry for thermocouple measurements and emissivity compensation for noncontact radiation sensors. Reference junction compensation automatically adjusts the measurement depending on the temperature at the junction between the thermocouple wire and the copper wire of the instrument's measuring circuit. Radiation-type temperature measurements require an emissivity compensator. This compensator makes a calibration adjustment by comparing measurement of the same target with an optical pyrometer or calibrated thermocouple. The emissivity compensator is adjusted to make the radiation pyrometer measurement indication agree with the reference calibration instrument.

Temperature control is the third major phase of a temperature-control system. A temperature controller must provide sufficient energy to satisfy process requirements, even though operating conditions vary. Variations include changes in process load, fuel characteristics, and ambient temperature. Thus, controller requirements are more stringent when process requirements are demanding and especially when operating conditions vary significantly.

Table 2 Limits of error for thermocouples

ANSI type	Type of thermocouple	Temperature range °C	Temperature range °F	Limits of error(a) Standard	Limits of error(a) Special
J	Iron and constantan	−190 to −75	−310 to −100	...	±2%
		−175 to 315	−100 to 600	±2 °C (±4 °F)	±1 °C (±2 °F)
		315–425	600–800	±2 °C (±4 °F)	±0.33%
		425–760	800–1400	±0.75%	±0.33%
K	Nickel, chromium and nickel, aluminum	0–275	32–530	±2 °C (±4 °F)	±1 °C (±2 °F)
		275–1260	30–2300	±0.75%	±0.38%
N	Nickel, chromium, silicon-nickel, silicon, magnesium	0–275	32–530	±2 °C (±4 °F)	±1 °C (±2 °F)
		275–1260	30–2300	±0.75%	±0.38%
T	Copper and constantan	−185 to −60	−300 to −75	...	±1%
		−100 to −60	−150 to −75	±2%	±1%
		−60 to 95	−75 to 200	±1 °C (±1.5 °F)	±0.5 °C (±0.75 °F)
		95–370	200–700	±0.75%	±0.38%
E	Nickel, chromium and constantan	0–315	32–600	±2 °C (±3 °F)	±1 °C (±2 °F)
		315–870	600–1600	±0.5%	±0.38%
S	Platinum, 10% rhodium and platinum	−15 to 540	0–1000	±1.5 °C (±2.5 °F)	±1 °C (±1.5 °F)
		540–1480	1000–2700	±0.25%	±0.15%
R	Platinum, 13% rhodium and platinum	−15 to 540	0–1000	±1.5 °C (±2.5 °F)	...
		540–1480	1000–2700	±0.25%	...
B	Platinum, 30% rhodium and platinum, 6% rhodium	870–1705	1600–3100	±0.5%	...

(a) When expressed as a percentage, the limit of error is a percentage of the temperature reading, not of the range. Source: Ref 1

In operation, the controller set point that represents the desired temperature is compared with the process or actual temperature. The stability of the controller and its sensitivity to the difference between desired and actual temperatures are critical. Based on this comparison, the controller regulates the energy flow to the process.

The two basic types of control are the two-position, or on-off, type, and the proportioning, or modulating, type. These two basic types exist in many variations. Generally, proportioning, or modulating, types are preferred as opposed to the on-off types because the proportioning types (regardless of the source of energy) permit much closer temperature control.

A typical digital ramp-soak, programmable, single-point temperature controller is shown in Fig. 2. The unit is capable of holding 49 heating/cooling cycles in memory, with each cycle containing up to 99 separate heating set points or heating/cooling steps. The controller can be configured for any thermocouple or RTD input with the control output configured for any of the output modes described earlier.

The various measurement and control-loop configurations must be configured properly to achieve optimum operation of the system. For example, if the temperature sensor, which may be only 1% of the instrumentation cost, is out of calibration or improperly located, the performance of the complete system is degraded. Poor input into even the most sophisticated temperature measurement and control systems will produce poor results.

Noncontact Temperature Instruments. Temperature is a measure of the ability to transfer heat. Heat may be transferred by conduction, convection (actually conduction to a fluid followed by translation), and by thermal radiation. Noncontact temperature sensors depend on the thermally generated electromagnetic radiation from a surface of a test object. Electromagnetic radiation is emitted from a heated body when electrons within the body change to a lower energy state. Both the intensity and the

Fig. 2 Ramp-soak, programmable, single-point temperature controller. Courtesy of Honeywell, Inc.

wavelength of the radiation depend on the temperature of the surface atoms or molecules. Radiation pyrometers such as the one shown in Fig. 3 are used as noncontacting thermometers for temperatures from 0 to 3200 °C (32 to 5800 °F). These instruments may be sighted to view target surfaces as small as 0.0254 mm (0.001 in.) in diameter and may be used to monitor and/or control furnace temperatures or the surface temperatures for induction, laser, or electron beam hardening. Radiometers and pyrometers are rugged, low-cost devices that can be used in an industrial environment for long-term process monitoring.

The advantage of radiation thermometers is that they yield accurate temperature measurements of the target without touching the test object, which is referred to here as a target because radiation thermometers must be sighted on the object of interest. Thus, radiation thermometers are preferred to contact thermometers (thermocouples, resistance thermometers, and so on) where the target is difficult to touch because it is moving, fragile, small, of small "thermal mass," hot, corrosive, or in a corrosive or protected environment.

Atmosphere Control

The purpose of atmosphere control is to maintain consistent levels of the various atmosphere constituents and to determine whether changes in those levels are required in order to produce a desired result under a given set of conditions. Controls are required for various heat treating operations that use a variety of different atmospheres. All methods of atmosphere control can effectively be divided into two groups: those involving control of the atmosphere once it is inside the furnace and those involving control of the atmosphere supply before it is introduced into the furnace. Such control is achieved through the use of atmosphere control devices.

For success in metallurgical atmosphere control, gas analysis instrumentation must be applied to the furnace and at the gas generator, if one

Fig. 3 An optical pyrometer. Courtesy of The Pyrometer Instrument Company

is used. This is true for endothermic as well as exothermic atmosphere applications. Analyzers and associated control systems applied to a furnace should never be expected to correct deficiencies in the basic atmosphere produced by the gas generator. Instrumentation dedicated to the furnace is designed to provide information and control relative to operation of the furnace and to the product being processed. Likewise, instrumentation dedicated to the generator is designed to provide information and control relative to operation of the generator and to the atmosphere produced by it.

Atmosphere Sensors and Control Systems

The most common sensors available for the wide variety of furnace atmospheres used can be categorized into three major types: oxygen probe, dew point, and infrared. These sensors can effectively be used to control endothermic, exothermic, nitrogen-methanol, nitrogen-hydrocarbon, and nitrogen-hydrogen-type atmospheres. Infrared control is most often used for endothermic and exothermic generator operation and, in some cases, for exothermic furnace atmospheres as well. Nitrogen-hydrogen atmospheres are typically controlled by dew point. Direct oxygen or dew point monitoring devices are also often used for applications such as bell furnace annealing.

The oxygen probe is an in situ type device; that is, it directly samples the atmosphere being measured. The electrical signal generated by an oxygen probe is directly proportional to the carbon potential of the atmosphere. Infrared control measures a sample drawn from the furnace. Infrared control is usually used to measure carbon monoxide and/or carbon dioxide levels. Dew point control also analyzes a sample drawn external to the furnace.

The oxygen probe is based in theory on a hot ceramic electrochemical cell. The probe will respond to oxygen, hydrogen, carbon monoxide, water, and carbon dioxide and thus can determine the oxidization potential of a gas. The output of the oxygen probe is a direct measurement of the oxidation potential of the atmosphere at the process temperature of the furnace. Therefore, when the probe temperature is close to the furnace temperature, the response of the probe is a direct indication of whether the atmosphere will oxidize or reduce steel, provided the composition of the atmosphere with regard to the proportions of carbon gases and hydrogen is known. Under such conditions, the probe will give a reliable indication of the oxidation/reduction situation for all furnace temperatures.

The oxygen probe is a closed-end tube usually constructed of lime-stabilized zirconia or yttria-stabilized material for temperatures up to 1600 °C (2900 °F). The typical oxygen probe construction is shown in Fig. 4. When such a probe is subjected to elevated temperatures, the nonporous sheath material acts as a solid electrolyte that permits the passage of oxygen ions when the inner and outer surfaces are subjected to atmospheres of different oxygen partial pressures—for example, a reference gas such as air, because its oxygen content is constant at 20.9% by volume at sea level, and the process furnace atmosphere, respectively. The electromotive force (emf) thus generated, and measured via the electrodes attached to the sheath, is related directly to, and provides an accurate quantification of, atmosphere characteristics in terms of its oxidizing/reducing, or, in some endothermic-atmosphere applications, carburizing/decarburizing tendencies (carbon potential) at a known temperature (Fig. 5). Oxygen probes made by different manufacturers have slightly different voltage (emf) outputs for a given carbon level and temperature due to the method of probe construction.

Oxygen probe systems are used extensively for control of furnaces used for hardening, carburizing, and carbonitriding. A single-point programmable controller such as is shown in Fig. 6, can provide up to 6 cycles in memory with 3 different carbon levels each on a variable timer. The unit has all the oxygen probe relationships of % carbon (mV) versus temperature for all the different probe manufacturers stored in memory. The unit can be configured for any one of the many control output modes.

Because of the fast response rate of probes, on/off control systems utilizing solenoid valves for regulating propane, natural gas, or liquid enrichment are sometimes adequate for batch furnaces. The in situ probe exposed directly to the furnace atmosphere controls carbon potential by adjusting the furnace carbon level to the control instrument set-point value. The oxygen probe supplies to the control station an electrical signal related to the carbon potential. High- and low-deviation contacts are ad-

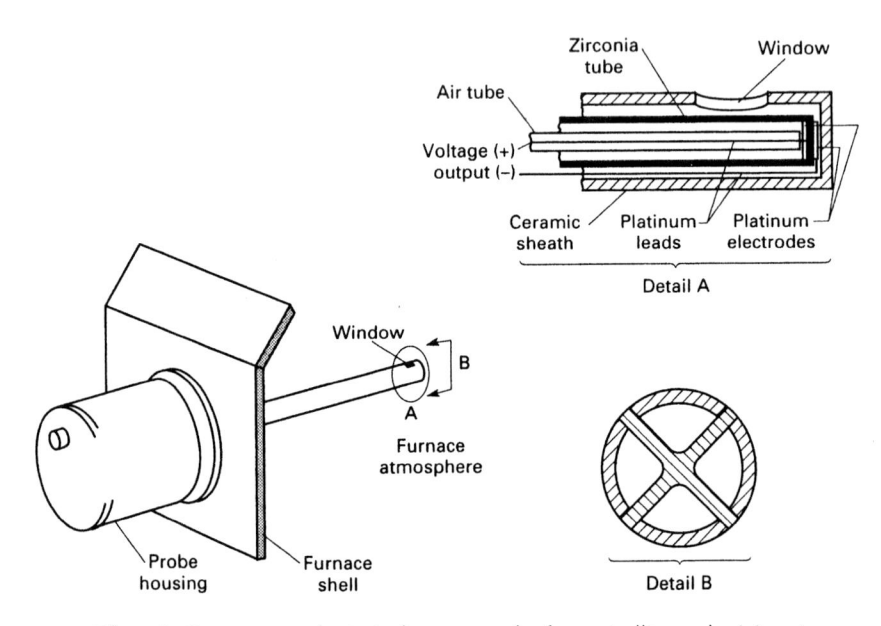

Fig. 4 Components of a typical oxygen probe for controlling carburizing atmospheres. Detail B shows construction of X-cap tip. Source: Ref 3

justed throughout the control instrument and either contact is made by the electrical signal from the probe. The low-deviation contact controls the solenoid valve that supplies enriching gas or liquid to the furnace. The high-deviation contact can be made to control the solenoid valve to add air or an oxidizer to the furnace. Carbon-concentration reproducibility of ±0.02% is frequently achieved using systems of this type, provided the cycle, temperature, and furnace conditions remain constant.

Endothermic generators also can be controlled by use of the oxygen probe. In this case, because of the difficulty of placing an in situ probe in the retort of the generator, the probe is used to sample atmosphere from the output line of the generator. The electrical signal from the probe is then wired to the control instrument that activates a solenoid valve or proportional valve located on the air-bypass line to the mixer. Thus, the air/gas ratio is automatically adjusted to give the desired properties of the endothermic gas.

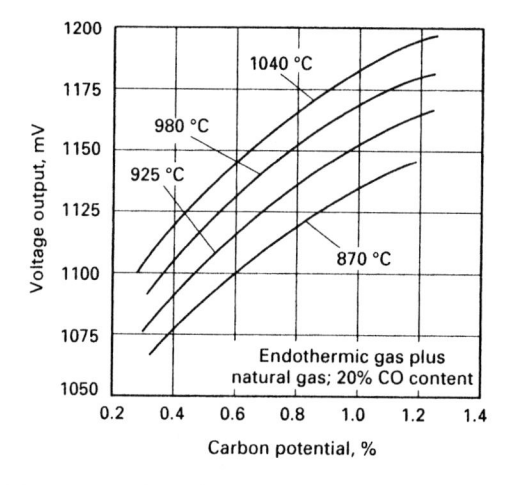

Fig. 5 Voltage across electrodes of a typical oxygen probe as a function of carbon potential at four temperatures, for endothermic gas enriched with natural gas and containing 20% CO. Source: Ref 3

Fig. 6 Programmable digital carbon controller. Courtesy of Honeywell, Inc.

It is good practice to regularly compare the carbon potential determined by a control system to the actual carbon potential of the atmosphere as determined by another method such as through shim analysis. A shim is a thin low-carbon metal sample that, when placed in the furnace, will quickly carburize to a level equal to the furnace carbon potential. Adjusting the control system to match the carbon potential indicated by a shim test can improve the accuracy of the entire system. Details on the shim analysis procedure can be found in Ref 4.

Dew Point Instrument. Use of a dew point measuring device is another method of monitoring or controlling the carbon potential of a furnace atmosphere. It can also be used to determine the moisture content of any given atmosphere.

One type of dew point instrument uses an aluminum oxide sensor to generate an electrical signal that is proportional to the moisture content of the sample stream. The sensor consists of an aluminum-base material with an aluminum oxide etched on its surface. The oxide is then covered with a thin permeable metal layer (Fig. 7). The inner aluminum base and the outer metal layer form two electrodes in what is basically a capacitor.

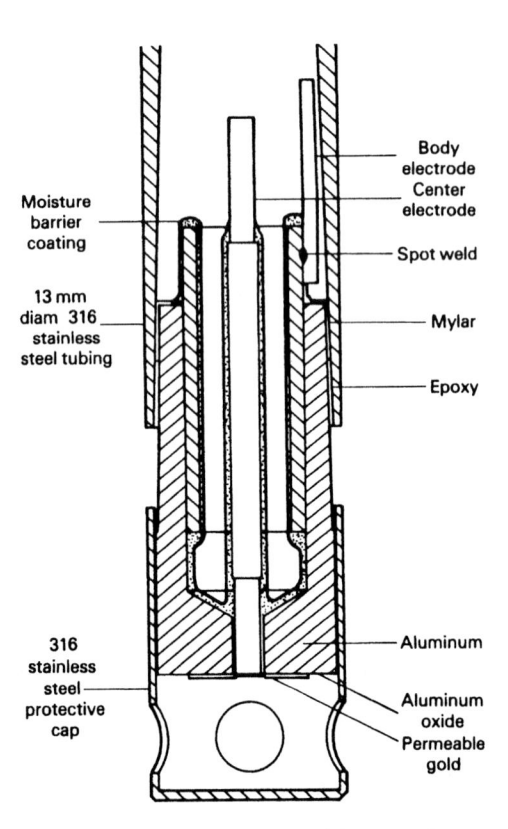

Fig. 7 The aluminum oxide sensor, a type of dew point instrument, measures water content of an atmosphere by change in capacitance between two electrodes. Source: Ref 5

Moisture passes through the outer metal layer and is absorbed on the oxide. This changes the capacitance of the entire assembly proportionally to the moisture content in the atmosphere.

Infrared analyzers are based on the principle that any compound present in the furnace atmosphere mixture will absorb infrared energy in proportion to its weight in the mixture. The wavelengths absorbed are different for each compound. Elemental gases such as hydrogen and oxygen do not absorb infrared radiation and therefore cannot be measured by this method. Infrared analyzers are normally used to measure carbon monoxide, carbon dioxide, and methane.

Integrated Control Systems

Until the mid-1980s, most heat treating equipment relied on relays, timers, manually operated push buttons, and dedicated process control instruments for the control and the sequencing of heat treating equipment operations. This meant that the equipment operator was usually very heavily involved in ensuring that several aspects of the process were followed. While these furnaces may have produced consistent loads of product with a reasonable reliability, the level of operator attention and skill required often affected the final results. In addition, the reliability of mechanical relays, timers, and push buttons exposed to dust, humidity, and other environmental plant factors was somewhat unpredictable and subject to frequent failure. A well-planned and executed preventative maintenance program was mandatory to keep malfunctions and furnace downtime to a minimum.

The development of the programmable logic controller (PLC) offered a solution to these problems. The PLC was originally designed to replace relays, timers, and other hardwired logic control systems and to basically simplify the management of several individual control instruments. Programmable logic controllers are often used to retrofit and update older heat treating equipment.

One of several PLCs available for the control of all functions, temperatures, time cycles, and atmosphere constituents and their control for an atmosphere batch furnace is shown in Fig. 8. This model can be programmed for more than 50 complete and distinct heat treating cycles.

Often the PLC is interfaced with a personal computer to enhance the data storage and diagnostic capabilities of the overall system. Programmable logic controllers have proved to be reliable, extremely powerful, and very practical for use as control devices on even the most complex heat treat processing systems. There are several manufacturers of PLCs and most furnace equipment makers can provide this control option as part of a standard control package.

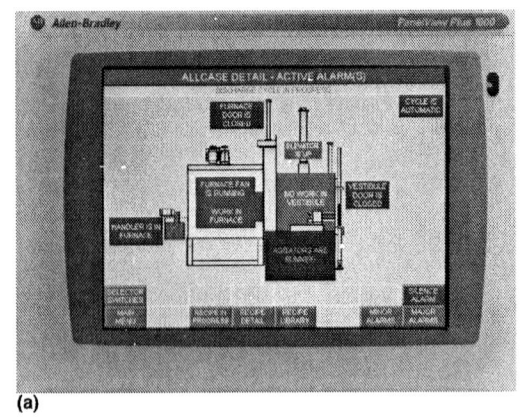

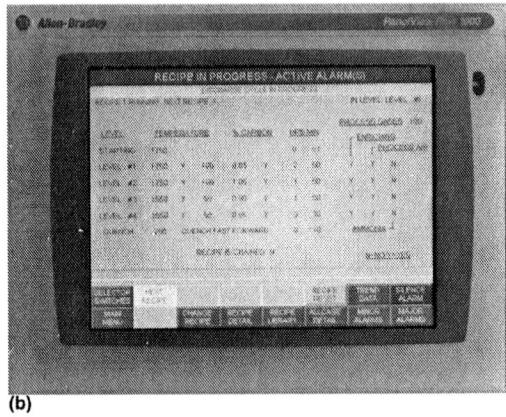

Fig. 8 Batch atmosphere furnace programmable logic controller (PLC) control system. (a) Furnace function status screen. (b) Furnace recipe control screen. Courtesy of Surface Combustion Inc.

REFERENCES

1. A.S. Tenney III, Temperature Control, *Heat Treating,* Vol 4, *ASM Handbook,* ASM International, 1991, p 529–541
2. H.E. Boyer, Chapter 4, *Practical Heat Treating,* 1st ed., American Society for Metals, 1984, p 59–90
3. M.S. Stanescu et al., Control of Surface Carbon Content in Heat Treating of Steel, *Heat Treating,* Vol 4, *ASM Handbook,* ASM International, 1991, p 573–586
4. Evaluation of Carbon Control in Processed Parts, *Heat Treating,* Vol 4, *ASM Handbook,* ASM International, 1991, p 587–600
5. J.G. Rowe, Furnace Atmosphere Control, *Heat Treating,* Vol 4, *ASM Handbook,* ASM International, 1991, p 568–572

SELECTED REFERENCE

- F.G. Lambating, Instrumentation and Controls, *Practical Heat Treating,* Course 42, Lesson 10, Materials Engineering Institute, ASM International, 1995

CHAPTER 6

Heat Treating of Carbon Steels

AS DISCUSSED IN CHAPTER 2, "Fundamentals of the Heat Treating of Steel," carbon steels are not strictly alloys of iron and carbon; manganese, silicon, phosphorus, and sulfur usually are present in small amounts. This condition has always led to some confusion regarding what carbon steels are, as well as what they are not.

Much study and effort has gone into developing a definition of carbon steels, as well as a list of compositions that are universally accepted by steel producers and fabricators. These efforts are summarized in Tables 1 to 6. The development and standardization of carbon steel compositions has been established by the American Iron and Steel Institute (AISI) and the Society of Automotive Engineers (SAE, or SAE International). These standards for carbon steels cover composition only. Hundreds of standards or codes established by other societies or agencies exist that cover steels or other alloys. Frequently, these standards place primary emphasis on mechanical properties rather than on composition. For this reason, it becomes difficult and sometimes dangerous to translate one specification to another. No attempt is made in this book to deal with compositions other than those established by AISI-SAE, except for the newer UNS system described next.

Table 1 Types and approximate percentages of identifying elements in standard carbon steels

Series designation	Description
10XX	Nonresulfurized, 1.00 manganese maximum
11XX	Resulfurized
12XX	Rephosphorized and resulfurized
15XX	Nonresulfurized, over 1.00 manganese maximum

Source: Ref 1

Unified Numbering System

The standard carbon and alloy grades established by AISI or SAE have been assigned designations in the Unified Numbering System (UNS) by ASTM International (ASTM E 527) and the SAE International (SAE J 1086). In the composition tables in this chapter, UNS numbers are listed along with their corresponding AISI-SAE numbers where available.

The UNS number consists of a single letter prefix followed by five numerals. The prefix letter G indicates standard grades of carbon or alloy steels, while the prefix letter H indicates standard grades that meet certain hardenability limits. The first four digits of the UNS designations usually correspond to the standard AISI-SAE designations, while the last digit

Table 2 Compositions of selected standard nonresulfurized carbon steels (1.0% Mn)

Steel designation AISI or SAE	UNS No.	C	Mn	P max	S max
1008	G10080	0.10 max	0.30–0.50	0.040	0.050
1012	G10120	0.10–0.15	0.30–0.60	0.040	0.050
1015	G10150	0.13–0.18	0.30–0.60	0.040	0.050
1018	G10180	0.15–0.20	0.60–0.90	0.040	0.050
1020	G10200	0.18–0.23	0.30–0.60	0.040	0.050
1022	G10220	0.18–0.23	0.70–1.00	0.040	0.050
1025	G10250	0.22–0.28	0.30–0.60	0.040	0.050
1029	G10290	0.25–0.31	0.60–0.90	0.040	0.050
1030	G10300	0.28–0.34	0.60–0.90	0.040	0.050
1035	G10350	0.32–0.38	0.60–0.90	0.040	0.050
1038	G10380	0.35–0.42	0.60–0.90	0.040	0.050
1040	G10400	0.37–0.44	0.60–0.90	0.040	0.050
1045	G10450	0.43–0.50	0.60–0.90	0.040	0.050
1050	G10500	0.48–0.55	0.60–0.90	0.040	0.050
1055	G10550	0.50–0.60	0.60–0.90	0.040	0.050
1060	G10600	0.55–0.65	0.60–0.90	0.040	0.050
1070	G10700	0.65–0.75	0.60–0.90	0.040	0.050
1080	G10800	0.75–0.88	0.60–0.90	0.040	0.050
1084	G10840	0.80–0.93	0.60–0.90	0.040	0.050
1090	G10900	0.85–0.98	0.60–0.90	0.040	0.050
1095	G10950	0.90–1.03	0.30–0.50	0.040	0.050

Source: Ref 1

Table 3 Compositions of standard nonresulfurized carbon steels (over 1.0% Mn)

Steel designation AISI or SAE	UNS No.	C	Mn	P max	S max
1513	G15130	0.10–0.16	1.10–1.40	0.040	0.050
1522	G15220	0.18–0.24	1.10–1.40	0.040	0.050
1524	G15240	0.19–0.25	1.35–1.65	0.040	0.050
1526	G15260	0.22–0.29	1.10–1.40	0.040	0.050
1527	G15270	0.22–0.29	1.20–1.50	0.040	0.050
1541	G15410	0.36–0.44	1.35–1.65	0.040	0.050
1548	G15480	0.44–0.52	1.10–1.40	0.040	0.050
1551	G15510	0.45–0.56	0.85–1.15	0.040	0.050
1552	G15520	0.47–0.55	1.20–1.50	0.040	0.050
1561	G15610	0.55–0.65	0.75–1.05	0.040	0.050
1566	G15660	0.60–0.71	0.85–1.15	0.040	0.050

Source: Ref 1

(other than zero) denotes some additional composition requirement such as lead or boron. The digit is sometimes a 6, which is used to designate steels that are made by the basic electric furnace with special practices. The AISI-SAE alloys steels in Chapter 7, "Heat Treating of Alloy Steels," are also covered by the UNS system.

Table 4 Compositions of standard carbon H-steels and standard carbon boron H-steels

Steel designation AISI or SAE	UNS No.	C	Mn	P max	S max	Si
Standard carbon H-steels						
1038H	H10380	0.34–0.43	0.50–1.00	0.040	0.050	0.15–0.30
1045H	H10450	0.42–0.51	0.50–1.00	0.040	0.050	0.15–0.30
1522H	H15220	0.17–0.25	1.00–1.50	0.040	0.050	0.15–0.30
1524H	H15240	0.18–0.26	1.25–1.75(a)	0.040	0.050	0.15–0.30
1526H	H15260	0.21–0.30	1.00–1.50	0.040	0.050	0.15–0.30
1541H	H15410	0.35–0.45	1.25–1.75(a)	0.040	0.050	0.15–0.30
Standard carbon boron H-steels						
15B21H	H15211	0.17–0.24	0.70–1.20	0.040	0.050	0.15–0.30
15B35H	H15351	0.31–0.39	0.70–1.20	0.040	0.050	0.15–0.30
15B37H	H15371	0.30–0.39	1.00–1.50	0.040	0.050	0.15–0.30
15B41H	H15411	0.35–0.45	1.25–1.75(a)	0.040	0.050	0.15–0.30
15B48H	H15481	0.43–0.53	1.00–1.50	0.040	0.050	0.15–0.30
15B62H	H15621	0.54–0.67	1.00–1.50	0.040	0.050	0.40–0.60

(a) Standard AISI-SAE H-steels with 1.75 manganese maximum are classified as carbon steels. Source: Ref 1

Table 5 Compositions of standard resulfurized carbon steels

Steel designation AISI or SAE	UNS No.	C	Mn	P max	S
1108	G11080	0.08–0.13	0.50–0.80	0.040	0.08–0.13
1110	G11100	0.08–0.13	0.30–0.60	0.040	0.08–0.13
1113	G11130	0.13 max	0.70–1.00	0.07–0.12	0.24–0.33
1117	G11170	0.14–0.20	1.30–1.60	0.040	0.08–0.13
1118	G11180	0.14–0.20	1.30–1.60	0.040	0.08–0.13
1137	G11370	0.32–0.39	1.35–1.65	0.040	0.08–0.13
1139	G11390	0.35–0.43	1.35–1.65	0.040	0.13–0.20
1140	G11400	0.37–0.44	0.70–1.00	0.040	0.08–0.13
1141	G11410	0.37–0.45	1.35–1.65	0.040	0.08–0.13
1144	G11440	0.40–0.48	1.35–1.65	0.040	0.24–0.33
1146	G11460	0.42–0.49	0.70–1.00	0.040	0.08–0.13
1151	G11510	0.48–0.55	0.70–1.00	0.040	0.08–0.13

Source: Ref 1

Table 6 Compositions of standard rephosphorized and resulfurized carbon steels

Steel designation AISI or SAE	UNS No.	C	Mn	P	S	Pb
1211	G12110	0.13 max	0.60–0.90	0.07–0.12	0.10–0.15	. . .
1212	G12120	0.13 max	0.70–1.00	0.07–0.12	0.16–0.23	. . .
1213	G12130	0.13 max	0.70–1.00	0.07–0.12	0.24–0.33	. . .
1215	G12150	0.09 max	0.75–1.05	0.04–0.09	0.26–0.35	. . .
12L14	G12144	0.15 max	0.85–1.15	0.04–0.09	0.26–0.35	0.15–0.35

Source: Ref 1

What Are Carbon Steels?

By the AISI classification, a steel is considered to be a carbon steel when:

- No minimum content is specified or required for aluminum, boron, chromium, cobalt, columbium, molybdenum, nickel, titanium, tungsten, vanadium, zirconium, or any other element added to obtain a desired alloying effect.
- The specified minimum for copper does not exceed 0.40%.
- The maximum specified content does not exceed the following limits: manganese, 1.65%; silicon, 0.60%; and copper, 0.60%.

Steels are sold to definite chemical limits, and each grade of plain carbon steel is assigned a code number that specifies its chemical composition. There are several coding systems, but the AISI-SAE system is the most comprehensive and widely used. It uses a code of four digits for all carbon steels. The first two always indicate the general type off steel. For instance, consider a steel identified as AISI 1040 grade. The digits 10 indicate it to be a plain carbon steel with a manganese content of no greater than 1.0%. The next two digits, 40, indicate the mean carbon content, in this case, 0.40%. Thus, when the code 1040 is seen, the reader knows the steel is a plain carbon grade containing 0.40% C plus or minus a few points. The steel described above is the first entry in Table 1. The 10XX means that it is a plain carbon steel—the XX indicates that the carbon content may vary.

Approximately 20 different compositions of the 10XX group are listed in Table 2. The complete list contains 47 compositions, but the 20 that are omitted are interspersed within the group shown and vary only slightly in carbon and/or manganese content. The large number of compositions that vary minimally are designated as such for the convenience of the steel producer and are of little interest to the heat treater. Steels designated as 1012, 1017, 1037, 1044, 1059, and 1086 are examples of compositions purposely omitted from Table 2. Some of these grades are produced only for wire rods and wire. From Table 2, it is evident that the entire carbon range of 0.08 to 0.95% is covered and in relatively small increments of carbon.

All of the grades listed in Table 2 are considered very low hardenability steels; none carry the "H" suffix, which is used to denote a specific hardenability that is guaranteed by the steel producer. Steel 1008 is incapable of developing much hardness even when heated above its transformation temperature and drastically quenched because there is not enough carbon present. On the other hand, 1905 steel can develop a hardness of 66 HRC when very thin sections are cooled quickly, or high hardnesses can be produced in the surface layers by drastic quenching. This practice is sometimes termed "shell hardening" because a very hard shell is produced on the outside of the workpiece, while the subsurface layers are relatively soft because they were not cooled fast enough to exceed the critical cooling rate. This condition is often useful in practical applications.

To provide higher hardenability, additional groups of carbon steels have been developed—higher manganese grades and/or boron treated grades. Compositions of these special grades, which are considered carbon steels, are listed in Tables 3 and 4.

Higher Manganese Carbon Steels

Because of differences in properties, and in particular, their response to heat treatment, the higher manganese carbon steels have been removed from the 10XX series and are assigned to the 15XX series. These steels are marketed as carbon steels, but because of their higher manganese content, they respond to heat treatment like some alloy steels. The same method of designation prevails for the 15XX series; that is, the digits 15 denote higher manganese (up to 1.65%), and the last two digits designate the mean carbon content. For example, steel 1551 has a carbon range of 0.45 to 0.56%.

In carbon steels, for each element specified, there is a permissible range of composition, or a maximum limit, rather than a single specified value. This is because it is difficult to melt to the exact desired chemical percentage of each element; some leeway must be allowed for practical purposes. Manganese, phosphorus, and sulfur appear in the composition of every commercial steel either as residual elements or alloys. In the case of manganese, the amounts present usually require that this element be purposely added. Phosphorus, sulfur, and silicon are useful in their own right in special applications.

Generally, however, they are undesirable and are considered impurities that have been carried along in the materials used to make up the steel. Because their complete removal is quite expensive, their presence is reduced to some nominal amount (0.04% maximum or less for phosphorus and sulfur), which renders them practically ineffective as far as the properties of the steel are concerned. They are then carried along in this reduced amount rather than eliminated entirely.

In studying Table 3, note that none of the eleven 15XX steels have the "H" suffix; therefore, no specific hardenability is guaranteed. Although the manganese ranges are generally higher compared with the 10XX steels (Table 2), there is a substantial variation in the manganese ranges for the steels shown in Table 3. In two instances (1524 and 1541), the upper boundaries of the manganese ranges reach the maximum limits that allow them to be considered carbon steels. This variation in manganese content causes a substantial difference in hardenability and thus in response to heat treatment.

Carbon H-Steels

A few carbon steels are available as H-grades. Compositions for six of these steels are listed in Table 4—two 10XX and four 15XX grades. In Table 4, note that 1524H and 1541H permit manganese contents up to 1.75%. These represent exceptions to the general description of carbon steels given previously. As a rule, when the manganese content exceeds 1.65%, a steel is considered an alloy steel.

For hardenability bands of all H-steels (see Ref 1 and 2), the data presented in Fig. 1 illustrate the positive effect of manganese on hardenability.

Figure 1(a) shows the hardenability for 1038H steel with a manganese range of 0.60 to 0.90%. Maximum hardness at the quenched end of the end-quench bar (see Chapter 3, "Hardness and Hardenability") approaches 60 HRC but drops drastically away from the end (commonly described as a steep hardenability curve). Therefore, this steel is considered to be very low in hardenability.

Referring to Fig. 1(b), the effect of manganese on hardenability is readily evident for 1541H. Maximum hardness shown for both boundaries of the band shows a slightly higher initial hardness for 1541H compared with 1038H, but this is due to the slightly higher carbon range for 1541H. The less steep curves that characterize the 1541H grade are the result of higher manganese content.

Boron-Treated Carbon Steels

Another approach to increasing the hardenability of carbon steels is by treatment with boron. This approach has become quite popular with fabricators as a means of obtaining greater hardenability without using more expensive alloy steels.

Because of the very small amount of boron used (0.0005 to 0.003%), steels containing boron to increase their hardenability are usually termed as boron treated, rather than regarding boron as an alloying element. Carbon steels that contain boron are identified by inserting the letter "B" between the second and third digits of the AISI number. For instance, a boron-treated 1541 steel would be written as 15B41.

Compositions for six standard grades of boron-treated carbon steels are listed in Table 4. However, other grades of carbon steels are available as boron-treated grades, usually by special order.

The profound effect of boron on the hardenability is demonstrated in Fig. 1(c). The two steels shown are essentially the same in composition except for the boron treatment. The effect of boron is quite evident on 15B41H. The precise mechanism of how boron (especially in such small amounts) accomplishes what it does is not entirely clear. However, it is well known that it slows the critical cooling rate; thus, it follows that hardenability increases.

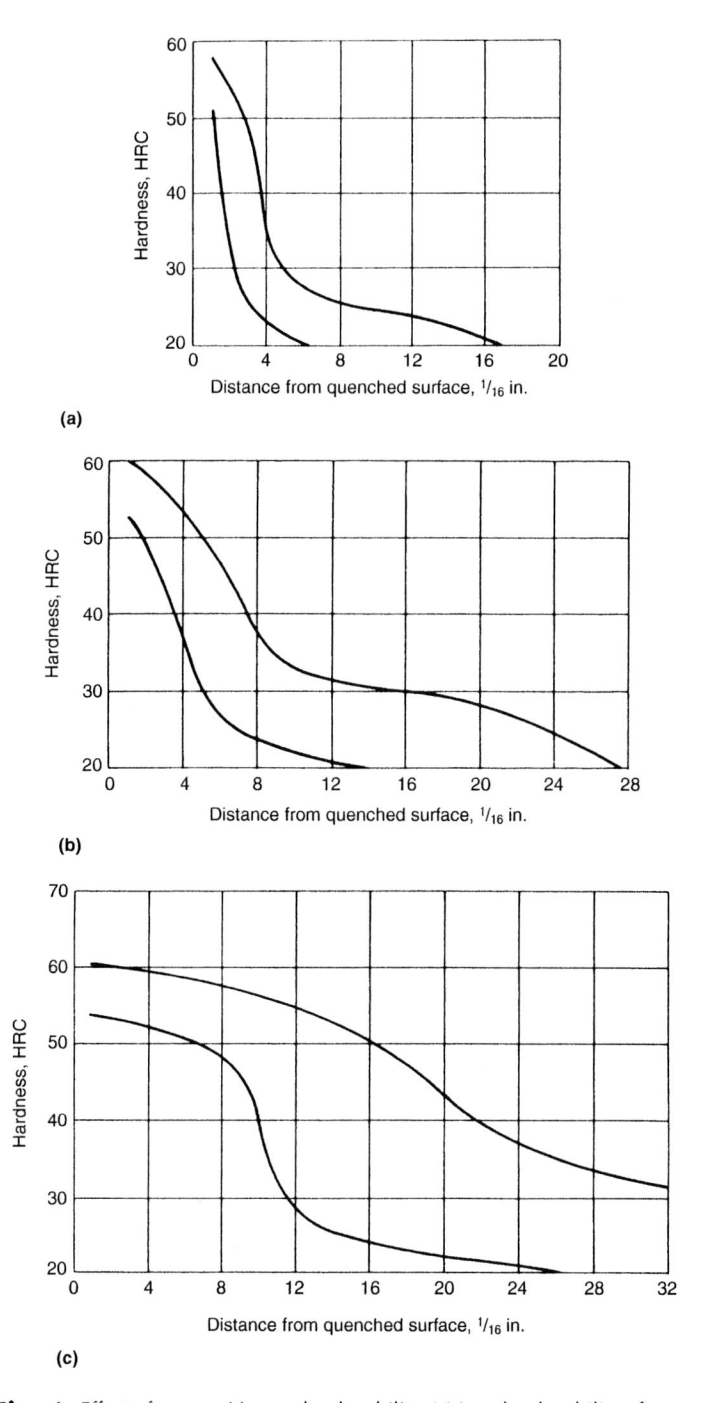

Fig. 1 Effect of composition on hardenability. (a) Low hardenability of a conventional 1038H carbon steel. Source: Ref 3. (b) Effect of manganese on hardenability of 1541H steel. Source: Ref 4. (c) Effect of manganese and boron on hardenability of 15B41H steel. Source: Ref 5

Free-Machining Carbon Steels

For the carbon steels discussed previously, sulfur and phosphorus were generally regarded as impurities; thus, it was desirable to keep their contents as low as possible. Free-machining carbon steels contain controlled amounts of these elements to improve machinability.

There are currently two different series of free-machining carbon steels. First is the series assigned AISI 11XX, referred to as resulfurized carbon steels. The carbon content ranges vary from 0.08 to 0.13% C for AISI 1110 to 0.48 to 0.55% C for AISI 1151. Several members of this series also contain a relatively high manganese content (as high as 1.65% for 1144), and sulfur contents as high as 0.33%. Phosphorus content for this series is, however, restricted to a maximum of 0.040%. Compositions for the standard resulfurized grades are given in Table 5.

A second series of free-machining carbon steels is identified as rephosphorized and resulfurized carbon steels and is coded as AISI 12XX.

All of the 12XX grades are low in carbon content (0.15% maximum). Also, the maximum manganese content is generally lower compared with the 11XX series. Sulfur content is also high in the 12XX steels (0.35% maximum for two grades), but the outstanding difference between the 11XX and 12XX series is that the steels of the latter series have been rephosphorized as well as resulfurized, for better machinability than can be provided by resulfurization alone. Compositions of the 12XX series are presented in Table 6.

Leaded Steels. The last composition shown in Table 6 also contains lead as indicated by the letter "L" between the second and third digits of the designation (12L14). While this is the only standard grade that contains lead and controlled amounts of sulfur and/or phosphorus, almost any of the free-machining grades, as well as many of the nonfree-machining grades (10XX and 15XX series), are available by special order with lead additions. Although lead additions do not provide the degree of improved machinability that can be provided by resulfurization and/or rephosphorization, neither is there as great a sacrifice of mechanical properties for lead additions compared with sulfur and/or phosphorus additions.

Effects of Free-Machining Additives on Properties and Heat Treating Procedures

To say the least, the free-machining additives are not beneficial to the mechanical properties of the workpieces into which these steels are made or to heat treating procedures. Free-cutting additives create a dispersion of tiny voids in the steel, which is undesirable and may lead to cracking during heat treatment. Therefore, in using any free-machining steel, the decision must be made as to whether the impairment of mechanical properties can be tolerated to gain better machinability, which is the sole benefit.

Effects of Sulfur. Additions of sulfur probably disturb the mechanical properties more than either of the other two additives. Figure 2 shows the polished and unetched surface of a resulfurized free-machining steel. Sulfur combines with iron and/or manganese to form the prominent inclusions. Because inclusions have no strength, each one creates a tiny void. They serve as chip breakers in machining applications and prevent buildup on edges of cutting tools, sometimes eliminating the need for cutting fluids. Inclusions are "built in" and are thus more effective in this area. Resulfurization has various effects on properties of finished workpieces—notably on impact strength and transverse tensile strength (opposite to rolling direction).

As a rule, the effects of sulfur on heat treating procedures is minimal; that is, the procedures for heat treating a resulfurized steel versus its non-resulfurized counterpart (same carbon content) are generally the same. There is one factor, however, that can be misleading to the heat treater if not clearly understood. In the higher sulfur grades such as 1144, sulfur combines with manganese, forming manganese sulfide, which impoverishes the matrix of its manganese and lowers hardenability. For example, 1144 steel may contain up to 1.65% Mn, which suggests that its hardenability may be about equal to that of 1541. However, this is not necessarily ture; 1144 has a very high sulfur range (0.24 to 0.33%) so that some of the manganese may be tied up by the sulfur.

Effects of phosphorus are similar to those just described for sulfur. Because phosphorus is not added in great amounts, its effect on mechanical properties is minimal and its effect on heat treating procedures is essentially nil.

Effects of Lead. Lead may be added to any of the carbon steels, as well as to alloy steels (see Chapter 7), to provide improved machinability.

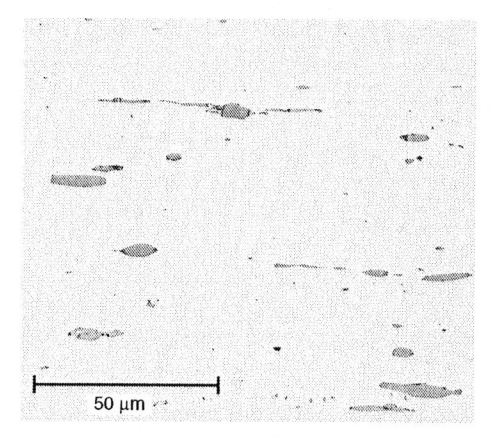

Fig. 2 Manganese sulfide inclusions in resulfurized free-machining steel. Unetched (as-polished). Original magnification: 500×. Source: Ref 6

However, the amount of improvement provided by the addition of 0.15 to 0.35% Pb is not nearly as great compared with sulfur and/or phosphorus additions. Likewise, lead additions are far less likely to impair mechanical properties compared with other additives, notably sulfur. Leaded steels are produced by adding fine lead shot to a stream of molten steel as it enters the ingot mold. Because lead is totally insoluble in steel, it supposedly remains as a very fine dispersion of almost submicroscopic particles in the solidified steel, if the leading procedure is properly performed. Lead segregations can and sometimes do occur, which can cause serious difficulty in heat treatment. Almost every heat treater who has treated quantities of leaded steel parts has had the disheartening experience of finding holes or porosity in the heat treated workpieces that were caused by segregations of lead that melted out during heat treatment.

General Precautions. None of the carbon steels that contain free-machining additives, and especially the resulfurized grades, should be considered for applications that require any appreciable amount of cold forming. Likewise, while free-machining steels are not totally unforgeable, they are prone to split in forging. Consequently, they should be considered only for forgings of very mild severity.

Classification for Heat Treatment

To simplify consideration of the treatments for various applications, the steels in this discussion are classified as follows: group I, 0.08 to 0.25% C; group II, 0.30 to 0.50% C; and group III, 0.55 to 0.95% C. A relatively few steels, such as 1025 and 1029, can be assigned to more than one group, depending on their precise carbon content.

Group I (0.08 to 0.25% C). The three principal types of heat treatment used on these low-carbon steels are (a) process treating of material to prepare it for subsequent operations; (b) treating of finished parts to improve mechanical properties; (c) case hardening, notably by carburizing or carbonitriding to develop a hard, wear-resistant surface. It is often necessary to process anneal drawn products between operations, thus relieving work strains in order to permit further working (Fig. 3). This operation normally is carried out at temperatures between the recrystallization temperature and the lower transformation temperatures. The effect is to soften by recrystallization of ferrite. It is desirable to keep the recrystallized grain size relatively fine. This is promoted by rapid heating and short holding time at temperature. A similar practice may be used in the treatment of low-carbon, cold headed bolts made from cold drawn wire. Sometimes, the strains introduced by cold working so weaken the heads that they break through the most severely worked portion under slight additional strain. Process annealing is used to overcome this condition. Stress relieving at about 540 °C (1000 °F) is more effective than annealing in retaining the normal mechanical properties of the shank of the cold headed bolt.

Heat treating frequently is used to improve machinability. The generally poor machinability of the low-carbon steels, except those containing sulfur or other additive elements, results principally from the fact that the proportion of free ferrite to carbide is high. This situation can be modified by putting the carbide into its most voluminous form—pearlite—and dispersing fine particles of this pearlite evenly throughout the ferrite mass. Normalizing commonly is used with success, but best results are obtained by quenching the steel in oil from 815 to 870 °C (1500 to 1600 °F). With the exception of steels containing a carbon content approaching 0.25%, little or no martensite is formed, and the parts do not require tempering.

Group II (0.30 to 0.50% C). Because of the higher carbon content, quenching and tempering become increasingly important when steels of this group are considered. They are the most versatile of the carbon steels because their hardenability can be varied over a wide range by suitable controls. In this group of steels, there is a continuous change from water-

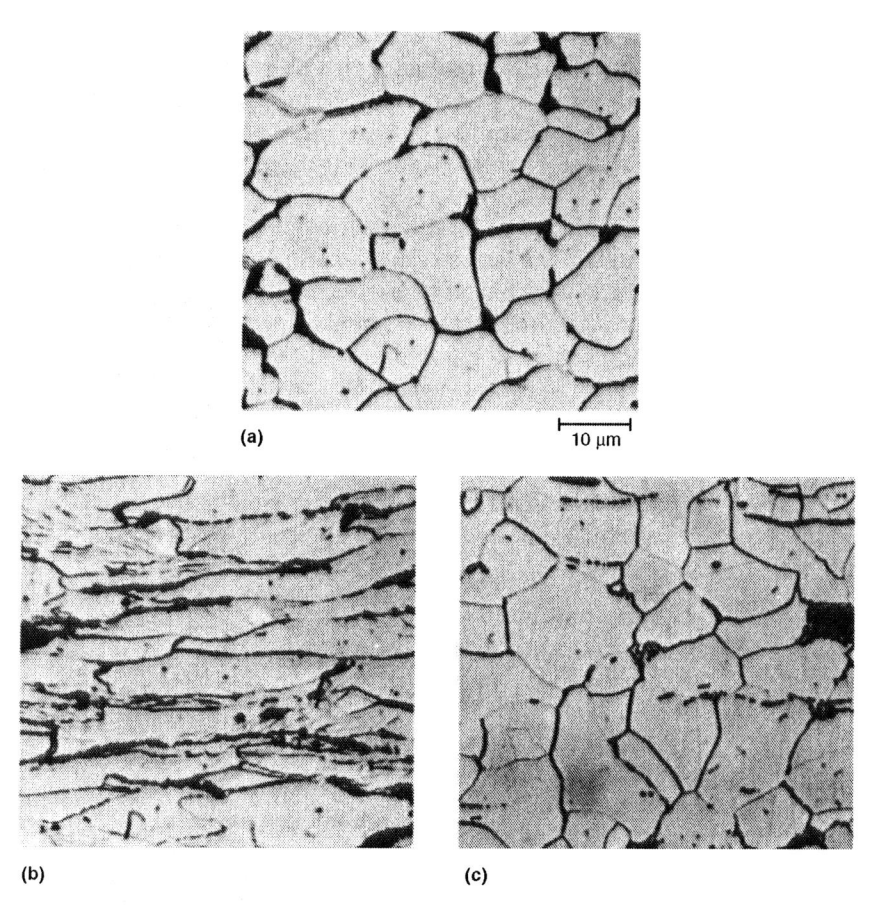

Fig. 3 Microstructure of 1008 steel. (a) At 1000× after slight cold reduction. (b) Same steel but after a 60% cold reduction. (c) Same steel but after process annealing at 595 °C (1100 °F). Source: Ref 7

hardening to oil-hardening types. Hardenability is very sensitive to changes in chemical composition, particularly to the content of manganese, silicon, and residual elements, as well as grain size. These steels are also very sensitive to changes in section size.

The medium-carbon steels should be either normalized or annealed before hardening in order to obtain the best mechanical properties after hardening and tempering. Parts made from bar stock are frequently given no treatment prior to hardening (the prior treatment having been performed at the steel mill), but it is common practice to normalize or anneal forgings.

These steels, whether hot finished or cold finished, machine reasonably well in bar stock form and are machined as received, except in the higher-carbon grades and small sizes that require annealing to reduce the as-received hardness. Forgings usually are normalized to improve machinability over that encountered with the fully annealed structure.

In hardening, the selection of quenching medium varies with steel composition, design of the part, hardenability of the steel, and the hardness desired in the finished part. Water is the quenching medium most commonly used because of the low hardenability of the steel grades. Caustic soda solution (5 to 10% NaOH) is used in some instances with improved results. It is faster than water and may produce better mechanical properties in all but light sections. It is hazardous, however, and operators must be protected against contact with it. Salt solutions (brine) are often successful used. They are not dangerous to operators, but their corrosive action on iron or steel parts or equipment is potentially serious. When the section is light or the properties required after heat treatment are not very high, oil quenching often is used. Finally, medium-carbon steels are readily case hardened by flame or induction hardening.

Group III (0.55 to 0.95% C). Forged parts made of these steels should be annealed for several reasons. Refinement of the forged structure is important in producing a high-quality, hardened product. The parts come from the forging operation too hard for cold trimming of the flash or for any machining operations. Ordinary annealing practice, followed by furnace cooling to about 595 °C (1100 °F) is satisfactory for most parts.

Hardening by conventional quenching is used on most parts made from steels in this group. However, special techniques are required at times. Both oil and water quenching are used; water is used for heavy sections.

Austempering commonly is used for processing these high-carbon grades; see the description of austempering at the end of this chapter.

Tools with cutting edges are sometimes heated in liquid baths to the lowest temperatures at which the part can be hardened and are then quenched in brine. The fast heating of the liquid bath plus the low temperature fail to put all of the available carbon into solution. As a result, the cutting edge consists of martensite containing less carbon than indicated by the chemical composition of the steel and containing many embedded particles of cementite. In this condition, the tool is at its maximum toughness relative to its hardness, and the embedded carbides promote long life of the cutting edge. Final hardness is 55 to 60 HRC. Steels in this group are also commonly hardened by flame or induction methods (see Chapter 9, "Flame and Induction Hardening").

Heat Treating Procedures for Specific Grades of Carbon Steels

Detailed heat treating procedures are presented subsequently for a representative group of carbon steels beginning with the very low carbon deep-drawing grade 1008. For greater details or precise procedures on specific steels that are not discussed in this section, see Ref 7.

1008 through 1019—Recommended Heat Treating Practice

The most common heat treatment applied to this steel is process annealing, which consists of heating to approximately 600 °C (1100 °F), which recrystallizes cold-worked structures. This process frequently is used as an intermediate anneal prior to further cold working (Fig. 3). Note in Fig. 3(a) that the grains are generally symmetrical because the steel has been subjected to only a slight amount of cold reduction. The low carbon content of 1008 is indicated by the small amount of pearlite (black areas). Figure 3(b) shows the same steel after severe cold reduction; note how the grains have become flattened and elongated. This is accompanied by a marked increase in hardness caused by cold working. Restoration of the low hardness and initial grain structure is accomplished by process annealing at 595 °C (1100 °F), as shown in Fig. 3(c). This treatment, and the results, are representative of the low-carbon grades that are used for cold forming applications.

Case Hardening. Hard, wear-resistant surfaces can be obtained on parts by carbonitriding (see Chapter 8). Depth of case developed depends on time and temperature. Carbonitride at 760 to 870 °C (1400 to 1600 °F). Atmosphere usually comprises an enriched carrier gas from an endothermic generator, plus about 2 to 5% anhydrous ammonia. Case depths usually range from about 0.08 to 0.25 mm (0.003 to 0.010 in.). Maximum surface hardness usually is obtained by oil quenching directly from the carbonitriding temperature. Cyanide or cyanide-free liquid salt baths may also be used to develop cases essentially the same as those obtained in the gas process. Temperatures, time at temperature, and quenching practice are also approximately the same.

Although the vast majority of parts case hardened by carbonitriding are not tempered, they can be rendered less brittle by tempering at 150 to 205 °C (300 to 400 °F) without appreciable loss of surface hardness.

1020—Recommended Heat Treating Practice

Normalizing. Heat to 925 °C (1700 °F). Air cool.

Annealing. Heat to 870 °C (1600 °F). Cool slowly, preferably in furnace.

Hardening. Can be case hardened by any one of several processes, which range from light case hardening, such as carbonitriding and the others described for grade 1008, to deeper case carburizing in gas, solid, or liquid mediums. Most carburizing is done in a gaseous mixture of methane combined with one of several carrier gases, using the temperature range of 870 to 955 °C (1600 to 1750 °F). Carburize for desired case depth with a 0.90 carbon potential. Case depth achieved is always a function of time and temperature. For most furnaces, a temperature of 955 °C (1750 °F) approaches the practical maximum without causing excessive deterioration in the furnace. With the advent of vacuum carburizing, temperatures up to 1095 °C (2000 °F) can be used to develop a given case depth in about one-half the time required at the more conventional temperature of 925 °C (1700 °F) (see Chapter 8).

Hardening after carburizing is usually achieved by quenching directly into water or brine from the carburizing temperature. After the desired carburizing cycle had been completed, the furnace temperature can be decreased or a lower temperature zone can be used for a continuous furnace to 845 °C (1550 °F) for a diffusion cycle. Quenching into water or brine then tempering at 150 °C (300 °F) follows.

1035—Recommended Heat Treating Practice

Normalizing, if required, is accomplished by heating to 915 °C (1675 °F) and cooling in still air.

Annealing. Heat to 870 °C (1600 °F). Furnace cool at a rate not exceeding 28 °C (50 °F) per hour to 650 °C (1200 °F).

Hardening. Austenitize at 855 °C (1575 °F). Quench in water or brine, except for sections under 6.35 mm (¼ in.), which may be oil quenched.

Tempering. As-quenched hardness should be approximately 45 HRC. Hardness can be adjusted downward by tempering (see curve in Fig. 4).

1045, 1045H—Recommended Heat Treating Practice

Normalizing. Heat to 900 °C (1650 °F). Cool in air.

Annealing. Heat to 845 °C (1550 °F). Cool in furnace at a rate not exceeding 28 °C (50 °F) per hour to 650 °C (1200 °F).

Hardening. Austenitize at 845 °C (1550 °F). Quench in water or brine. Oil quench sections under 6.35 mm (¼ in.) thick.

Tempering after Hardening. Hardness of at least 55 HRC, if properly austenitized and quenched. Hardness can be adjusted by tempering (see curve in Fig. 4).

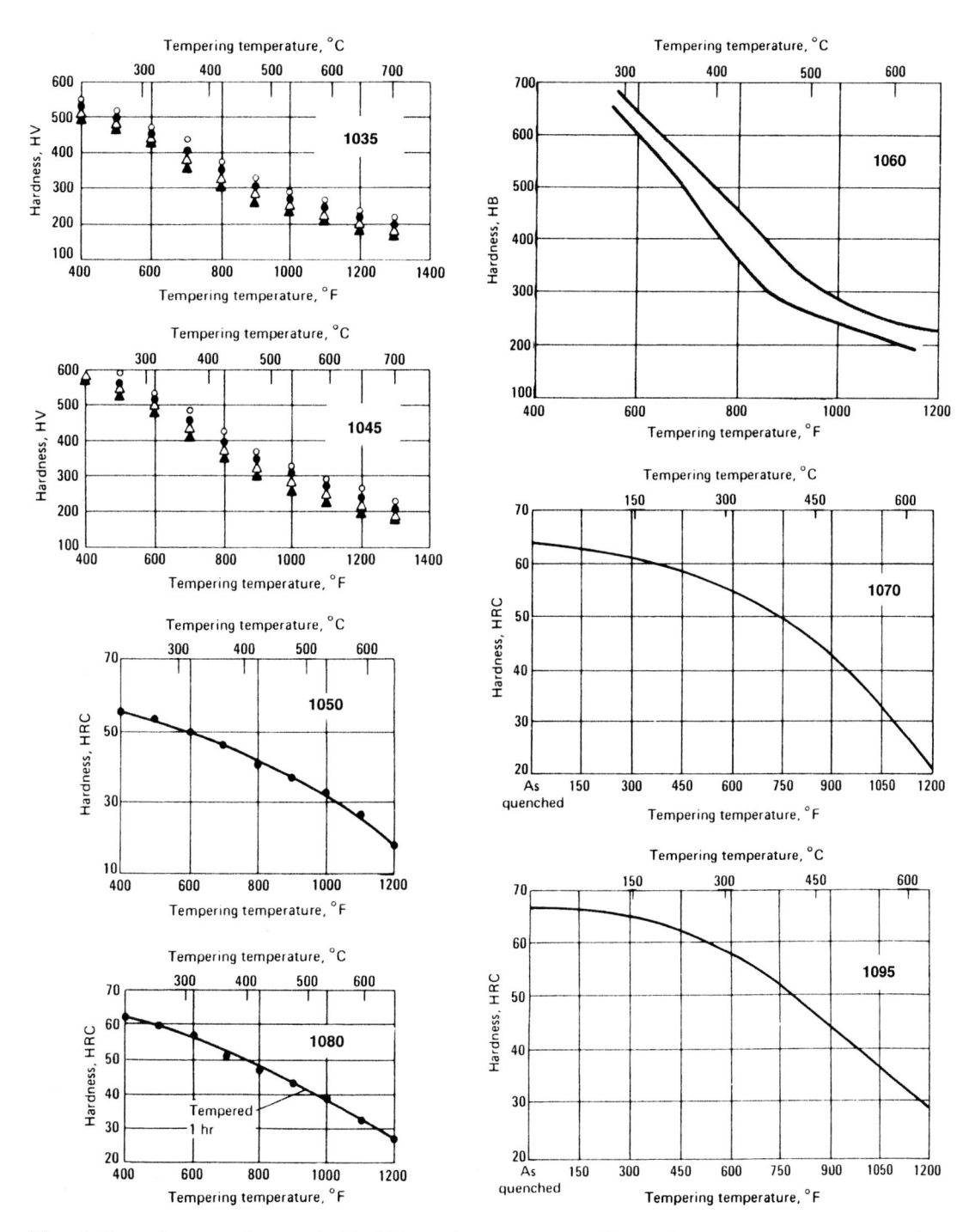

Fig. 4 Tempering curves for several of the higher-carbon 10XX series carbon steels. Specimens, 3.18 to 6.35 mm (1/8 to 1/4 in.) thick. ○: 10 min. ●: 1 hr. △: 4 hr. ▲: 24 hr. 1035: Tempered. 1045: Quenched. 1050 and 1080: Quenched. Tempered 1 hr. 1060: Specimen, 4.8 to 105 mm (3/16 to 4-1/8 in.) thick. Quenched. 1070 and 1095: Represent an average based on a fully quenched surface. Source: Ref 7

This steel and others having a carbon content within the range of approximately 0.35 to 0.55% usually are normalized followed by tempering. For thick sections, it is obviously impossible to exceed the critical cooling rate to form martensite. Thus, a compromise is to transform the microstructure by normalizing to fine pearlite (see time-temperature-transformation, TTT, curve in Chapter 2, also Fig. 5 in this chapter). The normalizing operation is then followed by tempering to relieve stresses and develop the required hardness and/or strength.

1050—Recommended Heat Treating Practice

Normalizing. Heat to 900 °C (1650 °F). Cool in air.

Annealing. Heat to 830 °C (1525 °F). Cool in furnace at a rate not exceeding 28 °C (50 °F) per hour to 650 °C (1200 °F).

Hardening. Heat to 830 °C (1525 °F). Quench in water or brine. Sections less than 6.35 mm (¼ in.) thick may be fully hardened by oil quenching. Normalize and temper as described for 1045.

Tempering. As-quenched hardness of 58 to 60 HRC can be adjusted downward by the proper tempering temperature as shown by the curve presented in Fig. 4.

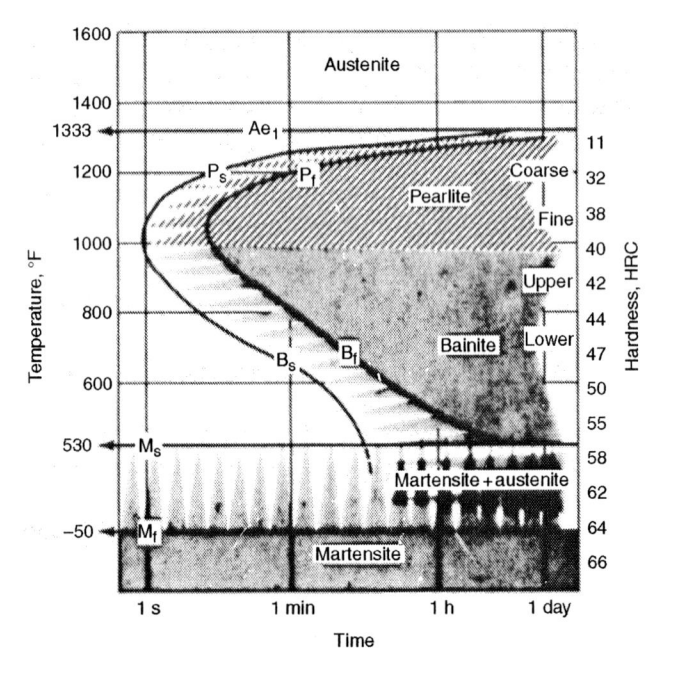

Fig. 5 Complete isothermal transformation diagram for 0.80% carbon steel. All of the transformation products are named. Bainite transformation takes place isothermally between 275 and about 525 °C (530 and about 975 °F). If austenite is rapidly cooled from above A_1 past the nose of the curve and to temperatures below 275 °C (530 °F), M_s martensite starts to form. As long as cooling continues, more martensite forms. Transformation of austenite to martensite is not complete until M_f is reached. Source: Ref 7

1060—Recommended Heat Treating Practice

Normalizing. Heat to 885 °C (1625 °F). Cool in air.
Annealing. Heat to 830 °C (1525 °F). Furnace cool to 650 °C (1200 °F) at a rate not exceeding 28 °C (50 °F) per hour.
Hardening. Heat to 815 °C (1500 °F). Quench in water or brine. Oil quench sections under 6.35 mm (¼ in.)
Tempering. As-quenched hardness from 62 to 65 HRC. This maximum hardness can be adjusted downward by proper tempering temperature (see curve in Fig. 4).
Austempering. Thin sections (typically springs) are austempered. Results in a bainitic structure and hardness of approximately 46 to 52 HRC. Austenitize at 815 °C (1500 °F). Quench in molten salt bath at 315 °C (600 °F). Hold at temperature for at least 1 h. Air cool. No tempering required. A more detailed discussion of the austempering practice is given at the end of this chapter.

1070—Recommended Heat Treatment

Normalizing. Heat to 885 °C (1625 °F). Cool in air.
Annealing. Heat to 830 °C (1525 °F). Furnace cool to 650 °C (1200 °F) at a rate not exceeding 28 °C (50 °F) per hour.
Hardening. Heat to 815 °C (1500 °F). Quench in water or brine. Oil quench sections under 6.35 mm (¼ in.) thick.
Tempering. As-quenched hardness of approximately 65 HRC. Hardness can be adjusted downward by proper tempering as demonstrated by the tempering curve in Fig. 4.
Austempering is also a practical treatment for 1070, using the same practice show for 1060.

1080—Recommended Heat Treating Practice

Normalizing. Heat to 870 °C (1600 °F). Cool in air.
Annealing. Heat to 815 °C (1500 °F). Furnace cool to 650 °C (1200 °F) at a rate not exceeding 28 °C (50 °F) per hour.
Hardening. Heat to 815 °C (1500 °F). Quench in water or brine. Oil quench sections under 6.35 mm (¼ in.) thick.
Tempering. As-quenched hardness of approximately 65 HRC. Hardness can be adjusted downward by proper tempering (see curve in Fig. 4).
Austempering. Grade 1080 is frequently subjected to austempering because of its wide usage, namely for small, thin flat springs. The same practice as outlined for 1060 is applicable for 1080.

1095—Recommended Heat Treating Practice

Normalizing. Heat to 855 °C (1575 °F). Cool in air.
Annealing. As is generally true for all high-carbon steels, bar stock supplied by mills in spheroidized condition. Annealed with structure of

fine spheroidal carbides in ferrite matrix. When parts are machined from bars in this condition, no normalizing or annealing required. Forgings should always be normalized. Anneal by heating at 800 °C (1475 °F). Soak thoroughly. Furnace cool to 650 °C (1200 °F) at a rate not exceeding 28 °C (50 °F) per hour. From 650 °C (1200 °F) to ambient temperature, cooling rate is not critical. This relatively simple annealing process will provide predominately spheroidized structure, desired for subsequent heat treating or machining.

Hardening. Heat to 800 °C (1475 °F). Quench in water or brine. Oil quench sections under 1.58 mm (3/16 in.) for hardening.

Tempering. As-quenched hardness as high as 66 HRC. Can be adjusted downward by tempering as shown by the curve in Fig. 4.

Austempering. Responds well to austempering (bainitic hardening). Austenitize at 800 °C (1475 °F). Quench in agitated molten salt bath at 315 °C (600 °F). Hold for 2 h. Cool in air.

12L14—Recommended Heat Treating Pratice

This grade represents one of the most free-machining grades that has ever been developed—resulfurized, rephosphorized, and leaded. This grade should never be considered for applications where forging or welding is involved; consequently, normalizing and annealing are seldom required.

Case hardening treatments such as carbonitriding are commonly used for parts made from 12L14 and other low-carbon free-machining grades (see steel 1008 and refer to Chapter 8).

1137—Recommended Heat Treating Practice

Normalizing. Not usually required. If necessary, heat to 900 °C (1650 °F). Cool in air.

Annealing. Heat to 885 °C (1625 °F). Furnace cool to 650 °C (1200 °F) at a rate not exxceeding 28 °C (50 °F).

Hardening. Heat to 845 °C (1550 °F). Quench in water or brine. For full hardness, oil quench sections not exceeding 9.52 mm (3/8 in.) thick.

Tempering. As-quenched hardness approximately 45 HRC. Hardness can be adjusted downward by tempering as shown in Fig. 6.

Strengthening by Cold Drawing and Stress Relieving. Grade 1137 bars and other medium-carbon resulfurized steels are frequently strengthened to desirable levels without quench hardening. Increase the draft during cold drawing by 10 to 35% above normal. Stress relieve by heating at approximately 315 °C (600 °F). Produces yield strenths of up to 690 MPa (100 ksi) or higher in bars up to about 19.05 mm (3/4 in.) diam. Machinability is very good.

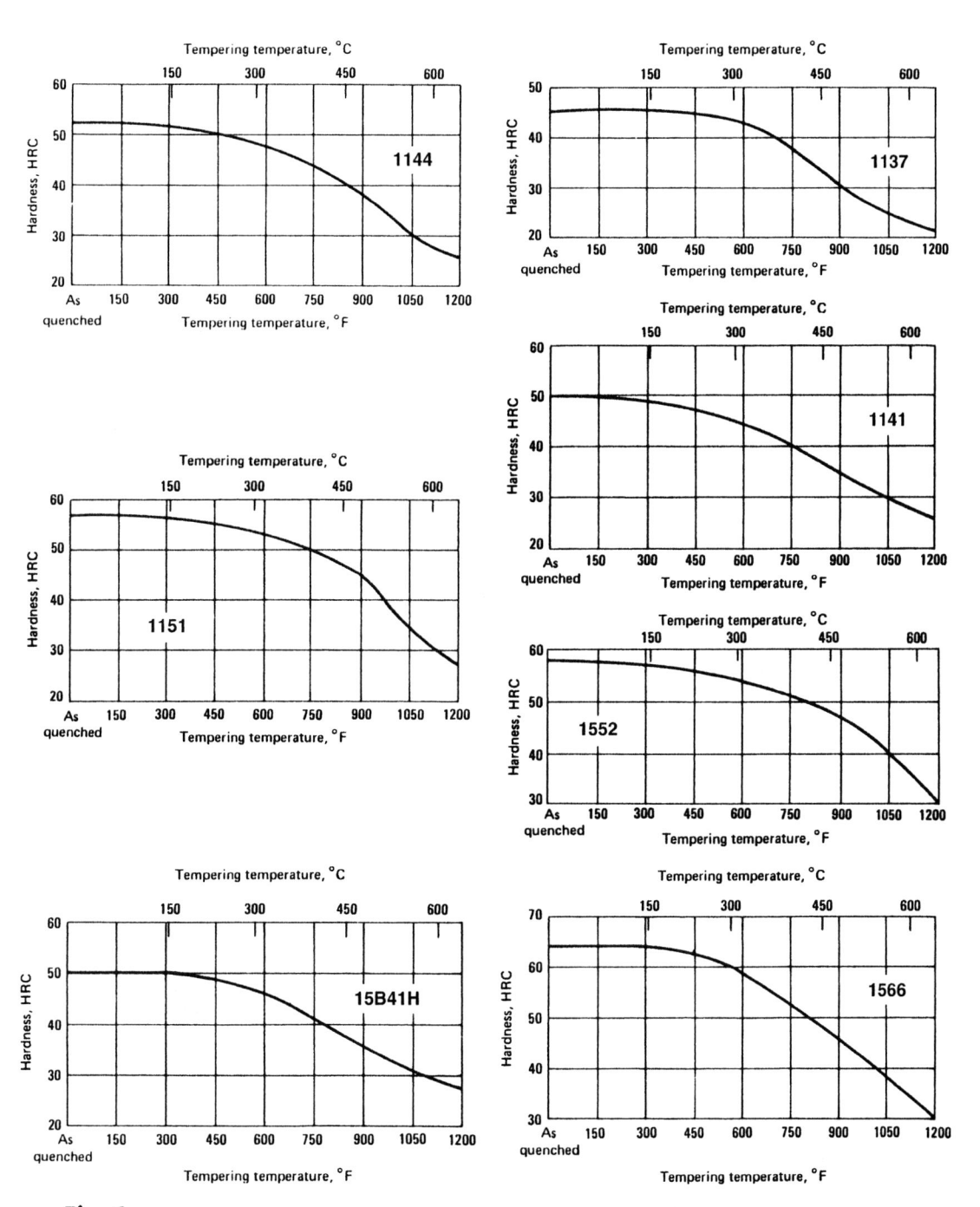

Fig. 6 Tempering curves for several of the higher-carbon grades of the 11XX and 15XX series. Source: Ref 7

1141—Recommended Heat Treating Practice

Normalizing. Not usually required. If necessary, heat to 885 °C (1625 °F). Cool in air.

Annealing. Usually purchased by fabricator in condition for machining. If required may be annealed by heating to 845 °C (1550 °F). Cool in furnace to 650 °C (1200 °F), or lower at a rate nor exceeding 28 °C (50 °F) per hour.

Hardening. Austenitize at 830 °C (1525 °F). Quench in water or brine. For full hardness, oil quench sections less than 9.52 mm (3/8 in.) thick.

Tempering. Depending on precise carbon content, as-quenched hardness usually 48 to 52 HRC. Hardness can be reduced by tempering (see curve in Fig. 6).

Grade 1141 is also amenable to strengthening by cold drawing and stress relieving as described previously for 1137.

1144—Recommended Heat Treating Practice

This grade is generally considered a special-purpose steel. As shown in Table 5, it has a very high sulfur content, equal to free-machining 1213, thus permitting extremely fast machining with heavy cuts. Machined finishes are unusually good. High sulfur content reduces transverse impact and ductility. Can be drawn by heavy drafts at elevated temperature, 370 °C (700 °F), which results in relatively high strength and high hardness, up to 35 HRC. Machinability is excellent. Widely used for producing machined parts put in service without heat treatment. Can be purchased in the cold drawn condition and heat treated. Depending on precise carbon content, as-quenched and fully hardened 1144 should be approximately 52 to 55 HRC. Never used where forging or welding is involved.

Normalizing. Seldom required. If necessary, heat to 885 °C (1625 °F). Cool in air.

Annealing. Seldom necessary. May be annealed by heating 845 °C (1550 °F). Furnace cool to 650 °C (1200 °F) at a rate not exceeding 28 °C (50 °F) per hour.

Hardening. Austenitize at 830 °C (1525 °F). Quench in water or brine. For full hardness, oil quench sections less than 9.52 mm (3/8 in.) thick.

Tempering. Depending on precise carbon content, as-quenched hardness usually 52 to 55 HRC. Hardness can be reduced by tempering, as shown in Fig. 6.

1151—Recommended Heat Treating Practice

Normalizing. Seldom used. If necessary, heat to 900 °C (1600 °F). Cool in air.

Annealing. Seldom used. If necessary, heat to 870 °C (1550 °F). Cool in furnace at a rate not exceeding 28 °C (50 °F) per hour to 650 °C (1200 °F).

Hardening. Austenitize at 830 °C (1525 °F). Quench in water or brine. Oil quench sections under 6.35 mm (¼ in.) thick.

Tempering after Hardening. Hardness of at least 55 HRC, if properly austenitized and quenched. Hardness can be adjusted by tempering, as shown in Fig. 6.

1522, 1522H—Recommended Heat Treating Practice

Normalizing. Heat to 925 °C (1700 °F). Air cool.

Annealing. Heat to 870 °C (1600 °F). Cool slowly, preferably in the furnace.

Hardening. Can be case hardened by any one of several processes, from light case hardening by carbonitriding and salt bath nitriding described for grade 1008 to deeper case carburizing in gas, solid, or liquid media. See carburizing process described for grade 1020. For 1522H, use carburizing temperature of 900 to 925 °C (1650 to 1700 °F). Use oil for cooling medium. Because of higher hardenability when compared with 1020, thicker sections can be oil quenched for full hardness. Refer to case hardening procedure given for grade 1020 and also Chapter 8.

Tempering. Temper at 150 °C (300 °F), or higher for 1 h, if some sacrifice of hardness can be tolerated.

15B41H—Recommended Heat Treating Practice

Normalizing. Heat to 900 °C (1650 °F), and cool in air.

Annealing. Heat to 830 °C (1525 °F). Furnace cool at a rate not exceeding 28 °C (50 °F) per hour to 650 °C (1200 °F).

Hardening. Heat to 845 °C (1550 °F). Should be regarded as an alloy steel in quenching from austenitizing temperature because of the boron addition, which greatly increases hardenability compared with conventioned 1541 steel (see Fig. 1c). Therefore, oil quenching is usually practiced for parts made from 15B41.

Tempering after Hardening. As-quenched hardness is generally around 52 HRC, which can be reduced as desired by tempering (see Fig. 6).

Tempering after Normalizing. For large sections, normalize by conventional practice. Results in structure of fine pearlite. Temper up to about 540 °C (1000 °F). Mechanical properties attained by this treatment are not equal to those obtained by quenching and tempering but are much better than the coarse pearlitic structure that results from annealing.

1552—Recommended Heat Treating Practice

Normalizing. Heat to 900 °C (1650 °F). Cool in air.

Annealing. Heat to 830 °C (1525 °F). Cool in furnace at a rate not exceeding 28 °C (50 °F) per hour to 650 °C (1200 °F).

Hardening. Heat to 830 °C (1525 °F). Because of higher hardenability provided by the high manganese content, oil quench heavier sections for full hardness. When induction hardening, use the least severe quench that will produce full hardness. This minimizes the possibility of quench cracking.

Tempering. As-quenched hardness of 58 to 60 HRC can be reduced by tempering, as shown in Fig. 6.

1566—Recommended Heat Treating Practice

This grade is simply a slightly higher manganese version of 1070. Therefore, it is used for purposes that parallel the application of 1070, namely, springs and a large variety of hand tools including saws and hammers, or similar applications requiring more hardenability than can be provided by 1070.

Normalizing. Heat to 885 °C (1625 °F). Cool in air.

Annealing. Heat to 830 °C (1525 °F). Furnace cool to 650 °C (1200 °F) at a rate not exceeding 28 °C (50 °F) per hour.

Hardening. Heat to 815 °C (1500 °F). Lessen severity of quench to avoid cracking compared with quenching of 1070.

Tempering. Maximum hardness near 65 HRC can be reduced by tempering, as shown in Fig. 6.

Austempering. Thin sections (typically springs) are commonly austempered, resulting in bainitic structure and a hardness range of approximately 46 to 52 HRC. Austenitize at 815 °C (1500 °F). Quench in molten salt bath at 315 °C (600 °F). Hold at temperature for 1 h. Air cool. No tempering is required.

Martensitic and Nonmartensitic Structures

Throughout this book, it is stressed that the ideal condition when hardening an entire part made from the subject steel is to attain a 100% martensitic structure from the quench. This may be an objective of quenching, but when one examines the TTT curve presented in Chapter 2 and a similar curve in Fig. 5, it is apparent that 100% martensite is impossible to obtain in carbon steels except in very thin sections. To do so, quenching must be very severe to exceed rapid cooling rates.

Therefore, in practice, the objective of forming 100% martensite in the as-quenched parts is not achieved except in very thin sections. As section thickness increases, the resulting microstructure becomes more of a mixture. However, even in quenching very heavy sections, formation of coarse pearlite can be avoided (near the top of Fig. 5), thus forming fine pearlite, probably some bainite and perhaps some martensite, depending on the pattern of the cooling curve attained. Note that the hardness of the quenched product gradually decreases (right side of Fig. 5) from martensite to coarse pearlite; that is, the lower the temperature where transformation takes place, the harder the quenched product.

Some massive parts, such as large shaft forgings made from steels similar to 1045 or 1050, are not quenched at all but are simply normalized by heating to about 900 °C (1650 °F) and by cooling in still air. They are then reheated (tempered) to approximately 540 °C (1000 °F).

As a rule, such treatments result in a structure of fine pearlite, which has good mechanical properties even though martensite is not formed.

Tempering of Quenched Carbon Steels

All parts made from carbon steels (or from quench-hardened steels) should be tempered immediately after cooling to near room temperature by quenching. This prevails for through-hardened, as well as case hardened, parts (carburized or carbonitrided specifically). It is, however, far more important that through-hardened parts are tempered; this becomes increasingly so as the carbon content increases. Many carburized or carbonitrided parts are never tempered, but this is not considered good practice.

The need for tempering also increases as the amount of martensite formed increases. Martensite is formed when carbon atoms are trapped in the iron lattice. When a high-carbon martensite is formed, the iron lattice actually has a tetragonal shape (instead of strictly cubic). In this condition, it is metastable and extremely brittle (fragile). The structure is known as white martensite because of the way it responds to etchants when prepared for microscopic examination. This metastable martensite changes to some degree, even at room temperature in a matter of time. The change from tetragonal to cubic martensite can be accomplished quickly, however, by heating. At approximately 85 °C (185 °F), the change begins, generally known as the first stage of tempering. This heating procedure brings about a marked decrease in brittleness without decreasing hardness as measured by indentation. Before heat treating developed from an art to a science, toolmakers heated and quenched tools in boiling water, making them far less brittle.

Although 85 °C (185 °F) does represent the beginning of tempering, a minimum tempering temperature of at least 150 °C (300 °F) is nearly standard because hardness is not lost from this tempering temperature (Fig. 4, 6).

Curves that show the relationship of tempering temperature and hardness for 14 different carbon steels are presented in Fig. 4 and 6. All hardness readings were taken at room temperature after cooling from the tempering operation. Also, all data are based on thin sections wherein 100% (or near 100%) martensite is present. Depending also on the precise carbon contents within the allowable ranges, the as-quenched hardness may be slightly higher or lower than those shown. As the amount of nonmartensitic products increases, the as-quenched hardness is lower, and the entire curve drops slightly.

In examining Fig. 4 and 6, note that the hardness for any given grade drops gradually, which is characteristic for all carbon steels. This condition results from a gradual decomposition of the martensite as tempering temperature increases.

Selection of tempering temperature depends entirely on the mechanical requirements, but there are two simple rules that should always be observed in establishing a tempering temperature. A tempering temperature should be used that is as high as can be tolerated, considering the resulting hardness, and the need for a specific hardness in service. A tempering temperature should always be used that is at least as high as, and preferably slightly higher, than the maximum service temperature to which the parts will be exposed in service.

Austempering of Steel (Ref 8)

Austempering is the isothermal transformation of a ferrous alloy at a temperature below that of pearlite formation and above that of martensite formation (Fig. 5). Steel is austempered by being:

- Heated to a temperature within the austenitizing range, usually 790 to 870 °C (1450 to 1600 °F)
- Quenched in a bath maintained at a constant temperature, usually in the range of 260 to 400 °C (500 to 750 °F)
- Allowed to transform isothermally to bainite in this bath
- Cooled to room temperature, usually in still air

The fundamental difference between austempering and conventional quenching and tempering is shown schematically in Fig. 7.

A conventional heating, quenching, and tempering cycle is shown at the left. Following the black lines down from Ae_3, note that the lines miss

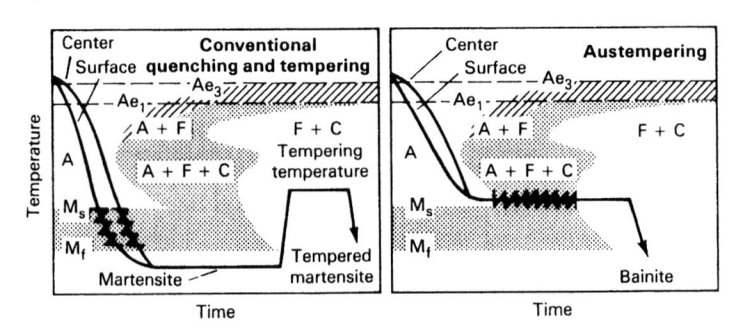

Fig. 7 Comparison of time-temperature-transformation cycles for conventional quenching and tempering and for austempering. Source: Ref 8

the nose of the TTT curve, continue straight through the martensite formation zone M_s to M_f), then remain at constant temperature until transformation is complete, then reheating for tempering, holding time and cooling from the tempering temperature. Thus, the left side of Fig. 7 represents a conventional austenitizing, quenching, and tempering cycle, wherein 100% martensite would be formed.

On the right side of Fig. 7, the workpiece is quenched past the nose of the TTT curve, but cooling is arrested at a temperature just above the martensite zone, the workpiece is then allowed to transform isothermally, after which it is cooled in air Under these conditions, the transformation is to bainite rather than martensite (Fig. 5). Tempering is not required.

As rule, parts that have been austempered have greater toughness at the same hardness level compared with quenched and tempered counterparts. Other advantages of the austempering process are:

- Reduced distortion, which lessens subsequent machining time, stock removal, and cost
- The shortest overall time cycle to through harden within the hardness range of 35 to 55 HRC, with resulting savings in energy and capital investment

Quenching Mediums. Molten salt is the quenching medium most commonly used in austempering because (a) it transfers heat rapidly; (b) it virtually eliminates the problem of a vapor phase barrier during the initial stage of quenching; (c) its viscosity is uniform over a wide range of temperature; (d) its viscosity is low at austempering temperatures (near that of water at room temperature), thus minimizing dragout losses; (e) it remains stable at operating temperatures and is completely soluble in water, thus facilitating subsequent cleaning operations; and (f) the salt can be easily recovered from wash waters so that there is no discharge to drain.

Formulations and characteristics of two typical salt quenching baths are given in Table 7. The high-range salt is suitable for austempering only, whereas the wide-range salt may alternatively be used for austempering or as a quenching medium.

Selection of Grade for Austempering. The selection of steel for austempering must be based on transformation characteristics as indicated in

Table 7 Compositions and characteristics of salts used for austempering

	High range	Wide range
Sodium nitrate	45 to 55%	0 to 25%
Potassium nitrate	45 to 55%	45 to 55%
Sodium nitrite	...	25 to 35%
Melting point (approx.)	220 °C (430 °F)	150 to 165 °C (300 to 330 °F)
Working temperature range	260 to 595 °C (500 to 1100 °F)	175 to 540 °C (345 to 1000 °F)

Source: Ref 8

TTT diagrams. Three important considerations are: the location of the nose of the TTT curve and the time available for bypassing it, the time required for complete transformation of austenite to bainite at the austempering temperature, and the location of the M_s point.

As indicate in Fig. 5, 1080 carbon steel possesses transformation characteristics that provide it with limited suitability for austempering. Cooling from the austenitizing temperature to the austempering bath must be accomplished in about 1 s to avoid the nose of the TTT curve, and thus prevent transformation to pearlite during cooling, Depending on the temperature, isothermal transformation in the bath is completed within a period ranging from a few minutes to about 1 h. Because of the rapid cooling rate required, austempering of 1080 can be successfully applied only to thin sections of about 5 mm (0.2 in.) maximum.

Alloy steels, because of their greater hardenability, are often better suited for austempering than are carbon steels (see Chapter 7).

Applications of Austempering. This process is substituted for conventional quenching and tempering for either or both of two reasons: (a) to obtain improved mechanical properties (particularly higher ductility or notch toughness at a given high hardness), and (b) to decrease the likelihood of cracking and distortion. In some applications, austempering may be less expensive than conventional quenching and tempering. This is most likely when small parts are treated in an automated setup wherein conventional quenching and tempering comprise a three-step operation—that is, austenitizing, quenching, and tempering. Austempering requires only two processing steps: austenitizing and isothermal transformation in an austempering bath.

The range of austempering applications generally encompasses parts fabricated from bars of small diameter or from sheet or strip of small cross section. austempering is particularly applicable to thin-section carbon steel parts requiring exceptional toughness at a hardness near 50 HRC.

REFERENCES

1. H. Chandler, Ed., Carbon & Alloy Steels Introduction, *Heat Treater's Guide: Practices and Procedures for Irons and Steels,* 2nd ed., ASM International, 1995, p 111–120
2. Hardenability Curves, *Properties and Selection: Irons, Steels, and High-Performance Alloys,* Vol 1, *ASM Handbook,* ASM International, 1990, p 485–570
3. H. Chandler, Ed., Nonresulfurized Carbon Steels (1000 Series), *Heat Treater's Guide: Practices and Procedures for Irons and Steels,* 2nd ed., ASM International, 1995, p 172
4. H. Chandler, Ed., High Manganese Carbon Steels (1500 Series), *Heat Treater's Guide: Practices and Procedures for Irons and Steels,* 2nd ed., ASM International, 1995, p 250

5. H. Chandler, Ed., High Managanese Carbon Steels (1500 Steels), *Heat Treater's Guide: Practices and Procedures for Irons and Steels,* 2nd ed., ASM International, 1995, p 253
6. B.L. Bramfitt and S.J. Lawrence, Metallography and Microstructures of Carbon and Low-Alloy Steels, *Metallography and Microstructures,* Vol 9, *ASM Handbook,* ASM International, 2004, p 608–626
7. H.E. Boyer, Chapter 5, *Practical Heat Treating,* 1st ed., American Society for Metals, 1984, p 91–120
8. J.R. Keough, W.J. Laird, and A.D. Godding, Austempering of Steel, *Heat Treating,* Vol 4, *ASM Handbook,* ASM International, 1991, p 152–63

SELECTED REFERENCES

- C.R Brooks, *Principles of the Heat Treatment of Plain Carbon and Low Alloy Steels,* ASM International, 1996
- G.E. Totten and M.A.H. Howes, Ed., *Steel Heat Treatment Handbook,* Marcel Dekker, Inc., 1997

CHAPTER 7

Heat Treating of Alloy Steels

IT IS LOGICAL for one to assume that alloy steels are steels that contain alloying elements other than carbon. While this assumption is technically correct, it is not consistent with the nomenclature generally used by the metalworking community.

Stainless steels, a majority of the tool steels, and many special-purpose steels are highly alloyed but generally are designated for use in specific application, such as heat-resistant grades.

In general, and for purposes of this discussion, alloy steels refer to those steel grades for which specific alloy ranges have been established by American Iron and Steel Institute-Society of Automotive Engineers (AISI-SAE). The standard alloy steel grades established by AISI-SAE have also been assigned designations in the Unified Numbering System (UNS). The UNS is described in Chapter 6, "Heat Treating of Carbon Steels."

What Are Alloy Steels?

A steel is considered to be an alloy steel when the maximum of the range given for the content of alloying elements exceeds one or more of the following limits:

- Manganese: 1.65%
- Silicon: 0.60%
- Copper: 0.60%

A steel is also classified as an alloy steel when a difinite range or a definite minimum quantity of any of the following elements is specified or required within recognized limits:

- Aluminum
- Boron

- Chromium (up to 3.99%)
- Cobalt
- Molybdenum
- Nickel
- Niobium
- Titanium
- Tungsten
- Vanadium
- Zirconium

In addition to these elements, any other element added to obtain a desired effect is characterized as an alloying element.

Designation. Alloy steels also are identified by the AISI-SAE coding system, using the same general procedure described for carbon and resulfurized steels. The first two digits of the four-numeral series for the various grades of alloy steel and their meaning are given in Table 1.

The last two digits of the four-numeral series are intended to indicate the approximate middle of the carbon range. It is necessary, however, to deviate from this rule and to interpolate numbers in the case of some carbon ranges and for variations in manganese, sulfur, chromium, or other elements.

When boron is specified, the letter B is inserted after the first two numbers; for example, 41B30. When lead is specified, the letter L is inserted after the first two numbers; for example, 41L30.

Many alloy steels are specialty steels uniquely suited for certain applications. For instance, 52100 is used almost exclusively for antifriction bearings, and the 9200 series alloys are used for springs and applications

Table 1 Numerical designations of AISI and SAE grades of alloy steels

Series designation	Type and approximate percentages of identifying elements
13XX	Manganese, 1.75
40XX	Molybdenum, 0.20 or 0.25
41XX	Chromium, 0.50, 0.80, or 0.95; molybdenum 0.12, 0.20, or 0.30
43XX	Nickel 1.83, chromium 0.50 or 0.80, molybdenum 0.25
44XX	Molybdenum 0.53
46XX	Nickel 0.85 or 1.83, molybdenum 0.20 or 0.25
47XX	Nickel 1.05, chromium 0.45, molybdenum 0.20 or 0.35
48XX	Nickel 3.50, molybdenum 0.25
50XX	Chromium 0.40
51XX	Chromium 0.80, 0.88, 0.93, 0.95, or 1.00
5XXXX	Carbon 1.04, chromium 1.03 or 1.45
61XX	Chromium 0.60 or 0.95, vanadium 0.13 or 0.15 min
86XX	Nickel 0.55, chromium 0.50, molybdenum 0.20
87XX	Nickel 0.55, chromium 0.50, molybdenum 0.25
88XX	Nickel 0.55, chromium 0.50, molybdenum 0.35
92XX	Silicon 2.00
50BXX	Chromium 0.25 or 0.50
51BXX	Chromium 0.80
81BXX	Nickel 0.30, chromium 0.45, molybdenum 0.12
94BXX	Nickel 0.45, chromium 0.40, molybdenum 0.12

Note: B denotes boron steel. Source: Ref 1

where shock resistance is a factor. In most instances, however, a single composition of steel serves a wide range of applications, for example, the 4340, 8640, and 8740 grades.

H-Steels. Many of the standard grades of alloy steels are guaranteed to meet the requirements of the AISI-SAE hardenability bands. Such steels carry the suffix letter H, for example, 4140H. Therefore, if a steel must meet standard hardenability requirements, it should be so specified.

Restricted hardenability (RH) steels are standard grades of alloy steels that are guaranteed to meet an even more restricted (narrower) hardenability band than H-steels.

AISI-SAE Compositions. Compositions of 58 standard alloy steels are listed in Table 2. In many instances, there are only minor variations among grades. For this reason, all of them are not listed in this chapter. Approximately one-half (which is representative) of the standard alloy steel compositions are listed in Table 2. For a complete list, see Ref 1 and 2. Compositions of standard boron alloy steels are listed in Table 3.

Free-Machining Alloy Steels. As a rule, alloy steels are used instead of carbon steels to improve certain mechanical properties, either as produced or through heat treatment. Therefore, for most applications, users hesitate to specify alloy steels that contain free-machining additives. At best, these additives lower steel quality to some extent, thus partially defeating the purpose of using alloy steels. There are, however, notable exceptions. Consequently, most alloy steels are available with additions of lead or sulfur by special order.

Purposes Served by Alloying Elements

As a rule, the alloys contained in AISI-SAE alloy steels include manganese and silicon (over specified amounts), nickel, chromium, molybdenum, and vanadium, but in a variety of combinations. Boron, because it is used in such small quantities, is not considered an alloy. Other alloying elements such as copper, cobalt, tungsten, and titanium are not usually specified in alloy steels but are used in stainless and tool steels (see Chapters 10, "Heat Treating of Stainless Steels," and 11, "Heat Treating of Tool Steels").

Although there may be several reasons for using certain alloy steels as opposed to carbon steels of the same carbon content, hardenability is by far the most common reason. Hardenability controls the mechanical properties that can be obtained by heat treatment.

Although a few alloy steels contain sufficient quantities of alloying elements to provide some resistance to heat and/or corrosion compared with their carbon steel counterparts, they are not usually selected because of these properties. Only a few of the alloy grades contain a total alloy content of near 5%. For the most part, their total alloy content is substantially less than 5%.

Table 2 Compositions of standard alloy steels

Steel designation AISI or SAE	UNS No.	C	Mn	P max	S max	Si	Ni	Cr	Mo
1330	G13300	0.28–0.33	1.60–1.90	0.035	0.040	0.15–0.30	...	...	...
1335	G13350	0.33–0.38	1.60–1.90	0.035	0.040	0.15–0.30	...	...	...
1340	G13400	0.38–0.43	1.60–1.90	0.035	0.040	0.15–0.30	...	...	...
1345	G13450	0.43–0.48	1.60–1.90	0.035	0.040	0.15–0.30	...	...	...
4023	G40230	0.20–0.25	0.70–0.90	0.035	0.040	0.15–0.30	...	...	0.20–0.30
4024	G40240	0.20–0.25	0.70–0.90	0.035	0.035–0.050	0.15–0.30	...	...	0.20–0.30
4027	G40270	0.25–0.30	0.70–0.90	0.035	0.040	0.15–0.30	...	...	0.20–0.30
4028	G40280	0.25–0.30	0.70–0.90	0.035	0.035–0.050	0.15–0.30	...	...	0.20–0.30
4037	G40370	0.35–0.40	0.70–0.90	0.035	0.040	0.15–0.30	...	...	0.20–0.30
4047	G40470	0.45–0.50	0.70–0.90	0.035	0.040	0.15–0.30	...	...	0.20–0.30
4118	G41180	0.18–0.23	0.70–0.90	0.035	0.040	0.15–0.30	...	0.40–0.60	0.08–0.15
4130	G41300	0.28–0.33	0.40–0.60	0.035	0.040	0.15–0.30	...	0.80–1.10	0.15–0.25
4137	G41370	0.35–0.40	0.70–0.90	0.035	0.040	0.15–0.30	...	0.80–1.10	0.15–0.25
4140	G41400	0.38–0.43	0.75–1.00	0.035	0.040	0.15–0.30	...	0.80–1.10	0.15–0.25
4142	G41420	0.40–0.45	0.75–1.00	0.035	0.040	0.15–0.30	...	0.80–1.10	0.15–0.25
4145	G41450	0.43–0.48	0.75–1.00	0.035	0.040	0.15–0.30	...	0.80–1.10	0.15–0.25
4147	G41470	0.45–0.50	0.75–1.00	0.035	0.040	0.15–0.30	...	0.80–1.10	0.15–0.25
4150	G41500	0.48–0.53	0.75–1.00	0.035	0.040	0.15–0.30	...	0.80–1.10	0.15–0.25
4161	G41610	0.56–0.64	0.75–1.00	0.035	0.040	0.15–0.30	...	0.70–0.90	0.25–0.35
4320	G4320	0.17–0.22	0.45–0.65	0.035	0.040	0.15–0.30	1.65–2.00	0.40–0.60	0.20–0.30
4340	G43400	0.38–0.43	0.60–0.80	0.035	0.040	0.15–0.30	1.65–2.00	0.70–0.90	0.20–0.30
E4340	G43406	0.38–0.43	0.65–0.85	0.025	0.025	0.15–0.30	1.65–2.00	0.70–0.90	0.20–0.30
4615	G46150	0.13–0.18	0.45–0.65	0.035	0.040	0.15–0.30	1.65–2.00	...	0.20–0.30
4620	G46200	0.17–0.22	0.45–0.65	0.035	0.040	0.15–0.30	1.65–2.00	...	0.20–0.30
4626	G46260	0.24–0.29	0.45–0.65	0.035	0.040	0.15–0.30	0.70–.100	...	0.15–0.25
4720	G47200	0.17–0.22	0.50–0.70	0.035	0.040	0.15–0.30	0.90–1.20	0.35–0.55	0.15–0.25
4815	G48150	0.13–0.18	0.40–0.60	0.035	0.040	0.15–0.30	3.25–3.75	...	0.20–0.30
4817	G48170	0.15–0.20	0.40–0.60	0.035	0.040	0.15–0.30	3.25–3.75	...	0.20–0.30
4820	G48200	0.18–0.23	0.50–0.70	0.035	0.040	0.15–0.30	3.25–3.75	...	0.20–030
5117	G51170	0.15–0.20	0.70–0.90	0.035	0.040	0.15–0.30	...	0.70–0.90	...
5120	G51200	0.17–0.22	0.70–0.90	0.035	0.040	0.15–0.30	...	0.70–0.90	...
5130	G51300	0.28–0.33	0.70–0.90	0.035	0.040	0.15–0.30	...	0.80–1.10	...
5132	G51320	0.30–0.35	0.60–0.80	0.035	0.040	0.15–0.30	...	0.75–1.00	...
5135	G51350	0.33–0.38	0.60–0.80	0.035	0.040	0.15–0.30	...	0.80–1.05	...
5140	G51400	0.38–0.43	0.70–0.90	0.035	0.040	0.15–0.30	...	070–0.90	...
5150	G51500	0.48–0.53	0.70–0.90	0.035	0.040	0.15–0.20	...	0.70–0.90	...
5155	G51550	0.51–0.59	0.70–0.90	0.035	0.040	0.15–0.30	...	0.70–0.90	...
5160	G51600	0.56–0.64	0.75–1.00	0.035	0.040	0.15–0.30	...	0.70–0.90	...
E51100	G51986	0.98–1.10	0.25–0.45	0.025	0.025	0.15–0.30	...	0.90–1.15	...
E52100	G52986	0.98–1.10	0.25–0.45	0.025	0.025	0.15–0.30	...	1.30–1.60	...
6118	G61180	0.16–0.21	0.50–0.70	0.035	0.040	0.15–0.30	...	0.50–0.70	0.10–0.15 V
6150	G61500	0.48–0.53	0.70–0.90	0.035	0.040	0.15–0.30	...	0.80–1.10	0.15 V min
8615	G86150	0.13–0.18	0.70–0.90	0.035	0.040	0.15–0.30	0.40–0.70	0.40–0.60	0.15–0.25
8617	G86170	0.15–0.20	0.70–0.90	0.035	0.040	0.15–0.30	0.40–0.70	0.40–0.60	0.15–0.25
8620	G86200	0.18–0.23	0.70–0.90	0.035	0.040	0.15–0.30	0.40–0.70	0.40–0.60	0.15–0.25
8622	G86220	0.20–0.25	0.70–0.90	0.035	0.040	0.15–0.30	0.40–0.70	0.40–0.60	0.15–0.25
8625	G86250	0.23–0.28	0.70–0.90	0.035	0.040	0.15–0.30	0.40–0.70	0.40–0.60	0.15–0.25
8627	G86270	0.25–0.30	0.70–0.90	0.035	0.040	0.15–0.30	0.40–0.70	0.40–0.60	0.15–0.25
8630	G86300	0.28–0.33	0.70–0.90	0.035	0.040	0.15–0.30	0.40–0.70	0.40–0.60	0.15–0.25
8637	G86370	0.35–0.40	0.75–1.00	0.035	0.040	0.15–0.30	0.40–0.70	0.40–0.60	0.15–0.25
8640	G86400	0.38–0.43	0.75–1.00	0.035	0.040	0.15–0.30	0.40–0.70	0.40–0.60	0.15–0.25
8642	G86420	0.40–0.45	0.75–1.00	0.035	0.040	0.15–0.30	0.40–0.70	0.40–0.60	0.15–0.25
8645	G86450	0.43–0.48	0.75–1.00	0.035	0.040	0.15–0.30	0.40–0.70	0.40–0.60	0.15–0.25
8655	G86550	0.51–0.59	0.75–1.00	0.035	0.040	0.15–0.30	0.40–0.70	0.40–0.60	0.15–0.25
8720	G87200	0.18–0.23	0.70–0.90	0.035	0.040	0.15–0.30	0.40–0.70	0.40–0.60	0.20–0.30
8740	G87400	0.38–0.43	0.75–1.00	0.035	0.040	0.15–0.30	0.40–0.70	0.40–0.60	0.20–0.30
8822	G88220	0.20–0.25	0.75–1.00	0.035	0.040	0.15–0.30	0.40–0.70	0.40–0.60	0.30–0.40
9260	G92600	0.56–0.64	0.75–1.00	0.035	0.040	1.80–2.20	...	...	...

Source: Ref 1

Effects of Specific Elements

In many instances, the effects of small amounts of two or more alloying elements used together are greater than larger amounts of a single element (most notable on hardenability).

Manganese is an important alloy in steel for several reasons and is present in virtually all steels in amounts of 0.30% or more. Manganese is a carbide former and has a marked effect on slowing the gamma-to-alpha transformation; therefore, it increases hardenability. Further, manganese in steel is important because of its ability to counter hot shortness, that is, the tendency to tear when being hot formed. In steel, iron and sulfur combine to form a sulfide that has a relatively low melting point. When the steel freezes, this sulfide solidifies in the grain boundaries. On subsequent reheating for rolling or forging, it tears or breaks apart. If manganese is added, it combines preferentially with the sulfur and forms a manganese sulfide of higher melting point, which by its distribution and nature eliminates hot shortness.

Silicon, to a very mild extent, retards critical cooling rate, thereby increasing hardenability. Silicon is used as a deoxidizer in steelmaking and therefore exists in small quantities—usually 0.15 to 0.30% as shown in Table 2. Silicon, as an alloy, is not extensively used in alloy steels although it has some strengthening effect. Silicon structural steels have seen considerable use. Silicon also improves shock resistance and is used where impact is a problem. It also is used as an alloy in spring steels (note grade 9260 in Table 2).

Nickel has a marked effect on the transformation of austenite by shifting the nose of the time-temperature-transformation (TTT) curves to the right, thus increasing hardenability. It also lowers the gamma-to-alpha transformation temperature to the point at which steel may become austenitic at room temperature when large amounts of nickel are used.

Nickel is a very versatile alloy. It provides an added degree of uniformity to quenched steels and strengthens the unquenched or annealed steels. It toughens ferritic-pearlitic steels, especially at low temperature,

Table 3 Compositions of standard boron (alloy) steels

Steel designation AISI or SAE	UNS No.	C	Mn	P max	S max	Si	Ni	Cr	Mo
50B44	G50441	0.43–0.48	0.75–1.00	0.035	0.040	0.15–0.30	...	0.40–0.60	...
50B46	G50461	0.44–0.49	0.75–1.00	0.035	0.040	0.15–0.30	...	0.20–0.35	...
50B50	G50501	0.48–0.53	0.75–1.00	0.035	0.040	0.15–0.30	...	0.40–0.60	...
50B60	G50601	0.56–0.64	0.75–1.00	0.035	0.040	0.15–0.30	...	0.40–0.60	...
51B60	G51601	0.56–0.64	0.75–1.00	0.035	0.040	0.15–0.30	...	0.70–0.90	...
81B45	G81451	0.43–0.48	0.75–1.00	0.035	0.040	0.15–0.30	0.20–0.40	0.35–0.55	0.08–0.15
94B17	G94171	0.15–0.20	0.75–1.00	0.035	0.040	0.15–0.30	0.30–0.60	0.30–0.50	0.08–0.15
94B30	G94301	0.28–0.33	0.75–1.00	0.035	0.040	0.15–0.30	0.30–0.60	0.30–0.50	0.08–0.15

Source: Ref 1

and gives good fatigue resistance. Nickel greatly increases corrosion and oxidation resistance. The amount of nickel used in alloy steels is usually less than 1.0%; the 43XX and 48XX grades are exceptions.

Chromium affects the properties of steel in a number of ways. It has a marked effect in slowing the rate of transformation of austenite; that is, it greatly increases the hardenability of any steel. Additionally, relatively large percentages of chromium greatly increase oxidation and corrosion resistance. However, because the amount of chromium used in alloy steels is 2.0% or less, its principal function for the AISI-SAE alloy steels is to increase hardenability.

Molybdenum, like chromium, greatly slows the gamma-to-alpha transformation; thus, it also increases hardenability. Molybdenum is, however, far more potent than chromium as indicated by the small amounts (generally less than 0.40%) used. Most benefit of molybdenum is achieved when used with nickel and/or chromium.

Vanadium is a strong deoxidizing agent that promotes fine grain size. In larger amounts (see Chapter 11), vanadium is a powerful carbide former that slows the gamma-to-alpha transformation, thus increasing hardenability. The amount of vanadium used in alloy steels is not, however, sufficient to form carbide. Consequently, it functions only as a grain refiner. As may be noted from Table 1, vanadium is specified only in the 61XX steels and even then in very small amounts.

Effects of Alloying on Hardenability

While alloying elements are sometimes used alone, they frequently are used in combinations of two or more elements because it has been found that the use of smaller amounts of two or more elements is more effective than large amounts of a single element. For instance, chromium is generally used with nickel, molybdenum, or some other element. Regardless of the type of amount of the alloy added, the general principles of austenite formation and transformation remain the same, although the reference points, such as the critical temperatures and the time and temperature of transformation, vary with the type and the amount of alloying element.

The main reason for using an alloy steel as opposed to its carbon steel counterpart is to obtain greater hardenability. There are many cases where alloy steels are really wasted; that is, they are used because a poorly informed user assumes that alloy steels are "better steels" than carbon steels. This is not necessarily true. One steel is "better" than another only because it is better suited to a specific application. In the annealed condition, all other factors being equal, mechanical properties of alloy steels are only slightly higher than carbon steels with the same carbon content. This condition prevails for heat treated parts that have very thin sections. However, as section thickness increases, the necessity for more hardenability becomes more important.

Relative Hardenability for Alloy Grades

As indicated from the wide variation in alloy content of the steels listed in Table 2, an equally wide variation in hardenability exists among the alloy grades. The alloy grades with the lowest hardenability are generally the 40XX series. For example, the hardenability of 4037H (shown in Fig. 1a) is little more, if any, than 1038H carbon steel.

(a)

(b)

(c)

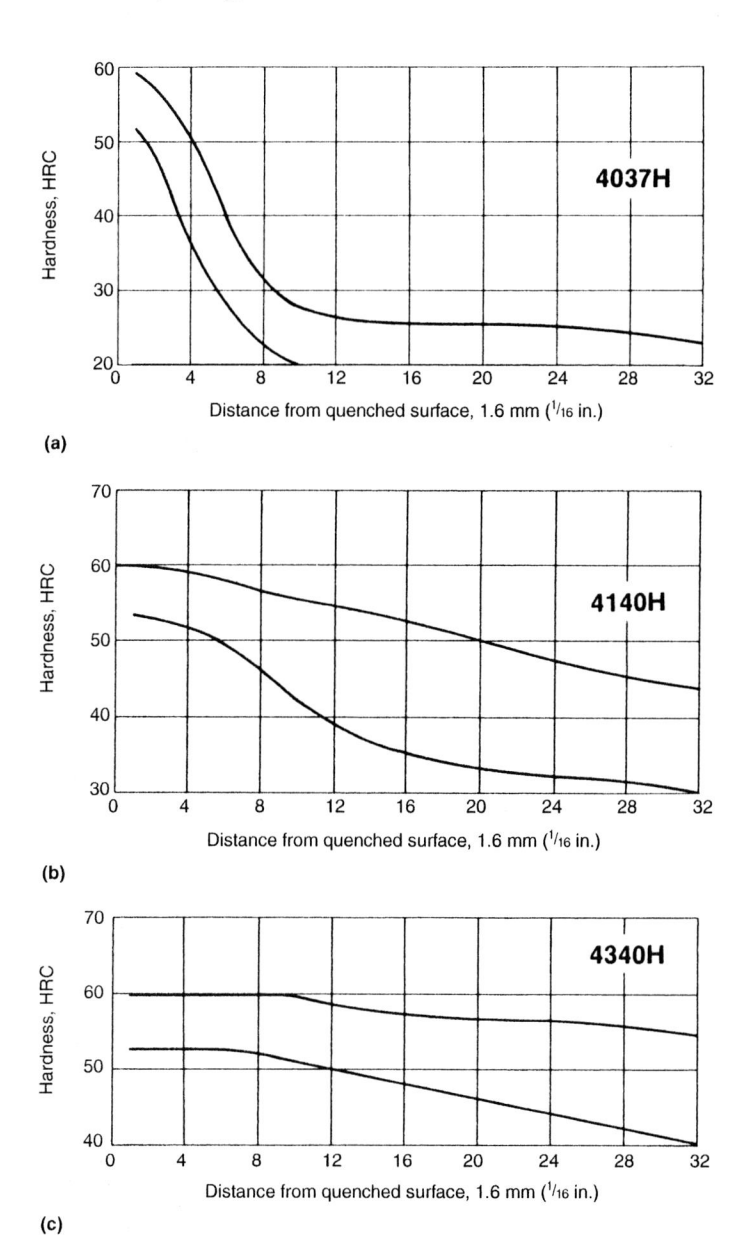

Fig. 1 Effect of total content on hardenability of three alloy steels (see text). Source: Ref 3

Hardenability bands for three widely used alloy steels are presented in Fig. 1. The band for 4037H (Fig. 1a) represents the minimum hardenability for the alloy grades; the molybdenum addition of 0.20 to 0.30% is the only element that can be considered an alloy by the established definitions.

The hardenability band for 4140H, a very popular alloy grade, is shown in Fig. 1(b). The marked effect of the chromium addition is obvious, and the even more marked effect of three alloys (nickel, chromium, and molybdenum) is shown in Fig. 1(c), the hardenability band for 4340H. This grade generally is considered the highest hardenability alloy steel. In Fig. 1(c), the upper boundary of the hardenability band is almost a straight line, which indicates that the hardenability of 4340H is approaching the hardenability of an air-hardening steel. As a practical note, a 150 mm (6 in.) round of 4340H steel can be austenitized and quenched in oil to produce at least 50% transformation to martensite to almost the center of the section.

The three hardenability bands in Fig. 1 clearly demonstrate how important it is to carefully consider section size in selecting an alloy grade to be heat treated.

Effect of Boron on Hardenability

While only a few standard grades of boron-treated steels are listed by AISI (Table 3), other grades that are treated with boron commonly are available by special order.

The marked effect of boron on hardenability is best demonstrated by comparing two alloy steels that otherwise have essentially the same composition (see Fig. 2 for a comparison of the hardenability bands for 5160H and 51B60H). The use of boron-treated alloy steels offers greater hardenability without using the more highly alloyed and more expensive grades of steels.

Heat Treating Procedures

Techniques used for heat treating the alloy steels are not significantly different from those used for carbon steels described in Chapter 6. All alloy steels with carbon contents that do not exceed approximately 0.25% are heat treated by one of the case hardening processes. Alloy steels of higher carbon contents are austenitized by heating above the upper transformation temperature, followed by quenching to near room temperature, and finally tempering to the desired hardness level.

Major variations involved in heat treating alloy steels compared with carbon steels are:

- Temperatures used for normalizing, annealing, and austenitized are generally at least 14 to 42 °C (25 to 75 °F) higher than for carbon steels of similar carbon content. There are some exceptions—notably the high-nickel grades like 4820.
- As alloy content increases, the annealing cycle becomes more complicated—principally, cooling from the annealing temperature must be much slower, or programmed for the specific alloy (see Ref 3) and required structure.
- Severe quenching mediums, such as brine or water, are rarely used for alloy steels, largely because the higher hardenability of alloy steels does not require rapid cooling rates. Alloy steels are also more susceptible to cracking from severe quenching than carbon steels. Procedures for heat treating four specific, but widely different, alloy steels

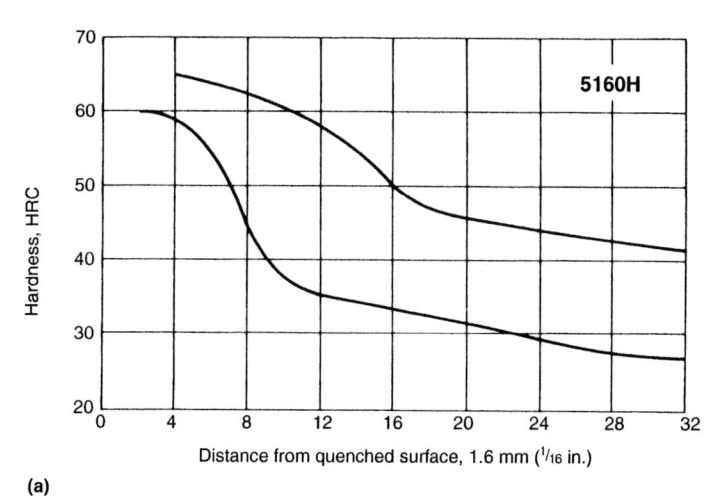

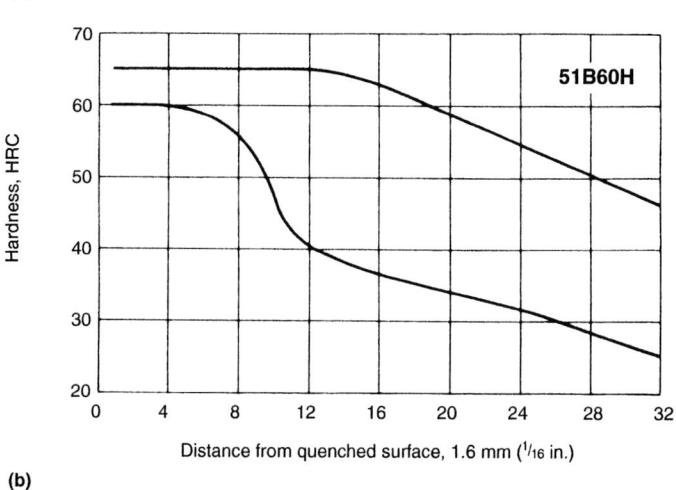

Fig. 2 Effect of boron on hardenability of 5160H alloy steel. Source: Ref 3

4037, 4037H—Recommended Heat Treating Practice

Normalizing. Heat to 870 °C (1600 °F), and cool in air.

Annealing. For a predominantly pearlitic structure, heat to 845 °C (1550 °F), cool rapidly to 745 °C (1370 °F), then at a rate not exceeding 11 °C (20 °F) per hour to 630 °C (1170 °F) or heat to 845 °C (1550 °F), cool rapidly to 660 °C (1225 °F) and hold for 5 h. For a predominantly spheroidized structure, heat to 760 °C, (1400 °F), and cool from 745 °C (1370 °F) to 630 °C (1170 °F), at a rate not exceeding 6 °C (10 °F) per hour, or heat to 760 °C (1400 °F), cool rapidly to 660 °C (1225 °F) and hold for 8 h.

Hardening. Heat to 845 °C (1550 °F) and quench in oil.

Tempering. Reheat to the temperature required to provide the desired hardness (see Fig. 3a).

4140, 4140H—Recommended Heat Treating Practice

Normalizing. Heat to 870 °C (1600 °F), and cool in air.

Annealing. For a predominantly pearlitic structure, heat to 845 °C (1550 °F) and cool to 755 °C (1390 °F) at a fairly rapid rate, then cool from 755 °C (1390 °F) to 665 °C (1230 °F) at a rate not exceeding 14 °C (25 °F) per hour; or heat to 845 °C (1550 °F), cool rapidly to 675 °C (1250 °F), and hold for 5 h. For a predominantly spheroidized structure, heat to 750 °C (1380 °F), and cool to 665 °C (1230 °F) at a rate not exceeding 6 °C (10 °F), per hour; or heat to 750 °C (1380 °F), cool fairly rapidly to 675 °C (1250 °F), and hold for 9 h.

Hardening. Austenitize at 855 °C (1575 °F), and quench in oil.

Tempering. Reheat after quenching to obtain the required hardness (see Fig. 3b).

4340, 4340—Recommended Heat Treating Practice

Normalizing. Heat to 870 °C (1600 °F), and cool in air.

Annealing. For a predominantly pearlitic structure (not usually preferred for this grade), heat to 830 °C (1525 °F), cool rapidly to 705 °C (1300 °F), then cool to 565 °C (1050 °F) at a rate not exceeding 8 °C (15 °F) per hour; or heat to 830 °C (1525 °F), cool rapidly to 650 °C (1200 °F), and hold for 8h.

For a predominantly spheroidized structure, heat to 750 °C (1380 °F), cool rapidly to 705 °C (1300 °F), then cool to 565 °C (1050 °F) at a rate not exceeding 3 °C (5 °F) per hour; or heat to 750 °C (1380 °F), cool rapidly to 650 °C (1200 °F), and hold for 12 h. A spheroidized structure is usually preferred for both machining and heat treating.

Chapter 7: Heat Treating of Alloy Steels / 135

Hardening. Austenitize at 845 °C (1550 °F) and quench in oil. Thin sections may be fully hardened by air cooling.

Tempering. In common with all high-hardenability steels, 4340H is susceptible to quench cracking. Before parts reach ambient temperature (38 to 49 °C, or 100 to 120 °F), they should be placed in the tempering

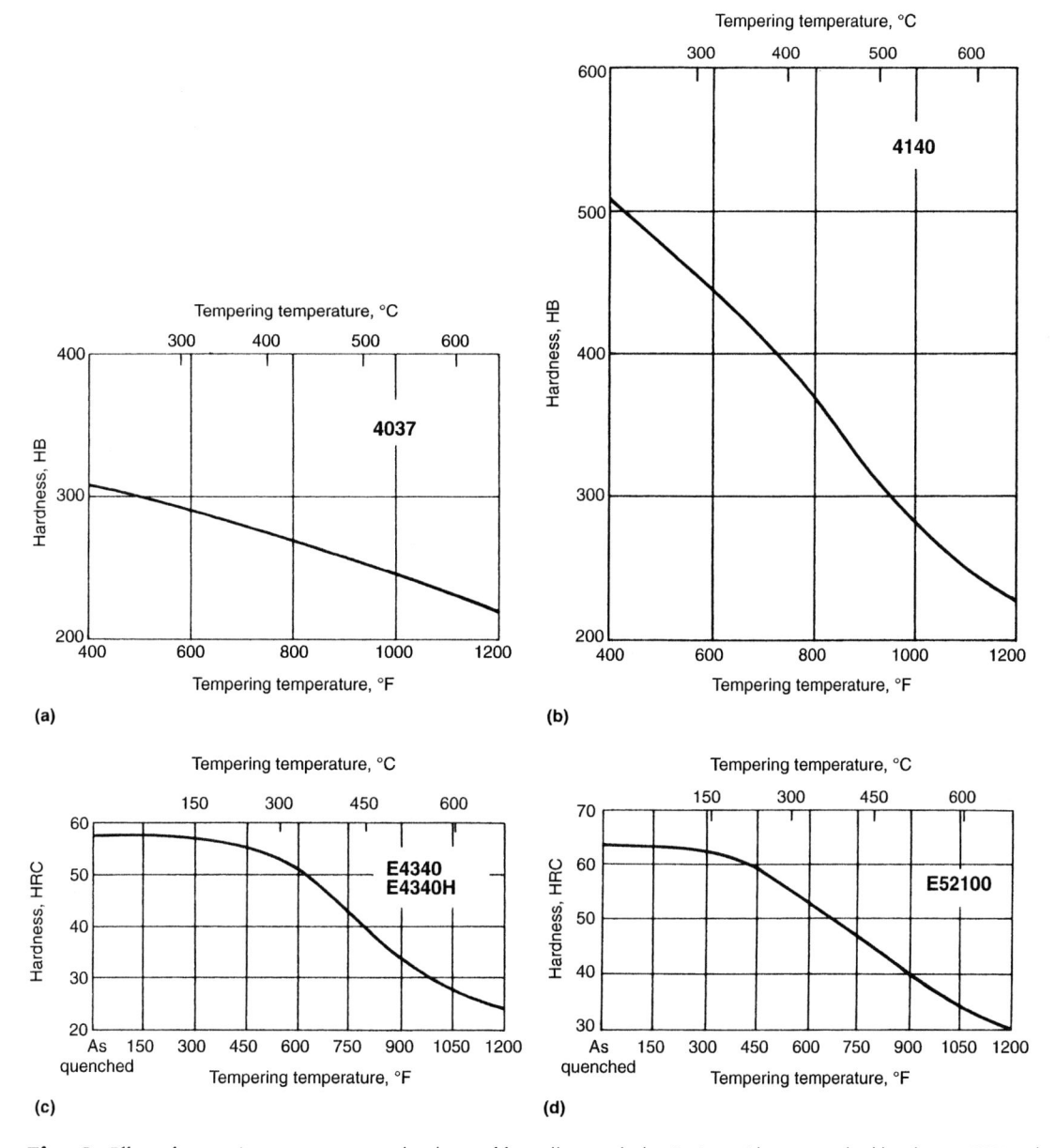

Fig. 3 Effect of tempering temperature on hardness of four alloy steels, beginning with as-quenched hardness. 4037 and 4140: Normalized at 870 °C (1600 °F). Quenched from 845 °C (1550 °F) and tempered at 55 °C (100 °F) intervals in 13.72 mm (0.540 in.) rounds. Tested in 12.83 mm (0.505 in.) rounds. E4340, E4340H, and E52100: Represent an average based on a fully quenched structure. Source: Ref 3

furnace. Tempering temperature depends on the desired hardness or combination of mechanical properties (see Fig. 3c).

E52100—Recommended Heat Treating Practice

Normalizing. Heat to 885 °C (1625 °F), and cool in air.

Annealing. For a predominantly spheroidized structure that is generally desired for machining as well as heat treating, heat to 795 °C (1460 °F) and cool rapidly to 750 °C (1380 °F), then continue cooling to 675 °C (1250 °F) at a rate not exceeding 6 °C (10 °F) per hour; or as an alternative technique, heat to 795 °C (1460 °F) cool rapidly to 690 °C (1275 °F) and hold for 16 h.

Hardening. Austenitize at 845 °C (1550 °F) in a neutral salt bath or in a gaseous atmosphere with a carbon potential of near 1.0%, and quench in oil.

Tempering. After quenching, parts should be tempered as soon as they have uniformly reached near ambient temperature; 38 to 49 °C (100 to 120 °F) is ideal. Because of the high carbon content, parts must be tempered to at least 120 °C (250 °F) to convert the tetragonal martensite to cubic martensite. Most commercial practice calls for tempering at 150 °C (300 °F), which does not reduce the as-quenched hardness to any significant amount. When a reduction in hardness from the as-quenched value of approximately two points HRC can be tolerated, a tempering temperature of 175 °C (350 °F) is recommended. Sometimes E52100 is subjected to higher tempering temperatures, with an accompanying loss of hardness (see Fig. 3d).

Effects of Tempering

From the data presented in Fig. 3, it can be seen that the as-quenched hardness of alloy steels is a function of carbon content, reaching a maximum of about 65 HRC for E52100. There is, however, some difference in the rate at which hardness decreases which increasing tempering temperature for alloy steels, as compared with carbon steels. This simply indicates the effects of alloy upon softening by increasing temperature; note especially the hardness versus tempering curve for 4340, 4340H in Fig. 3.

Austempering and Martempering Treatments

Austempering. Many grades of alloy steels are well suited to austempering—better suited, in many instances, than carbon steels. The thicker sections of alloy steels can be successfully austempered. Alloy steels that are frequently austempered include 5150, 6150, 50B60, and 51B60 For a description of austempering, see Chapter 6.

Martempering. The term "martempering" describes an elevated-temperature quenching procedure aimed at reducing cracks, distortion, or residual stresses. It is not a tempering procedure, as the name implies, and is more properly termed "marquenching."

Martempering of steel consists of (1) quenching from the austenitizing temperature into a hot fluid medium (hot oil, molten salt, molten metal, or a fluidized particle bed) at a temperature usually above the martensitic range (M, point); (2) holding in the quenching medium until the temperature throughout the steel is substantially uniform; and then (3) cooling (usually in air) at a moderate rate to prevent large differences in temperature between the outside and the center of the section. Formation of martensite occurs fairly uniformly throughout the workpiece during cooling to room temperature, thereby avoiding formation of excessive amounts of residual stress. The microstructure after martempering is essentially primary martensite, which is untempered and brittle for most applications. After being air cooled to room temperature, martempered parts are tempered in the same manner as if they had been coventionally quenched.

Figure 4 shows the significant difference between conventional quenching (a) and martempering (b) The principal advantage of martempering lies in the reduced thermal gradient between surface and center as the part is quenched to the isothermal temperature and then is air cooled to room temperature. Residual stresses developed during martempering are lower than those developed during conventional quenching because the greater thermal variations occur while the steel is in the relatively plastic austenitic condition and because final transformation and thermal changes occur throughout the part at approximately the same time. Martempering also reduces or eliminates susceptibility to cracking.

Generally, any steel that has sufficient hardenability to harden by oil quenching can be successfully martempered. It must be emphasized that martempering does not in any way eliminate the need for tempering (Fig. 4).

Modified martempering differs from "standard" martempering only in that the temperature of the quenching bath is below the M_s point (Fig. 4c). The lower temperature increases the severity of quenching. This is important for steels of lower hardenability that require faster cooling to harden to sufficient depth, or when the M_s is high and some bainite is detrimental to the finished part. Thus, modified martempering is applicable to a greater range of steel compositions than is the standard process.

Although hot oil is invariably the quenchant employed for modified martempering at 175 °C (350 °F) and lower, molten nitrate-nitrite salts (with water addition and agitation) are effective at temperatures as low as 175 °C (350 °F). Due to their higher heat-transfer coefficients, molten salts offer some metallurgical and operational advantages. For more details on these special heat treating processes, see Ref 4.

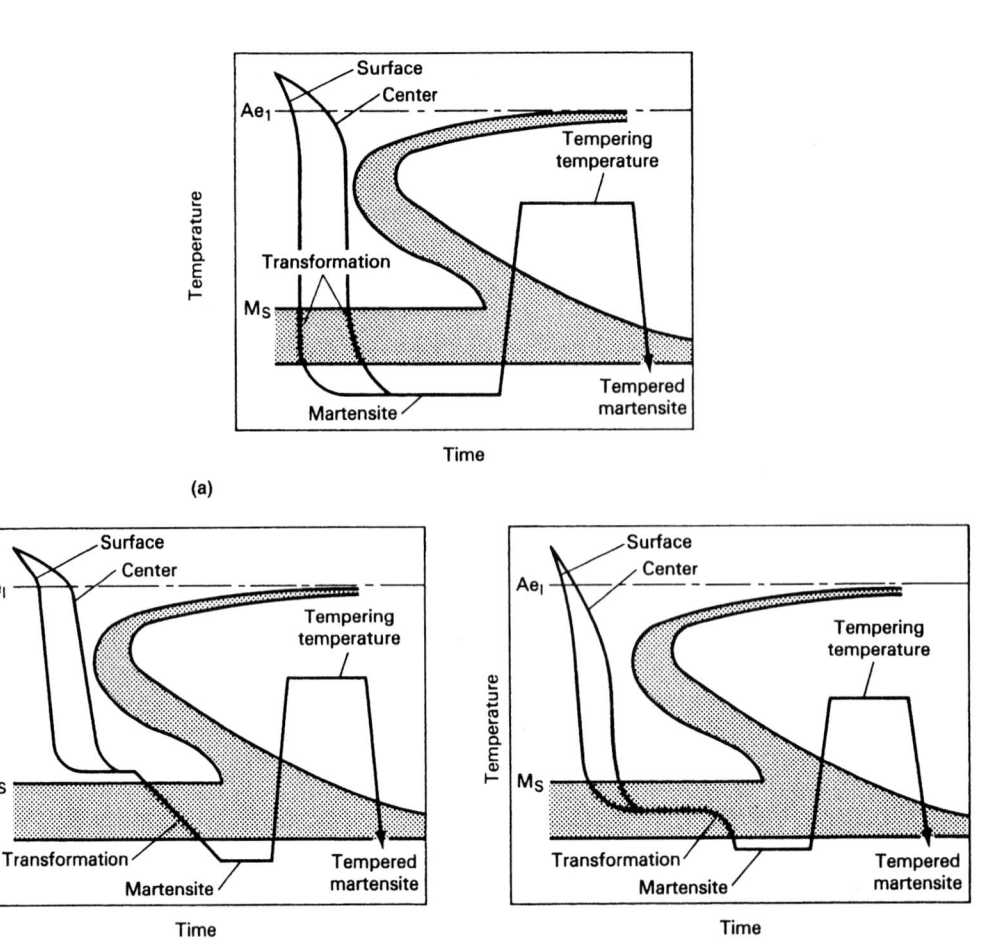

Fig. 4 Time-temperature-transformation diagrams with superimposed cooling curves showing quenching and tempering. (a) Conventional process. (b) Martempering. (c) Modified martempering. Source: Ref 4

REFERENCES

1. H. Chandler, Ed., Carbon & Alloy Steels Introduction, *Heat Treater's Guide: Practices and Procedures for Irons and Steels,* 2nd ed., ASM International, 1995, p 111–120
2. Classification and Designation of Carbon and Low-Alloy Steels, *Properties and Selection: Irons, Steels, and High-Performance Alloys,* Vol 1, *ASM Handbook,* ASM international, 1990, p 140–194
3. H. Chandler, Ed., Alloy Steels (1300 through 9700 series), *Heat Treater's Guide: Practices and Procedures for Irons and Steels,* 2nd ed., ASM International, 1995, p 226–511
4. H. Webster and W.J. Laird, Martempering of Steel, *Heat Treating,* Vol 4, *ASM Handbook,* ASM International, 1991 p 137–151

SELECTED REFERENCES

- C.R. Brooks, *Principles of the Heat Treatment of Plain Carbon and Low Alloy Steels,* ASM International, 1996
- G.E. Totten and M.A.H. Howes, Ed., *Steel Heat Treatment Handbook,* Marcel Dekker, Inc. 1997

CHAPTER **8**

Case Hardening of Steel

CASE HARDENING—the production of parts that have hard, wear-resistant surfaces, but with softer and/or tougher cores—can be accomplished by two distinct methods. One approach is to use a grade of steel that already contains sufficient carbon to provide the required surface after heat treatment. The surface areas requiring the higher hardness are then selectively heated and quenched. The second method is to use a steel that is not normally capable of being hardened to the desired degree, then alter the composition of the surface layers by "diffusion" so that it either can be hardened or, in some instances, becomes hard during processing. The first approach is discussed in Chapter 9, "Flame and Induction Hardening." This chapter is confined to the discussion of the second approach—hardening processes that involve changes in surface composition.

Classification of Case Hardening

Because each general type of case hardening that involves composition changes by diffusion may incorporate a "family" of several types, precise classification becomes difficult. However, for most practical purposes, case hardening treatments can be broadly classified into four groups: carburizing, carbonitriding, nitriding, and nitrocarburizing. Of these processes, carburizing is by far the most widely used. Table 1 compares the characteristics of case hardening processes.

Carburizing Processes

The primary object of carburizing is to provide a hard, wear-resistant surface with surface residual compressive stresses that results in an improvement in the useful life of ferrous engineering components. In carburizing, austenitized ferrous metal is brought into contact with an environment of sufficient carbon potential to cause absorption of carbon at the surface and, by diffusion, to create a carbon concentration gradient between the surface and the interior of the metal. Two factors may control the rate of carburizing—the carbon-absorption reaction at the surface and diffusion of carbon into the metal.

Carburizing is done at elevated temperature, generally in the range of 850 to 950°C (1550 to 1750 °F). However, temperatures as low as 790 °C (1450 °F) and as high as 1090 °C (2000 °F) have been used. Although the carburizing rate can be greatly increased at temperatures above about 950 °C (1750 °F), most furnace equipment has an increasingly limited life, causing the carburizing temperature to be limited.

Carburizing may be done in a gaseous environment (gas carburizing), a liquid salt bath (liquid carburizing), or with all the surfaces of the workpiece covered with a solid carbonaceous compound (pack carburizing). Regardless of the process, the objective of carburizing is to start with a relatively low-carbon steel (0.20% C) and increase the carbon content in the surface layers, resulting in a high-carbon, hardenable steel on the outside with a gradually decreasing carbon content to the underlying layers. Process control to keep the desired carbon level is an important part of any carburizing process. There is a limited discussion of atmosphere carbon control in Chapter 5, "Instrumentation and Control of Heat Treating Processes" (see also Ref 2). Generally, although not always, a carbon content of approximately 0.80 to 0.90% is desired. Of the various carburizing processes, gas carburizing is the most widely used.

Table 1 Typical characteristics of case hardening treatments

Process	Nature of case	Process temperature, °C (°F)	Typical case depth	Case hardness, HRC
Carburizing				
Pack	Diffused carbon	815–1090 (1500–2000)	125 μm–1.5 mm (5–60 mils)	50–63(a)
Gas	Diffused carbon	815–980 (1500–1800)	75 μm–1.5 mm (3–60 mils)	50–63(a)
Liquid	Diffused carbon and possibly nitrogen	815–980 (1500–1800)	50 μm–1.5 mm (2–60 mils)	50–65(a)
Vacuum	Diffused carbon	815–1090 (1500–2000)	75 μm–1.5 mm (3–60 mils)	50–63(a)
Nitriding				
Gas	Diffused nitrogen, nitrogen compounds	480–590 (900–1100)	125 μm–0.75 mm (5–30 mils)	50–70
Salt	Diffused nitrogen, nitrogen compounds	510–565 (950–1050)	2.5 μm–0.75 mm (0.1–30 mils)	50–70
Ion	Diffused nitrogen, nitrogen compounds	340–565 (650–1050)	75 μm–0.75 mm (3–30 mils)	50–70
Carbonitriding and nitrocarburizing				
Gas	Diffused carbon and nitrogen	760–870 (1400–1600)	75 μm–0.75 mm (3–30 mils)	50–65(a)
Liquid (cyaniding)	Diffused carbon and nitrogen	760–870 (1400–1600)	2.5–125 μm (0.1–5 mils)	50–65(a)
Ferritic nitrocarburizing	Diffused carbon and nitrogen	565–675 (1050–1250)	2.5–25 μm (0.1–1 mil)	40–60(a)

(a) Requires quench from austenitizing temperature. Source: Ref 1

Gas Carburizing

In gas carburizing, the carbon source usually is introduced into an essentially noncarburizing or weakly carburizing (0.40% C) carrier gas. In general, gas carburizing is more effective than pack or liquid carburizing with deeper and higher carbon content cases obtained more rapidly.

Gas carburizing is more economical and more adaptable for mass production than the other carburizing processes. The economy is brought about because a specified case depth may be achieved more rapidly in gas carburizing and it does not involve the extensive handling labor of the other methods.

Carburizing Gases. Natural gas, which is largely methane (CH_4), is the most common source of carbon for gas carburizing. Where natural gas supplies are unreliable or varies in composition, propane may be used also as a carbon source. However, undiluted natural gas or propane is much too rich to use directly so that only small quantities (5 to 20% by volume) are mixed with a carrier gas.

The carrier gas is usually an externally produced gas such as endothermic gas or a mixture of nitrogen and cracked methanol, which essentially has the same composition as endothermic gas (see section on furnace atmospheres in Chapter 4, "Furnaces and Related Equipment for Heat Treating"). The results obtained with the various types of carrier gases are essentially the same. Forced fan circulation serves to distribute the atmosphere evenly throughout the furnace and also provides temperature uniformity.

Drip Carburizing. As an alternative form of gas carburizing, liquid hydrocarbons are also extensively used as sources of carbon. These liquids are usually proprietary compounds that range in composition from pure hydrocarbons, such as dipentene or benzene, to oxygenated hydrocarbons, such as alcohols, glycols, or ketones. This process is sometimes called "drip carburizing" and should not be confused with liquid carburizing to be discussed later.

A liquid normally is fed in droplet form to a target plate in the furnace where it volatilizes almost instantaneously. The liquid dissociates thermally to provide a carburizing atmosphere. Forced fan circulation serves to distribute the atmosphere evenly throughout the furnace and also provides temperature uniformity. The liquid flow can be adjusted manually, although it can be a part of an automatically controlled system that provides almost any desired carbon potential.

Furnaces. A wide variety of conventional furnaces including pit, rotary, batch/integral quench, and continuous furnaces are used for gas carburizing (see illustrations of furnaces in Chapter 4). There are no specific limitations on furnace size. Selection of furnace type depends largely on workpiece shape and size, total production required, and production flow, as for any heat treating process.

Characteristics of a Carburized Case. It is appropriate at this point to acquaint the reader with a few fundamentals regarding a gas carburized case. These generalizations also apply to cases produced in solid (pack carburized) or liquid bath carbonaceous environments.

First, it must be understood that carburizing is a distinctly separate operation; that is, it is the process of diffusing carbon into the steel surface to make it hardenable and does not necessarily incorporate the hardening phase of the total operation. However, in most gas and liquid (salt bath) carburizing operations, the workpiece is quenched from an elevated temperature following the carburizing cycle. Therefore, in this specific section, it is considered that the carburized work is also hardened in a continuous cycle.

A second important consideration is the nature of a carburized case, which is often poorly understood by design and process engineers. At a temperature of approximately 925 °C (1700 °F), a steel surface is extremely active, and if the carbon content of its environment is higher than that of the steel, the steel absorbs carbon to achieve equilibrium with the environment. However, if the carbon potential of the environment is lower than that of the steel, the steel loses carbon to its environment (decarburization). However, equilibrium conditions prevail only at the steel surface, and as distance from the surface increases, carbon concentration decreases gradually to the original carbon content of the steel. This is illustrated in Fig. 1, which relates carbon content with distance from the surface. For the conditions given in Fig. 1 for an 8620H steel, the carbon content of the case decreases as distance from the surface increases. For example, the surface carbon is shown as 1.0%, but even at a depth of 0.5 mm (0.020 in.), the carbon concentration has decreased to about 0.80%. This drop in carbon content below the surface clearly illustrates a very important fact: finishing operations (grinding or other) must be planned

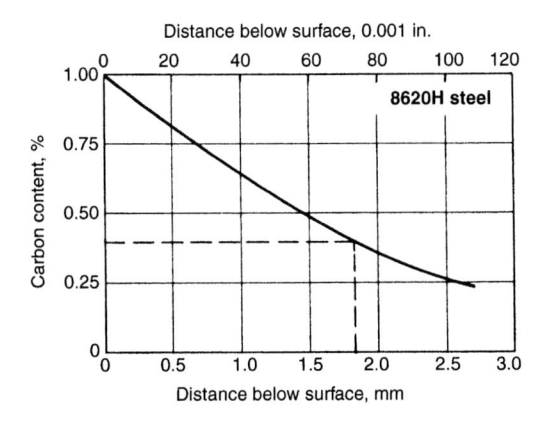

Fig. 1 Relationship of carbon concentration with distance from the surface. Source: Ref 3

carefully because too much stock removal in finishing removes the most valuable part of the case. A general rule is to plan operations so that no more than 10% (per side) of the case is removed during finishing.

Total Case versus Effective Case. Figure 1 also illustrates the principle of case depth. Total case depth (regardless of the carburizing method) refers to the very end point of the case—where the carbon content drops to the original level of the base metal (0.20% C in Fig. 1). In Fig. 1, this occurs at approximately 2.5 mm (0.10 in.). Because the point where the carbon content just reaches the base carbon level is very difficult to measure for case hardened alloy steels, total case depth is primarily used for carburized or carbonitrided plain carbon steels having shallow cases, under 0.5 mm (0.020 in.).

Effective case depth is generally more easily and more accurately measured than total case depth. It is thus more widely used by industry. It is generally agreed that effective case depth is the point below the surface where the case hardness drops to 50 HRC.

In Fig. 1, the intersection of the dashed lines shows the effective case, which ends at approximately 1.91 mm (0.075 in.) and at the 0.40% C level. However, the carbon level necessary to achieve 50 HRC may be 0.30 to 0.60% C depending on the severity of the quenchant used. The total case depth of any carburized case is a function principally of time and temperature, as discussed subsequently. Surface carbon concentration also has a minor effect on case depth.

Effect of Temperature. The maximum rate at which carbon can be added to steel is limited by the rate of diffusion of carbon in austenite. This diffusion rate increases greatly with temperature; the rate of carbon addition at 925 °C (1700 °F) is about 40% greater than at 870 °C (1600 °F). For example, in Table 2, at 870 °C (1600 °F), a total case depth of 0.64 mm (0.025 in.) is attained in 2 h, whereas a total case depth of 0.89 mm (0.035 in.) is developed in the same length of time at 925 °C (1700 °F). It is mainly for this reason that the most common temperature range for gas carburizing is 925 to 955 °C (1700 to 1750 °F). These temperatures

Table 2 Values of total case depth for various times and temperatures

Time, t, h	Case depth, mm (in.) after carburizing at		
	870 °C (1600 °F)	900 °C (1650 °F)	925 °C (1700 °F)
2	0.64 (0.025)	0.76 (0.030)	0.89 (0.035)
4	0.89 (0.035)	1.07 (0.042)	1.27 (0.050)
8	1.27 (0.050)	1.52 (0.060)	1.80 (0.071)
12	1.55 (0.061)	1.85 (0.073)	2.21 (0.087)
16	1.80 (0.071)	2.13 (0.084)	2.54 (0.100)
20	2.01 (0.079)	2.39 (0.094)	2.84 (0.112)
24	2.18 (0.086)	2.62 (0.103)	3.09 (0.122)
30	2.46 (0.097)	2.95 (0.116)	3.48 (0.137)
36	2.74 (0.108)	3.20 (0.126)	3.81 (0.150)

Source: Ref 4

permit a reasonably rapid carburizing rate without excessive deterioration of furnace equipment, particularly of heat-resistant alloys. There has been a trend toward raising the carburizing temperature to 980 °C (1800 °F) for certain deep-case requirements. For shallow-case carburizing in which case depth must be kept within a specified narrow range or distortion is critical, lower temperatures are frequently used. Data concerning several carburizing variables are presented in Fig. 2, but the effect of temperature is evident in interrelating the four sets of graphs that include case depth data for carburizing at 870 °C (1600 °F), 900 °C (1650 °F), 925 °C (1700 °F), and 955 °C (1750 °F), respectively.

Effect of Time. The depth of case achieved versus time is by no means a straight-line relationship, but it is more nearly like an inverse square relationship. Referring again to Table 2, it can be seen that the gain in

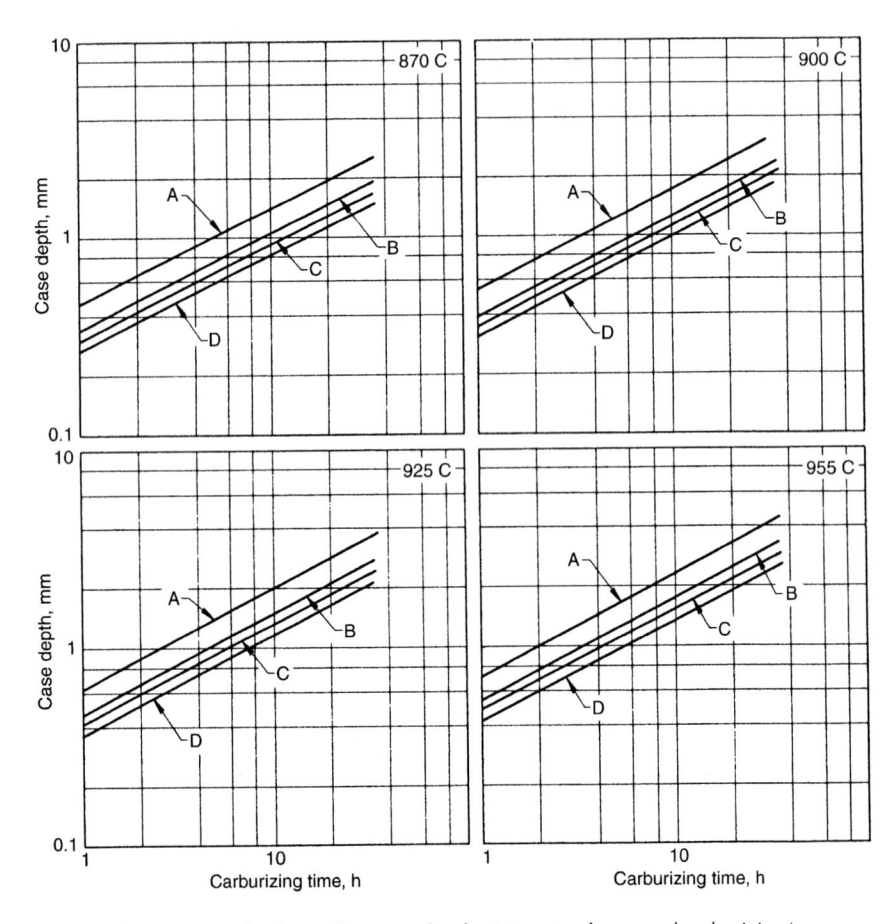

Fig. 2 Case depth as a function of carburizing time for normal carburizing (no diffusion cycle) of low-carbon and certain low-alloy steels. Curve A: Total case depth. Curve B: Effective case depth for surface carbon content of 1.1% to saturation. Curve C: Effective case depth for surface carbon content of 0.8 to 0.9%. Curve D: Effective case depth for surface carbon content of 0.7 to 0.8%. Source: Ref 5

case depth with increasing time at the carburizing temperature is a matter of rapidly diminishing returns. For example, at any of the three temperatures shown, to double the case depth obtained in 2 h, the time is approximately quadrupled. This fact is also borne out in the temperature-case depth-time data shown in Fig. 2. Note that as a matter of convenience, the time data are plotted on a logarithmic scale. Formulas are available for more precise calculations of case depth (Ref 6).

Control of Carbon Concentration. The means for determining and controlling the carbon potential of the gas is quite complex and is better described in Chapter 5 and Ref 2. However, several systems have been devised for this type of control so that all an operator needs to do is set the instrument for the desired surface carbon content and the gas mixture is controlled accordingly.

Vacuum Carburizing

Vacuum carburizing is a nonequilibrium, boost-diffuse type carburizing process (boost step to increase carbon content of austenite; diffusion step to provide gradual case/core transition). The steel being processed is austenitized in a rough vacuum, carburized in a partial pressure of hydrocarbon gas, diffused in a rough vacuum, and then quenched in either gas or oil. Compared with atmosphere carburizing, vacuum carburizing offers excellent uniformity and repeatability because of the high degree of process control possible with vacuum furnaces, potentially less distortion, and potentially reduced cycle times when the higher process temperatures possible with vacuum furnaces are used.

The process is carried out at pressures less than atmospheric—6.7 kPa to 40 kPa (50 to 300 torr)—and the carburizing temperatures may range from 900 to 1100 °C (1650 to 2000 °F) but usually are 980 to 1050 °C (1800 to 1925 °F). The atmosphere gas usually consists solely of enriching gas; nitrogen also may be used as a carrier gas. Generally, graphite heating elements and fixtures are used instead of metals in furnace construction. Therefore, higher carburizing temperatures (and, thus, shorter cycle times) may be used with vacuum carburizing than for other gas carburizing methods. The guidelines for boost/diffuse cycles for vacuum carburizing are shown in Appendix C, "Boost/Diffuse Cycles for Carburizing."

Plasma (Ion) Carburizing. An alternative vacuum process for carburizing offering faster carbon diffusion rates, better case uniformity, and less part distortion than either atmosphere or conventional vacuum carburizing is plasma (ion) carburizing. A complete explanation of this process is given in Ref 7.

Liquid Carburizing

The term "liquid carburizing" should not be confused with drip carburizing. Liquid carburizing is a method of case hardening ferrous metal parts by holding them above their transformation temperature in a molten salt bath. The salt decomposes and releases carbon and sometimes nitrogen that diffuse into the work metal surfaces so that a high degree of hardness can be developed on quenching.

Many liquid carburizing baths contain some cyanide, which introduces both carbon and nitrogen into the case. It should be pointed out that processes using cyanide present waste treatment environmental problems and for that reason have declined in usage over the past couple of decades. Another type of liquid bath uses a special grade of carbon rather than cyanide as the source of carbon. This bath produces a case that contains only carbon as the hardening agent.

Cyanide-Type Baths. Light case and deep case are arbitrary terms that have been associated with liquid carburizing in baths containing cyanide. There is necessarily some overlapping of bath compositions for the two types of cases. In general, the two types are distinguished more by operating temperature than by bath composition. Hence, the terms *low temperature* and *high temperature* are preferred.

Both low-temperature and high-temperature baths are supplied in different cyanide contents to satisfy individual requirements of carburizing activity (carbon potential) within the limitations of normal dragout and replenishment. In many instances, compatible companion compositions are available for starting the bath or for bath makeup, and for the regeneration or maintenance of carburizing activity.

Low-temperature cyanide-type baths are those usually operated in the temperature range from about 845 to 900 °C (1550 to 1650 °F). Low-temperature baths are best suited to the formation of total case depths of 0.075 to 0.75 mm (0.003 to 0.030 in.) deep. They are accelerated cyanogen baths containing various combinations and amounts of the constituents listed in Table 3.

High-temperature cyanide-type baths (deep case) are usually operated in the temperature range of about 900 to 955 °C (1650 to 1750 °F). This

Table 3 Operating compositions of liquid carburizing baths

Constituent	Light case, low temperature 845–900 °C (1550–1650 °F)	Deep case, high temperature 900–955 °C (1650–1750 °F)
Sodium cyanide	10–23	6–16
Barium chloride	...	30–55(a)
Salts of other alkaline earth metals(b)	0–10	0–10
Potassium chloride	0–25	0–20
Sodium chloride	20–40	0–20
Sodium carbonate	30 max	30 max
Accelerators other than those involving compounds of alkaline earth metals(c)	0–5	0–2
Sodium cyanate	1.0 max	0.5 max
Density of molten salt	1.76 g/cm^3 at 900 °C (0.0636 lb/in.^{3} at 1650°F)	2.00 g/cm^3 at 925 °C (0.0723 lb/in.^{3} at 1700°F)

(a) Proprietary barium chloride-free deep-case baths are available. (b) Calcium and strontium chlorides have been employed. Calcium chloride is more effective, but its hygroscopic nature has limited its use. (c) Among these accelerators are manganese dioxide, boron oxide, sodium fluoride, and sodium pyrophosphate. Source: Ref 8

range may be extended somewhat, but at lower temperatures, the carbon penetration rate becomes undesirably slow. Also, at temperatures higher than about 955 °C (1750 °F), deterioration of the bath and equipment is markedly accelerated. The most important use of high-temperature baths is for producing case depths of about 0.5 to 2 mm (0.020 to 0.080 in.), although deeper cases are possible. The baths consist of cyanide and a major proportion of barium chloride, with or without supplemental acceleration from other salts of the alkaline earth metals.

Noncyanide Liquid Carburizing. Liquid carburizing can be accomplished in a bath that contains a special grade of carbon instead of cyanide as the source of carbon. In this bath, carbon particles are dispersed in the molten salt by mechanical agitation, which is achieved with one or more simple propeller stirrers.

Operating temperatures for the bath are generally higher than for cyanide-type baths. A range of about 900 to 955 °C (1650 to 1750 °F) is most commonly employed. Temperatures below about 870 °C (1600 °F) are not recommended and may even lead to decarburization of the steel. The case depths and carbon profiles produced are in the same range as that for high-temperature cyanide-type baths.

Control of Case Depth and Carbon Concentration. In liquid carburizing, depth of case depends on time and temperature, as has been discussed for gas carburizing. This is indicated in Fig. 3, which shows the effect of three different carburizing temperatures on case depth for 1020 steel for time periods of 2 to 40 h.

Carbon concentration is controlled principally by control of the salt composition. The carbon and/or nitrogen that diffuses into the steel comes from decomposition of the salt. Therefore, any salt bath requires a great deal of attention in terms of bailing out the sludge and replenishing with fresh salt—usually every 8 h shift, depending on the amount of production and, therefore, the amount of dragout.

Furnaces. In general, all of the salt bath furnaces discussed and illustrated in Chapter 4 are suitable for salt bath carburizing.

Advantages and Limitations of Liquid Carburizing. Two specific advantages of liquid carburizing are realized. Selective carburizing can be achieved in many applications without stop-off procedures. For example, to carburize and harden only one end of a long shaftlike member, the workpiece, by suitable fixturing, can be held so that only the end is immersed in the bath. This is not possible with other carburizing methods. A second advantage is the flexibility of liquid carburizing in simultaneously processing a variety of workpieces. They can vary in size and shape as well as case depth requirements for parts.

A disadvantage of liquid carburizing is the requirement of washing after quenching. Another is that the salt adhering to the hot workpieces contaminates quenching mediums. Because of the problems associated with salt removal, liquid carburizing is not recommended for parts containing small holes, threads, or recessed areas that are difficult to clean. It should again be pointed out that processes using cyanide present waste treatment environmental problems and for that reason, cyanide-type processes have declined in usage over the past couple of decades.

Pack Carburizing

Pack carburizing is a process in which carbon monoxide derived from a solid compound decomposes at the metal surface into nascent carbon and carbon dioxide. The nascent carbon is then absorbed into the metal. The carbon dioxide resulting from this decomposition immediately reacts

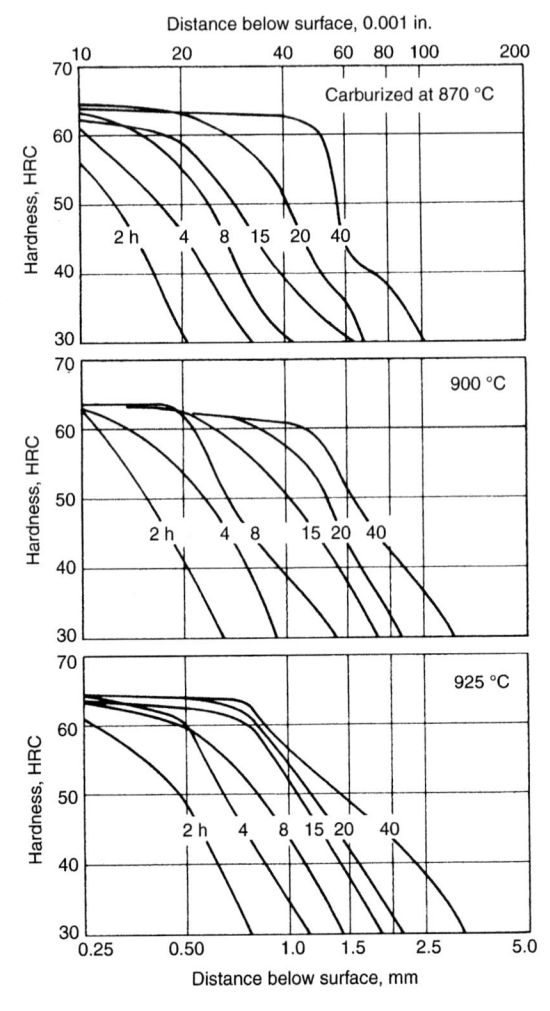

Fig. 3 Effects of time and temperature for liquid carburizing of 1020 steel. Source: Ref 9

with carbonaceous material present in the solid carburizing compound to produce fresh carbon monoxide. This reaction is enhanced by energizers or catalysts, such as barium carbonate ($BaCO_3$) and sodium carbonate (Na_2CO_3), that are present in the carburizing compound. These substances react with carbon to form additional carbon monoxide and an oxide of the energizing compound. The latter in turn reacts in part with carbon dioxide to re-form carbonate. Thus, in a closed system, the energizer is continuously being used and re-formed. Carburizing continues as long as enough carbon is present to react with the excess of carbon dioxide.

For the most part, pack carburizing has been replaced by other methods. However, there are special applications where pack carburizing is desired.

Furnaces. Pack carburizing is usually done in box batch-type furnaces. No prepared atmosphere is required. Typical furnaces used for this process are discussed in Chapter 4.

Processing. Parts to be processed are placed in steel or heat-resistant alloy boxes surrounded by a generous amount of the solid carburizing compound, usually purchased as a proprietary mixture. The boxes are then sealed with fire clay and heated to the carburizing temperature for the necessary time. Carburizing rates are generally lower for pack carburizing, but otherwise the same variables prevail. There are occasions where, at the end of the carburizing cycle, parts are removed from the boxes and directly quenched or reheated for hardening.

Carbon Potential and Gradient. The carbon potential of the atmosphere generated by the carburizing compound, as well as the carbon content obtained at the work surface, increases directly with an increase in the carbon monoxide to carbon dioxide ratio. Thus, more carbon is made available at the work surface by the use of energizers and carburizing materials that promote carbon monoxide formation.

The carbon concentration gradient of carburized parts is influenced principally by carbon potential, carburizing temperature, time, and the chemical composition of the steel.

Pack carburizing is normally performed in the temperature range of 815 to 955 °C (1500 to 1700 °F), although carburizing temperatures as high as 1095 °C (2000 °F) have been used.

The carburizing rate is more rapid at the start of the cycle and gradually diminishes as the cycle is extended as shown for gas carburizing.

Process Limitations. Principal disadvantages of pack carburizing as opposed to other carburizing are the additional labor required and the fact that the process is grossly inefficient in terms of energy consumption because of all the material (boxes and compound) that must be heated, which usually amounts to the weight of the workpieces or more. There is also the difficulty of direct quenching from the process, thus requiring the consumption of more energy for a separate reheating process.

Carbonitriding

Carbonitriding is a modified gas carburizing process, rather than a form of nitriding. The modification consists of introducing ammonia gas (commonly about 2 to 5%) into the carburizing atmosphere in order to add nitrogen to the case as it is being produced. Ammonia in the atmosphere dissociates to form nascent nitrogen at the work surface, and the nascent nitrogen diffuses into the steel simultaneously with carbon. Typically, carbonitriding is carried out at a lower temperature and for a shorter time than gas carburizing, in order to obtain a total case depth of about 0.8 mm (0.03 in.) maximum. Lower temperatures are possible in carbonitriding primarily because nitrogen is a powerful austenitizer; thus, the transformation temperature for any given steel is lowered. It also increases the hardenability of the case so that many plain low-carbon steels respond to carbonitriding and oil quenching that would not harden after being carburized and oil quenched.

Because of problems in disposing of cyanide-bearing wastes, carbonitriding is generally preferred to liquid cyaniding. In terms of case characteristics, carbonitriding differs from carburizing and nitriding in that carburized cases normally do not contain nitrogen and nitrided cases contain no added carbon, whereas carbonitrided cases contain both.

Applications. Carbonitriding is used mainly to impart a hard, wear-resistant but relatively thin case to a large variety of hardware items (mostly small sized) on a mass-production basis. Case depths usually range in thickness from 0.075 to 0.75 mm (0.003 to 0.030 in.). Figure 4 shows typical case hardness profiles for two steels.

A carbonitrided case has higher hardenability than a carburized case; consequently, by carbonitriding and quenching with either a carbon or very low-alloy steel, a hardened case can be produced with a less expen-

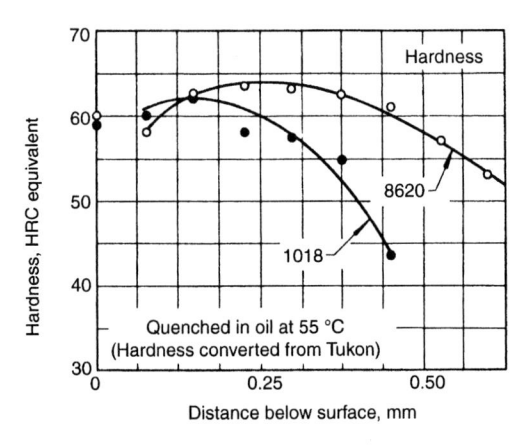

Fig. 4 Hardness-depth relationships for carbonitriding of one plain carbon and one alloy steel. Source: Ref 10

sive material. Because of the higher hardenability of carbonitrided cases, full hardness can often be obtained by oil quenching instead of quenching in water or brine as might be required for a conventional carburized case. Thus, there could be less workpiece distortion as a result of the gas carbonitriding process.

A carbonitrided case retains a higher hardness after tempering than a similarly carburized steel as shown in Fig. 5. This resistance to softening makes carbonitrided parts useful in engine or transmission parts operating for long periods at 110 to 165 °C (200 to 300 °F) above room temperature.

Steels for carbonitriding are generally of the lower-carbon variety (0.30% C max for plain carbon or alloy), although sometimes steels having a somewhat higher carbon content (0.30 to 0.35% C) are case hardened by carbonitriding.

Furnaces that are suited to gas carburizing are generally suitable for carbonitriding (see Chapter 4).

Furnace Atmospheres. Any atmosphere composition that can be used effectively for gas carburizing can be used for carbonitriding, with an ammonia addition. For more details on this process, see Ref 11 and 12.

Tempering. Carburized parts are generally tempered at 150 °C (300 °F) minimum for 1 h minimum or ½ h minimum for each inch of maximum cross section in order to improve mechanical properties. The actual tempering temperature used depends on the required surface hardness range. Tempering data on carburized and carbonitrided 8620H steel are presented in Fig. 5.

Carbonitrided parts may or may not be tempered depending on the application. The actual tempering temperature chosen depends on the specified surface hardness. A comparison of hardness values after carburizing or carbonitriding, hardening, and tempering of 16 mm (⅝ in.) diameter test pins made from 8620H and 1018 steel is also presented in Fig. 5.

Conventional Gas Nitriding

Gas nitriding is a case hardening process in which nitrogen is introduced into the surface of a ferrous alloy by holding the metal at a suitable temperature (below Ac_1 for ferritic steels) in contact with a nitrogenous gas, usually ammonia. At elevated temperature, the ammonia dissociates into its components according to the reaction:

$$2NH_3 \rightarrow 2N + 3H_2$$

The nitrogen, which is very active at the moment of decomposition of the ammonia gas, combines with the alloying elements in the steel to form nitrides. These nitrides form at the steel surface as a fine dispersion and impart extremely high hardness to the steel surface without the need for quenching.

Steels for Gas Nitriding. Because aluminum is the strongest nitride former of the common alloying elements, aluminum-containing steels (0.85 to 1.50% Al) yield the best nitriding results in terms of total alloy content. Chromium-containing steels can approximate these results if their chromium content is high enough (approximately 5% Cr). Plain carbon steels are not suited to gas nitriding because during the process, they form an extremely hard, brittle case that spalls ("flakes off") readily. Compositions of some typical aluminum-bearing steels are shown in Table 4.

Other steels that can be surface hardened by gas nitriding include medium carbon, chromium-containing alloy steels; hot work tool steels such as H11, H12, and H13; and most stainless steels. In conventional gas

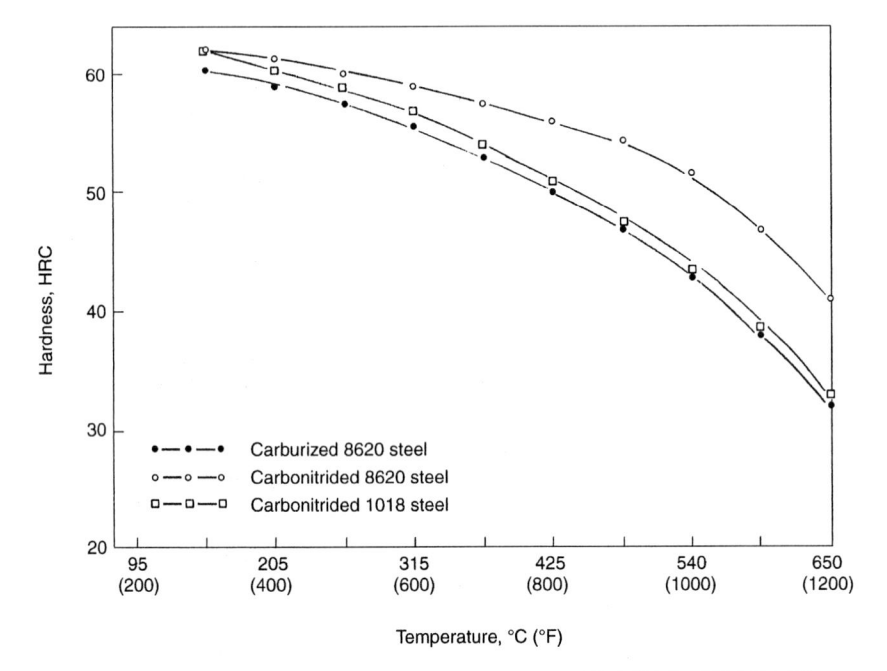

8620 carburized steel			8620 carbonitrided steel			1018 carbonitrided steel		
°C	°F	HRC	°C	°F	HRC	°C	°F	HRC
150	300	60.5	150	300	62.0	150	300	62.0
205	400	59.0	205	400	61.5	205	400	60.5
260	500	57.5	260	500	60.0	260	500	59.0
315	600	55.5	315	600	59.0	315	600	57.0
370	700	53.0	370	700	57.5	370	700	54.0
425	800	50.0	425	800	56.0	425	800	51.0
480	900	47.0	480	900	54.0	480	900	47.0
540	1000	43.0	540	1000	51.5	540	1000	43.5
595	1100	38.0	595	1100	47.0	595	1100	38.5
650	1200	32.0	650	1200	41.0	650	1200	32.5

Fig. 5 Tempering curves for carburized and oil-quenched and carbonitrided and oil-quenched steels. Source: J.L. Dossett, Midland Metal Treating Inc., Personal research, 1992

nitriding, best results can be obtained only by hardening and tempering the steels before nitriding (Table 4).

Furnaces. As a rule, gas nitriding is done in batch-type furnaces such as a pit or bell-type furnace (see Chapter 4). A major requirement is that the furnace can be tightly sealed to prevent infiltration of air.

Processing. Either a single- or a double-stage process may be used. In the single-stage process, a temperature range of about 495 to 525 °C (925 to 975 °F) is used, and the dissociation rate ranges from 15 to 30%.

The first stage of the double-stage process is, except for time, a duplication of the single-stage process. The second stage may proceed at the nitriding temperature employed for the first stage, or the temperature may be increased from 550 to 565 °C (1025 to 1050 °F), but at either temperature, the rate of dissociation in the second stage is increased from 65 to 85% (preferably 80 to 85%). Generally, an ammonia dissociator is necessary to obtain the required higher second-stage dissociation. In all cases, the furnace must be purged with ammonia or nitrogen so that the atmosphere contains no more than 5% of air before the nitriding cycle is started. Cycles may range from 10 to 50 h or more.

Case Characteristics. Nitrided cases are much harder (when correctly evaluated) than carburized cases, but they are shallower and have a hardness gradient not unlike a carburized case. Nitrided cases also have greater resistance to softening from heat compared with carburized cases.

Plasma (Ion) Nitriding

Plasma, or ion, nitriding is a method of surface hardening using glow discharge technology to introduce nascent (elemental) nitrogen to the surface of a metal part for subsequent diffusion into the material. In a vacuum, high-voltage electrical energy is used to form a plasma, through which nitrogen ions are accelerated to impinge on the workpiece. This ion bombardment heats the workpiece, cleans the surface, and provides active nitrogen. Ion nitriding provides better control of case chemistry and has other advantages, such as lower part distortion than conventional (gas) nitriding. A key difference between gas and ion nitriding is the mechanism

Table 4 Nominal composition and preliminary heat treating cycles for aluminum-containing low-alloy steels commonly gas nitrided

Steel			Composition, %								Austenitizing temperature(a)		Tempering temperature(a)	
SAE	AMS	Nitralloy	C	Mn	Si	Cr	Ni	Mo	Al	Se	°C	°F	°C	°F
...	...	G	0.35	0.55	0.30	1.2	...	0.20	1.0	...	955	1750	565–705	1050–1300
7140	6470	135M	0.42	0.55	0.30	1.6	...	0.38	1.0	...	955	1750	565–705	1050–1300
...	6475	N	0.24	0.55	0.30	1.15	3.5	0.25	1.0	...	900	1650	650–675	1200–1250
...	...	EZ	0.35	0.80	0.30	1.25	...	0.20	1.0	0.20	955	1750	565–705	1050–1300

Note: SAE: Society of Automotive Engineers; AMS: Aerospace Material Specification. (a) Sections up to 50 mm (2 in.) in diameter quenched in oil; larger sections may be water quenched. Source: Ref 13

used to generate nascent nitrogen at the surface of the work. Reference 14 provides additional information.

Salt Bath Nitriding

Steels that can be nitrided in a gaseous atmosphere may also be nitrided in molten salt (liquid nitriding) at essentially the same temperature—510 to 565 °C (950 to 1050 °F). As in liquid carburizing and cyaniding, the case hardening medium is molten cyanide. Unlike liquid carburizing and cyaniding, however, liquid nitriding is done below, instead of above, the transformation temperature range of the steel being treated.

Furnaces for salt bath nitriding are the same as those used for liquid carburizing (see Chapter 4).

Types of Salts. A typical commercial bath for liquid nitriding is composed of a mixture of sodium and potassium salts. The sodium salts, which constitute 60 to 70 wt% of the total mixture, consist of 96.5% NaCN, 2.5% Na_2CO_3, and 0.5% NaCNO. The potassium salts, 30 to 40 wt% of the mixture, consist of 96% KCN, 0.6% K_2CO_3, 0.75% KCNO, and 0.5% KCl. The operating temperature of this salt bath is 565 °C (1050 °F). Control of bath composition is critical.

Proprietary Salt Bath Processes. Aerated salt bath nitriding is a proprietary process that employs cyanide salts. In this process, measured amounts of air are pumped through the molten bath. The introduction of air provides agitation and stimulates chemical activity. The cyanide content of this bath, calculated as NaCN, is preferably maintained at about 50 to 60% of the total bath content, and the cyanate at 32 to 38%. The potassium content of the fused bath, calculated as elemental potassium, is between 10 and 30%, preferably about 18%. The potassium may be present as the cyanate or the cyanide, or both. The remainder of the bath is sodium carbonate.

This process produces a nitrogen-diffused case 0.3 mm (0.012 in.) deep on plain carbon or low-alloy steels in a 1.5 h cycle. The surface layer (0.005 to 0.013 mm, or 0.0002 to 0.0005 in., deep) of the case is composed of epsilon nitride (Fe_3N) and a nitrogen-bearing Fe_3C. Parts that are nitrided by the aerated bath process are machined to final dimensions before being nitrided.

Another proprietary aerated process for salt bath nitriding is capable of producing results similar to the process discussed above. The advantage of this process is that the salts do not contain cyanide.

Gaseous Nitrocarburizing

The term *nitrocarburizing* includes several proprietary names. In general, the results are similar to those obtained with aerated salt bath nitriding (sometimes called *salt bath nitrocarburizing*). A natural advantage of the gas process is the elimination of the need for salt removal, waste treatment, and disposal.

Ferritic nitrocarburizing treatments involve diffusion of both nitrogen and carbon into the surfaces of ferrous metals at temperatures below 675 °C (1250 °F), thus distinguishing it from carbonitriding, because nitrocarburizing is done with steels in their ferritic, rather than austenitic, phase.

The primary objective of such treatments is usually to improve antiscuffing characteristics of ferrous engineering components by providing the surface with a compound layer—really, a surface zone—exhibiting good wear/friction-resistant properties. In addition, fatigue characteristics can be considerably improved, particularly when nitrogen is retained in solid solution in the "diffusion zone" beneath the compound layer. This retention normally is achieved by quenching in oil or water from the treatment temperature. Corrosion resistance provided by the compound zone is an important secondary benefit.

Gaseous nitrocarburizing commonly employs sealed-quench batch furnaces of the same design used for carburizing and carbonitriding (see Chapter 4). Furnace operating temperatures are low enough and capable of holding very uniform temperatures to maintain steels in the ferritic condition. The atmosphere employed consists of ammonia diluted with a carrier gas. In one process, the atmosphere is formed from equal amounts of ammonia and endothermic gas. In another process, a typical atmosphere consists of 35% ammonia and 65% refined exothermic gas (nominally 97% nitrogen), which may be enriched with a hydrocarbon gas. High-purity nitrogen is used as a diluent in one of these processes. Gaseous nitrocarburizing is performed near 570 °C (1060 °F), a temperature below the austenite range for the iron-nitrogen system. Treatment times generally range from 1 to 5 h.

Plasma Nitrocarburizing

Plasma nitrocarburizing is a method of surface hardening employing vacuum and glow-discharge technology similar to plasma (ion) nitriding except that mixtures of hydrogen, nitrogen, and a carbon-bearing gas such as methane or carbon dioxide are used in the process. Additional information on nitriding and nitrocarburizing can be found in Ref 15 and 16.

REFERENCES

1. J.R. Davis, Ed., Process Selection Guide, *Surface Hardening of Steels: Understanding the Basics,* ASM International, 2002, p 1–15

2. M.S. Stanescu et al., Control of Surface Carbon Content in Heat Treating of Steel, *Heat Treating,* Vol 4, *ASM Handbook,* ASM International, 1991, p 573–586
3. *Carburizing and Carbonitriding,* American Society for Metals, 1977, p 21
4. H.E. Boyer, Chapter 10, *Practical Heat Treating,* American Society for Metals, 1984, p 180–195
5. *Carburizing and Carbonitriding,* American Society for Metals, 1977, p 68
6. G. Krauss, Surface Hardening, *Steels: Heat Treatment and Processing Principles,* ASM International, 1990, p 281–317
7. W.L. Grube and S. Verhoff, Plasma (Ion) Carburizing, *Heat Treating,* Vol 4, *ASM Handbook,* ASM International, 1991, p 352–362
8. A.D. Godding, Liquid Carburizing and Cyaniding, *Heat Treating,* Vol 4, *ASM Handbook,* ASM International, 1991, p 329–347
9. *Carburizing and Carbonitriding,* American Society for Metals, 1977, p 152
10. *Carburizing and Carbonitriding,* American Society for Metals, 1977, p 128
11. J.L. Dossett, Carbonitriding, *Heat Treating,* Vol 4, *ASM Handbook,* ASM International, 1991, p 376–386
12. J.L. Dossett, *Furnace Atmospheres Presentation,* ASM International Satellite Seminar, 1995
13. J.R. Davis, Ed., Nitriding, *Surface Hardening of Steels: Understanding the Basics,* ASM International, 2002, p 141–194
14. J.M. O'Brien and D. Goodman, Plasma (Ion) Nitriding, *Heat Treating,* Vol 4, *ASM Handbook,* ASM International, 1991, p 420–424
15. T. Bell, Gaseous and Plasma Nitrocarburizing, *Heat Treating,* Vol 4, *ASM Handbook,* ASM International, 1991, p 425–436
16. D. Pye, *Practical Nitriding and Ferritic Nitrocarburizing,* ASM International, 2003

CHAPTER **9**

Flame and Induction Hardening

CASE HARDENING OF STEEL (the subject of Chapter 8) deals exclusively with methods that involve changes in surface composition. Methods discussed in this chapter produce hard surfaces (commonly called cases) on ferrous metal parts (steels and cast irons) with no change in surface or base composition. While flame and induction hardening are the principal production methods used, other case hardening methods, including shell hardening and laser-beam and electron-beam hardening methods, are also used and are discussed briefly.

Hardened Zones in Low-Hardenability Steels

The use of low-hardenability steels, which is one of the oldest methods of producing a part with a hard surface and a relatively soft core without altering composition, is involved with each of these surface hardening methods. Although the part is heated throughout the section, a low-hardenability steel transforms to the austenitic state from softer products such as ferrite or pearlite so rapidly that only the outer layer is cooled quickly enough to form martensite. Figure 1 shows that only a narrow band of full hardness is obtained after heating and quenching. Because of the lesser mass involved, the depth of the hardened zone is greater on the smallest diameter section (see also Chapter 11, "Heat Treating of Tool Steels").

Because only the surface is hardened by these applied energy (thermal) processes, a loss of surface carbon, "decarburization," can cause decreased surface hardness in the areas affected. Decarburization generally is present, if not removed by machining, on all steel bar stock, forgings, and castings. Refer to Appendix B, "Decarburization of Steels," for a more detailed discussion of the problem of decarburization.

Shell Hardening in Molten Metal or Salt

Another method of hardening only the surface layers of a workpiece is by a technique commonly known as *shell hardening*. The workpiece is immersed either fully or selectively in a high-conductivity heating medium such as a fluidized bed or a salt bath, but only long enough to completely heat the outer layers. If the part is round (wheels or gears) and the requirement is that only the outer diameter (OD) be hardened, the part may be slowly turned so that only the OD portion is heated. The heating of the surface is followed by quenching. By this technique, the core portion is never heated sufficiently to form austenite. This technique is limited in use, largely by part design. Obviously, the part must have some appreciable minimum thickness or it would heat completely through. Cold working dies are typical examples of this application to surface hardening.

Flame Hardening

Flame hardening is a case hardening process in which a thin surface shell of a steel or cast iron part is heated rapidly to above the critical point of the ferrous material. After the grain structure of the shell has been austenitized, the part is quickly quenched, transforming the austenite to martensite while leaving the core of the part in its original state.

Flame hardening employs direct impingement of a high-temperature flame on the surface area to be hardened. The high-temperature flame is obtained by combustion of a mixture of a fuel gas with oxygen or air. Fuel gases commonly used are acetylene, methylacetylene propadiene (MAPP) gas, or propane. Fuel gas selection is based on the flame temperature desired. Different designs of flame heads (torches) are used depending on the application.

Using this method, the hardened layer may be varied from a very thin skin to depths of up to 6 mm (¼ in.). The depth of the hardened zone is controlled by the amount of time the flame head is allowed to heat the

Fig. 1 Cross section of three sizes of water-hardening tool steel (W1) after heating to 800 °C (1475 °F) and quenching in brine. Black rings indicate hardened zones (cases) (65 HRC). Cores range from 38 to 43 HRC. Source: Ref 1

steel. With short heating times, only a thin skin is made austenitic and hardened. With longer heating times, the heat penetrates to a greater depth and results in deeper hardening. All sorts of sections and sizes, as well as any accessible portion of a workpiece, can be hardened by this procedure. Likewise, quenching procedures may vary and are not necessarily the same as for the same steel heated and quenched from a furnace. This is because the quickly heated area receives a "mass quenching effect" from the mass of underlying cold metal. In many instances where the heated zone is shallow and the mass is relatively large, liquid quenching is not required to develop full hardness. Air blast cooling may achieve the same effects.

Almost any new flame-hardening operation requires developmental work (cut and try) to arrive at the optimum procedure for a specific workpiece. Past experience with similar workpieces is always helpful in developing specific procedures.

Material for flame hardening can be any ferrous metal (steel or cast iron) that has the potential for being directly hardened by any other method.

Equipment choices for flame hardening range from a handheld torch to an elaborate mechanized and automated setup, depending mostly on the number of identical parts to be processed.

Methods of Flame Hardening. The principal procedures employed in flame hardening are spot or stationary, progressive, spinning, and combination progressive-spinning modes of operation. Selection of the most appropriate system depends on shape, size, and composition of the workpiece; the area to be surface hardened; required depth of the hardened zone; and the number of pieces to be hardened.

Figure 2(a) illustrates the spot or stationary method, consisting of heating specific areas with a suitable flame head and subsequently quenching. Both torch and workpiece are kept stationary, and only the area immediately under the torch is heated and hardened. This is the simplest of the four systems because mechanical equipment is not required (except perhaps a fixture and timing device to ensure uniform processing of each workpiece). If desired, however, the operation could be automated.

The progressive method is used to harden large areas that are beyond the scope of spot hardening. As illustrated in Fig. 2(b), the heating torch and integral quenching head travel along the face of the object to be hardened. The torch and quench head also may be kept stationary, and the work moved along under it. Speeds of about 75 to 300 mm/min (3 to 12 in./min) are commonly employed. Thus, the work surface is progressively heated and hardened as either the torch or the work moves along. This method of flame hardening can be applied to various workpiece shapes and sizes but is used most often for hardening of long flat areas such as machine tool ways.

162 / Practical Heat Treating: Second Edition

A distinct disadvantage of the progressive method for flame hardening of rounds is that it is virtually impossible to avoid formation of a narrow zone that is slightly softer than the rest of the hardened skin, which is left at the starting-stopping point. In high-carbon steels, cracking also may be encountered at this zone when the flame overlaps the previously hardened area. For these reasons, it is usually preferable to harden cylindrical workpieces by spinning, as shown in Fig. 3 and 4. Equipment for hardening by the progressive method commonly consists of one or more flame heads and a quenching means mounted on a carriage that runs on a track at a regulated speed.

The spinning method is commonly employed for hardening small rounds. The torch usually is held stationary and the work rotated at speeds of up to 100 rpm. The quenching water is turned off while the piece is being heated. The work is rotated until the surface reaches the desired temperature or until the heat penetrates to the desired depth. At this time, the flames are extinguished and the quenching water turned on, or the part

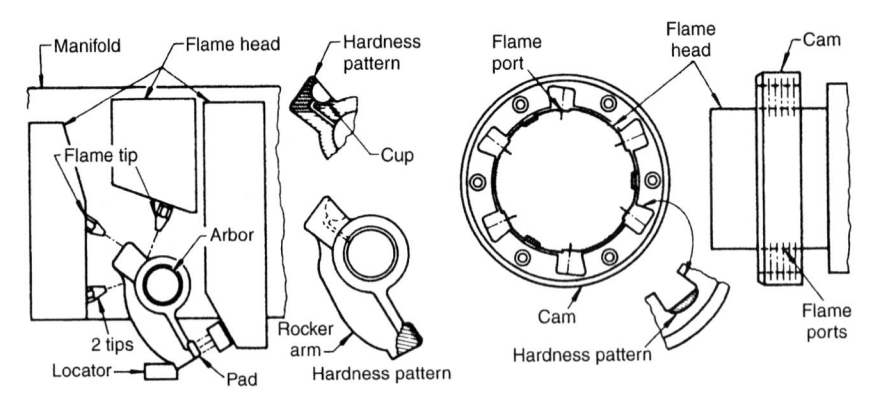

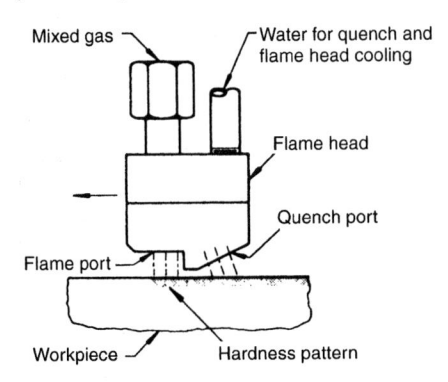

Fig. 2 Methods of flame heating. (a) Spot (stationary) heating of a rocker arm and the internal lobes of a cam; quench not shown. (b) Progressive method. Source: Ref 2

is dropped into a quenching tank. Sometimes multiple torches are used to ensure rapid and uniform heating. Three methods of spin hardening are illustrated in Fig. 3. In two of the methods, the workpiece is rotated. However, in the third application (shown at the bottom of Fig. 3), rotation of the flame head was more practical.

A combination of the progressive and spinning methods is illustrated in Fig. 4. Frequently used for hardening the surfaces of long cylinders, this method consists of moving a heating ring along the length of the cylinder at speeds of about 75 to 300 mm/min (3 to 12 in./min), followed by a quenching ring. The progressive spinning method differs from the

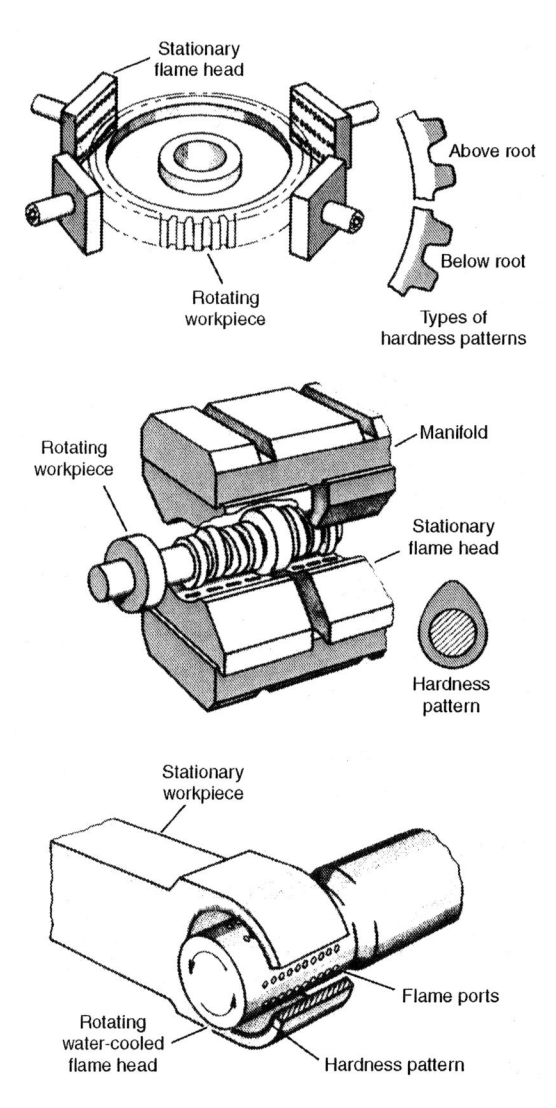

Fig. 3 Spinning methods of flame heating, in which (top center) the part rotates and (bottom) flame head rotates. Quench is not shown. Source: Ref 2

simple progressive method in that the work is spun while it is being heated. The rapidity of revolution depends on the size of the piece being hardened.

Multiple heating heads may be employed to ensure proper heating. In this method, as in the others, the depth of hardening depends on the heating time.

Quenching. In Fig. 2, 3, and 4, note that quenching facilities are "built in," which usually is the procedure for relatively small workpieces heat treated on a production basis. Larger workpieces may not require quenching.

In almost all instances, temperature-controlled water or a dilute polymer solution is used for spray quenching following flame heating. This is done for two reasons: spraying of oil presents a fire hazard, and the amount of water or polymer used can be closely controlled so that the workpiece is not fully quenched until it is completely cold. As mentioned in Chapter 2, "Fundamentals of the Heat Treating of Steel," drastic quenching is required only to exceed the critical cooling rate down to the lower temperature range, wherein cooling rate is no longer critical. The practical results are that steels (or cast irons) that might crack when water quenched by immersion can be safely hardened by the timed control of the water in a spray quench.

Tempering. Even though the workpiece is still quite warm when removed from a flame heating and quenching operation, there still is a need for tempering. Regardless of the system used to flame harden a workpiece, the part must be tempered after the hardening operation. Tempering prevents cracking by relieving some of the stress set up in hardening and imparts some degree of toughness to the hard case.

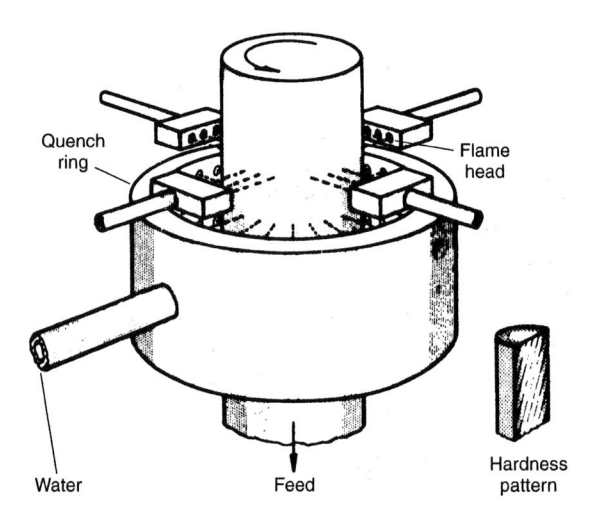

Fig. 4 Combination progressive-spinning method of flame hardening. Source: Ref 2

Induction Hardening

Electromagnetic induction is a very important and widely used method of heating a metal part. It can be used for several types of heat treating operations such as normalizing, annealing, heating for hardening, tempering, and heating of nonferrous metals. Additionally, this method of heating can also be used for melting, preheating for forging, and welding and brazing. The use of induction heating, as dealt with in this chapter, is confined principally to localized or selective heating for subsequent hardening (usually surface hardening). It should be noted, however, that induction hardening can be and often is used for through hardening parts.

The metallurgical principles and advantages of induction heating, compared with furnace heating, generally are:

- Ease of automation and control
- Reduced floor space requirements
- Quiet and cleaner operation
- Suitability for integration in a production line

Principles of Induction Heating

Any electrically conducting material can be heated by electromagnetic induction. As alternating current flows through the inductor, or work coil (which does not contact the workpiece), a highly concentrated, rapidly alternating magnetic field is established within the coil. The strength of this field depends primarily on the magnitude of the current flowing in the coil. The magnetic field thus established induces an electric potential in the part to be heated, and because the part represents a closed circuit, the induced voltage causes the flow of current. The resistance of the part to the flow of the induced current causes surface heating.

The depth of current penetration depends on workpiece permeability, resistivity, and the alternating current frequency. Because the first two factors vary comparatively little, the greatest variable is frequency. Depth of current penetration decreases as frequency increases. High-frequency current generally is used when shallow heating (thin case) is desired; intermediate and low frequencies are used in applications requiring deeper heating.

Most induction surface-hardening applications require comparatively high power densities and short heating cycles to restrict heating to the surface area. The principal metallurgical advantages that may be obtained by surface-hardening with induction are the same as for the other surface-hardening techniques. This would include the ability to harden lower hardenability ferrous materials and to have less distortion from processing.

The pattern of heating obtained by induction is determined by the (a) shape of the induction coil producing the magnetic field, (b) number of turns in the coil, (c) operating frequency, (d) alternating current power input, and (e) nature of the workpiece. Four examples of magnetic fields and induced currents produced by induction coils are shown in Fig. 5.

The rate of heating obtained with induction coils depends on the strength of the magnetic field to which the part is exposed. In the workpiece, this becomes a function of the induced currents and of the resistance to their flow.

The preceding paragraphs describe what happens when steel is heated by induction but do not explain why heat is developed in the workpiece. It is generally accepted that heat is developed and conducted to the interior in the following three ways.

One way is to place magnetic materials such as steel in a magnetic field, where the molecules tend to align themselves with the polarity of the field. If the current is reversed a great number of times per second, as in high-frequency alternating current, the molecules also tend to change their alignment a similar number of times. As a result, molecular friction is set up and heat is generated. When the steel is heated and becomes austenitic, it also becomes nonmagnetic. The heat generated by this effect then becomes negligible.

Another method of developing and conducting heat to the interior is by inducing current into a steel workpiece. The induced current has a tendency to swirl in much the same fashion as a pool of water swirls when stirred with a stick. The swirling effect of the current is known as eddy current, and it induces heat. The eddy current effect probably is the major source of heat, particularly when the steel has become nonmagnetic.

In a third method, heat is generated in the surface layers and carried to the interior by simple conduction. After the steel has been heated to the proper austenitizing temperature and required depth, quenching is accomplished by introducing a water spray or other suitable quenching fluid

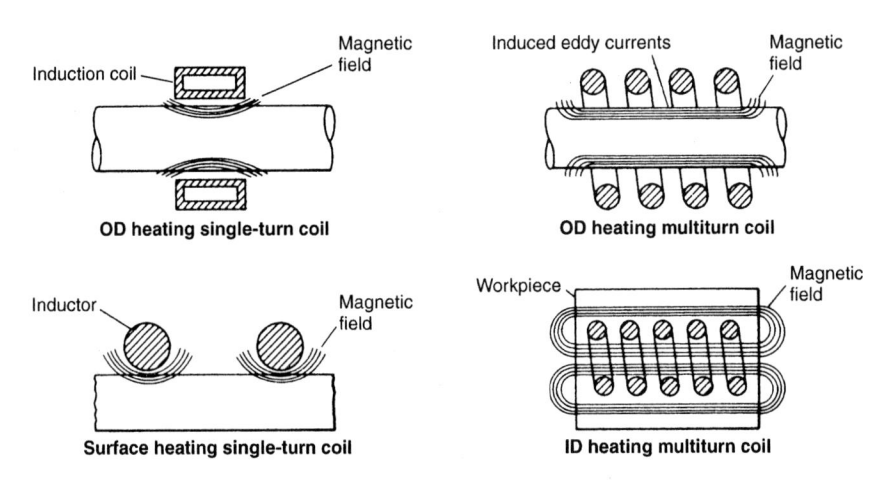

Fig. 5 Magnetic fields and induced currents produced by various induction coils. Source: Ref 3

through the spaces between the inductor coil or by dropping the workpiece from its heating position in the coil into a quenching tank. The depth of hardness penetration is controlled by the power input, the frequency used, and heating time in the inductor. Because the heating effect is very rapid, hardening only a thin skin without quenching is possible. In such instances, the thin hot skin is cooled by the large cold mass lying under it. The piece is then described as self-quenched or mass-quenched.

An induction heating operation in its simplest form is shown in Fig. 6. A small gear is manually placed within a multiple-turn coil (the inductor). The coil is made from copper tubing to permit water cooling of the coil when it is in operation. The outer ring is the quench ring, through which the quench media spray quenches the gear in a timed operation at the end of the heating cycle.

Equipment for Induction Hardening

The basic equipment for induction hardening consists of a power supply, controls, a means of matching the power supply with the load, an inductor, quenching equipment, cooling systems for the power supply and

Fig. 6 Induction hardening of a gear. The heating coils are surrounded by the spray quenching head. A thin surface layer of the gear will be heated, power turned off, and the spray turned on, resulting in a thin hard case. Source: Ref 1

quenchant, and a system for holding and positioning the workpieces on the coil. The part may range in degree of sophistication from a simple locator (as in Fig. 6) to a completely automated positioning and indexing system. The entire induction-hardening operation may be fully automated and controlled.

Power Supplies. The four most widely used power supplies are vacuum tube oscillators, motor generators, frequency multipliers, and solid-state (static) inverters.

The vacuum tube oscillator changes the power line voltage to 15,000 to 20,000 V and rectifies it to direct current. Then, by using a circuitry involving one or more vacuum tubes, capacitors, and coils, it generates high-voltage, high-frequency power at 200 to 450 kHz. Operating frequency is determined by workpiece size, inductor design, and load matching.

The motor generator consists of a conventional three-phase induction motor, which drives a high-frequency, single-phase alternator. The most common frequencies are 1 to 10 kHz. Frequency multipliers essentially are special transformers that triple the line frequency to single-phase 180 Hz. This may be tripled again to 540 Hz.

Frequency inverters use diodes and silicon-controlled rectifiers (SCRs). For the lower-frequency range (up to 10 kHz), the inverter is gradually becoming the main source of high-frequency power for two principal reasons: (a) equipment cost is lower than for motor-generator installations in terms of kW delivered, and (b) the inverter is more efficient because it does not use power when a workpiece is not being heated; the motor generator uses a significant amount of power when not heating a workpiece.

Medium-frequency power supplies using solid-state frequency conversions make up the vast majority of induction heating power supplies in use today. These systems have largely replaced motor-generator sets for medium-frequency conversion. As compared with motor generators, the basic advantages of solid-state power supplies are their improved efficiency, low initial cost and maintenance, and availability in a multitude of sizes and frequencies.

Frequencies, efficiencies, power ranges, and features for the four principal types of induction heating equipment are given in Table 1.

Inductors. For the most efficient heating, the work should be surrounded by the coil with a minimum air gap between the inductor and the work. An inductor that meets these requirements is shown in Fig. 7. This single-turn inductor contains separate internal chambers—one for cooling water and one for the quenchant. With suitable auxiliary equipment, this general design can be used for a variety of applications, including progressive surface hardening of long barlike products.

Despite the fact that greatest efficiency is attained when the inductor surrounds the work, this is not always feasible. For this reason, there are many other inductor designs including single- and multiple-turn types. There are even specially designed inductors for hardening internal surfaces such as cylinder bores. However, these generally are less efficient than the type shown in Fig. 7. A variety of coils and resulting heat patterns

Table 1 Characteristics of the four major power sources for induction heating

Power source	Frequency range	Power range	Efficiency, %	Features
Line frequency	60 Hz	100 kW to 100 MW	90–95	High efficiency; low cost; no complex equipment; deep current penetration
Motor-generator	500 Hz to 10 kHz	10 kW to 1 MW	75–85	Low sensitivity to ambient heat; low sensitivity to line surges; fixed frequency; low maintenance cost; spares not needed
Solid state	180 Hz to 50 kHz	1 kW to 2 MW	75–95	No standby current; high efficiency; no moving parts; needs protection outdoors; no warmup time; impedance matches changing loads
Vacuum tube	50 kHz to 10 MHz	1 kW to 500 kW	50–75	Shallow heating depth; localized heating; highest cost; impedance matches changing loads; lowest efficiency

Source: Ref 4

Fig. 7 An inductor with separate internal chambers for flow of quenchant and cooling water. Source: Ref 1

are shown in Fig. 8, which illustrates some of the attainable objectives with induction heating, including the heating of specific areas on flat surfaces (Fig. 8e).

Selection of Frequency

The application generally dictates frequency requirements, although some broad overlapping exists. In many instances, essentially the same results can be achieved with more than one frequency by using different power densities and heating times. For example, a frequency of 10 kHz at low power and long heating time could duplicate the case pattern as the heating of 3 kHz at higher power and shorter heating time; the same is true for 400 and 10 kHz. It should be pointed out that in these instances, the part distortion or the hardness profile across the case/core interface may not be the same.

In establishing processing cycles for surface hardening, the minimum depths that are considered practical for 3, 10, and 450 kHz are given in Table 2. Producing shallower depths than are required is expensive because it requires higher frequency and/or higher power—both of which increase the original equipment and operating costs.

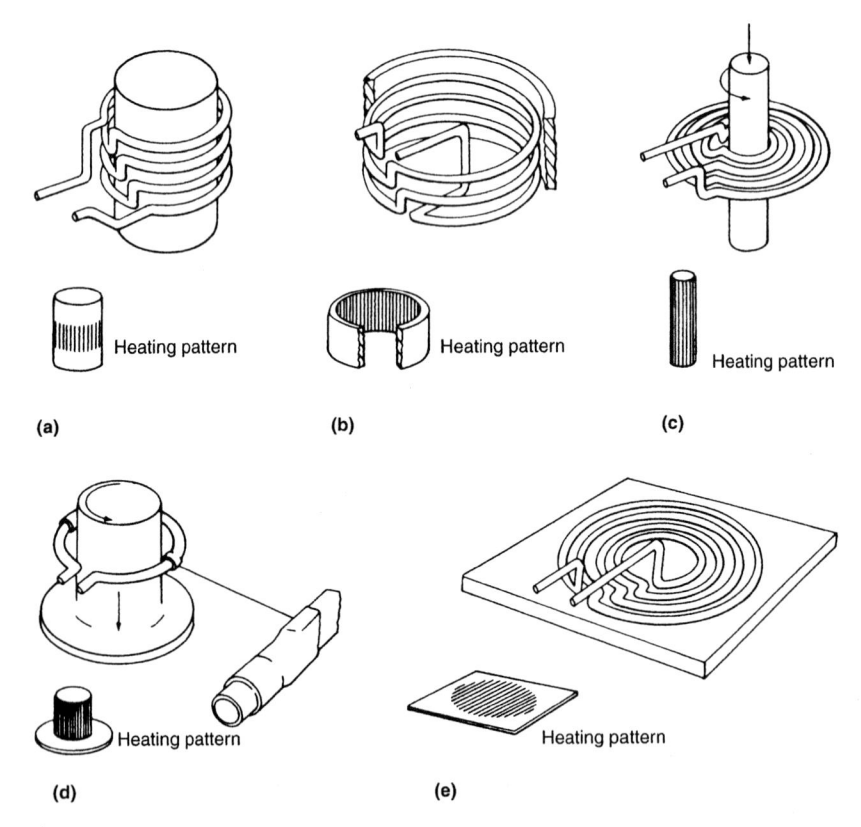

Fig. 8 Typical work coils for induction heating. Source: Ref 3

For most surface hardening applications, 10 or 3 kHz is satisfactory. Ten kHz provides a suitable contoured hardness pattern on cams and coarse-pitch gears. Finer pitch gears require 450 kHz to prevent through hardening. However, in some instances, because the roots are not heated effectively at the higher frequency, 10 kHz is used in a dual-frequency approach. The valleys are preheated with 10 kHz; 450 kHz is used for the tips. In all such applications, time must be kept short to limit heat flow by conduction.

Process Development

Development of optimum heating and cooling cycles for induction hardening of a specific workpiece is usually the result of some experimentation and sectioning of parts to verify case patterns and depths. For information that may be helpful in developing a new setup, see Ref 4 and 5. Reference to records of past experience with similar workpieces usually is the most helpful tool in establishing new cycles of heating and cooling.

Tempering of Induction-Hardened Parts

Induction-hardened parts generally must be tempered after the hardening operation. Tempering prevents cracking of higher hardenability ferrous parts by relieving some of the stress set up in hardening and imparts some degree of toughness to the hard case. The three most common tempering processes that are used are furnace tempering, induction tempering, and residual heat tempering.

Furnace tempering involves placing the entire workpiece into a batch or continuous tempering oven for a prescribed period time in order to achieve the desired surface hardness. Depending on specific hardenability and quenching conditions used, the tempering operation may need to be started within a very short time after hardening to prevent cracking of the parts.

Induction tempering may be accomplished by reheating the workpiece—after quenching and hardening—in the same coil but under lower power and different time conditions. Sometimes lower frequency (1 kHz) units are used at low power for induction tempering because of the greater depth of heating. Figure 9 shows tempering curves for both induction and furnace tempering for induction-hardened 1050 steel.

Table 2 Minimum hardness depths for production work in surface hardening

Frequency, Hz	Hardness depth(a), in.	Penetration of electrical energy(b), in.
3,000	0.060	0.035
10,000	0.040	0.020
450,000	0.020	0.003

(a) Approximate practical minimum depth of hardness in inches. (b) Approximate theoretical depth of penetration of electrical energy in inches. Source: Ref 1

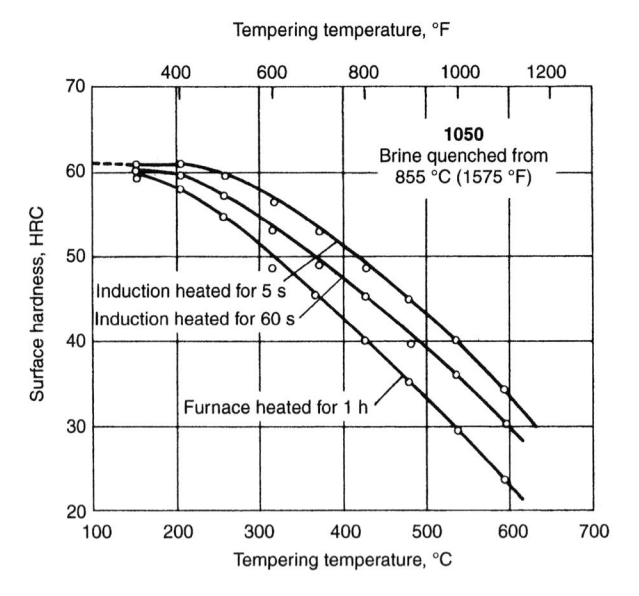

Fig. 9 Variation of hardness with tempering temperature for furnace and induction heating. Source: Ref 5

Other Surface Hardening Treatments

Electron-beam surface hardening treatments utilize the energy from an electron beam to heat the surface of a ferrous part to a typical hardened depth of 0.1 to 1.5 mm (0.004 to 0.060 in.). The rapid cooling for martensite formation occurs, after the energy transfer is completed, by self-quenching. The workpiece must be approximately 10 times the austenitized depth for effective self-quenching. This treatment, which causes minimum distortion, is used cost effectively on certain specialized applications.

Laser-surface-hardening treatments operate in a manner similar to electron-beam treatments except that the energy source is an industrial laser. Because steel has low infrared absorption, the workpieces may need to be coated with an absorptive material prior to treatment. References 6 and 7 provide more details on electron-beam and laser-surface-hardening processes.

REFERENCES

1. H.E. Boyer, Chapter 11, *Practical Heat Treating,* 1st ed., American Society for Metals, 1984, p 196–208
2. T. Ruglic, Flame Hardening, *Heat Treating,* Vol 4, *ASM Handbook,* ASM International, 1991, p 268–285
3. J.R. Davis, Ed., Surface Hardening by Applied Energy, *Surface Hardening of Steels: Understanding the Basics,* ASM International, 2002, p 237–274

4. P.A. Hassell and N.V. Ross, Induction Heat Treating of Steel, *Heat Treating,* Vol 4, *ASM Handbook,* ASM International, 1991, p 164–202
5. R.E. Haimbaugh, *Practical Induction Heat Treating,* ASM International, 2001, p 145
6. S. Schiller and S. Panzer, Electron Beam Surface Hardening, *Heat Treating,* Vol 4, *ASM Handbook,* ASM International, 1991, p 297–311
7. O.A. Sandven, Laser Surface Hardening, *Heat Treating,* Vol 4, *ASM Handbook,* ASM International, 1991, p 286–296

CHAPTER **10**

Heat Treating of Stainless Steels

THE TERM "STAINLESS STEELS" is a misnomer for two reasons: (a) none of the steels included in this grade are completely "stainless" when exposed to all possible atmospheric and corrosive media conditions; and (b) only a few are truly "steels" in terms of their response to heat treatments used for the carbon and alloy steels, as discussed in Chapters 6, "Heat Treating of Carbon Steels," and 7, "Heat Treating of Alloy Steels."

Stainless steels include approximately 60 standard compositions of corrosion- and/or heat-resistant iron-base alloys, in addition to more than 100 nonstandard compositions. Stainless steels contain a minimum of about 11% chromium. Few stainless steels contain more than 30% chromium or less than 50% iron.

What Are Stainless Steels?

Chromium is the basis for corrosion resistance in stainless steels. One or more additional elements may be used in conjunction with chromium for most grades of stainless steels, but chromium is the key element contributing to corrosion resistance. When chromium is added to iron in relatively small amounts (1 to 3%), a modest increase in corrosion resistance of the alloy is evident. However, as the amount of chromium approaches approximately 10%, a dramatic increase in corrosion resistance takes place; such an alloy is virtually impervious to rusting from almost any outdoor exposure to normal atmospheres, such as rain, humidity, and temperature variations. This is not necessarily true in salt-bearing or industrial atmospheres.

Therefore, a minimum requirement for a stainless steel is that it contain at least 11.0% chromium and that it is capable of resisting attack from normal atmospheric exposure. It is essential that both of these conditions are fulfilled for a steel to qualify as a stainless steel. Many highly alloyed iron-base alloys, such as certain tool steels, contain more than 11.0% chromium, but because of their high carbon content, they do not meet the minimum requirements for stainless steels.

The mechanism of corrosion resistance in stainless steels occurs when chromium alloys with iron and forms a transparent surface oxide that serves to protect the alloy. However, formation of this protective oxide is dependent on the chromium alloying with the ferrite (iron). Carbon has a high affinity for alloying with chromium and can, as carbon content increases, combine with most of the chromium. This action impoverishes the ferritic matrix and greatly reduces corrosion resistance because chromium carbide has little resistance to most corrosive media.

Classification of Wrought Stainless Steels

The American Iron and Steel Institute (AISI) has adopted standard designation numbers for approximately 60 grades of wrought stainless steels that are divided into four groups: austenitic, ferritic, martensitic, and precipitation-hardening grades. These terms generally are based on existing structures or on structures that can be attained in the steel by heat treatment. A fifth group of stainless steels—the duplex stainless steels—are supplied with a microstructure of approximately equal amounts of austenite and ferrite. Duplex stainless steels are not covered by the standard AISI groups.

Compositions of several prominent grades of each of the five groups are listed in Tables 1 to 5. Each table also includes the corresponding Unified Numbering System (UNS) designations (see Chapter 6 for a discussion of the UNS system). For more complete listings, see Ref 1 and 2.

In addition to the steels listed in Tables 1 to 5, there are more than 100 (perhaps many more) nonstandard compositions that are marketed under proprietary names. These nonstandard grades are produced in relatively small quantities; consequently, AISI does not list them as standard grades. In most instances, the nonstandard compositions have been developed to resist corrosion or heat, or both, in specific environments.

Austenitic Grades

Referring to Table 1, the austenitic grades carry identifying numbers of either 200 or 300, although most of the steels in Table 1 are 300 series alloys—the chromium-nickel grades. In the 200 series, some of the nickel has been replaced by manganese. The austenitic grades are used most widely in corrosive environments, although some grades (most notably, type 310) are used for elevated-temperature service up to 650 °C (1200 °F).

Because the austenitic grades do not change their crystal structure on heating, they do not respond to conventional quench-hardening treat-

Table 1 Chemical compositions of standard AISI austenitic stainless steels

UNS No.	Type/designation	C	Mn	Si	Cr	Ni	P	S	Other
S20100	201	0.15	5.5–7.5	1.00	16.0–18.0	3.5–5.5	0.06	0.30	0.25 N
S20200	202	0.15	7.5–10.0	1.00	17.0–19.0	4.0–6.0	0.06	0.30	0.25 N
S20500	205	0.12–0.25	14.0–15.5	1.00	16.5–18.0	1.0–1.75	0.06	0.30	0.32–0.40 N
S30100	301	0.15	2.0	1.00	16.0–18.0	6.0–8.0	0.045	0.03	...
S30200	302	0.15	2.0	1.00	17.0–19.0	8.0–10.0	0.045	0.03	...
S30215	302B	0.15	2.0	2.0–3.0	17.0–19.0	8.0–10.0	0.045	0.03	...
S30300	303	0.15	2.0	1.00	17.0–19.0	8.0–10.0	0.20	0.15 min	0.6 Mo(b)
S30323	303Se	0.15	2.0	1.00	17.0–19.0	8.0–10.0	0.20	0.06	0.15 min Se
S30400	304	0.08	2.0	1.00	18.0–20.0	8.0–10.5	0.045	0.03	...
S30403	304L	0.03	2.0	1.00	18.0–20.0	8.0–12.0	0.045	0.03	...
S30451	304N	0.08	2.0	1.00	18.0–20.0	8.0–10.5	0.045	0.03	0.10–0.16 N
S30500	305	0.12	2.0	1.00	17.0–19.0	10.5–13.0	0.045	0.03	...
S30800	308	0.08	2.0	1.00	19.0–21.0	10.0–12.0	0.045	0.03	...
S30900	309	0.20	2.0	1.00	22.0–24.0	12.0–15.0	0.045	0.03	...
S30908	309S	0.08	2.0	1.00	22.0–24.0	12.0–15.0	0.045	0.03	...
S31000	310	0.25	2.0	1.50	24.0–26.0	19.0–22.0	0.045	0.03	...
S31008	310S	0.08	2.0	1.50	24.0–26.0	19.0–22.0	0.045	0.03	...
S31400	314	0.25	2.0	1.5–3.0	23.0–26.0	19.0–22.0	0.045	0.03	...
S31600	316	0.08	2.0	1.00	16.0–18.0	10.0–14.0	0.045	0.03	2.0–3.0 Mo
S31620	316F	0.08	2.0	1.00	16.0–18.0	10.0–14.0	0.20	0.10 min	1.75–2.5 Mo
S31603	316L	0.03	2.0	1.00	16.0–18.0	10.0–14.0	0.045	0.03	2.0–3.0 Mo
S31651	316N	0.08	2.0	1.00	16.0–18.0	10.0–14.0	0.045	0.03	2.0–3.0 Mo; 0.10–0.16 N
S31700	317	0.08	2.0	1.00	18.0–20.0	11.0–15.0	0.045	0.03	3.0–4.0 Mo
S31703	317L	0.03	2.0	1.00	18.0–20.0	11.0–15.0	0.045	0.03	3.0–4.0 Mo
S32100	321	0.08	2.0	1.00	17.0–19.0	9.0–12.0	0.045	0.03	5 × %C min Ti
S34700	347	0.08	2.0	1.00	17.0–19.0	9.0–13.0	0.045	0.03	10 × %C min Nb
S34800	348	0.08	2.0	1.00	17.0–19.0	9.0–13.0	0.045	0.03	0.2 Co 10 × %C min Nb; 0.10 Ta
S38400	384	0.08	2.0	1.00	15.0–17.0	17.0–19.0	0.045	0.03	

(a) Single values are maximum values unless otherwise indicated. (b) Optional. Source: Ref 1

Table 2 Nominal chemical composition of standard AISI 400-series ferritic stainless steels

UNS No.	Type	C	Cr	Mo	Other
S42900	429	0.12	14.0–16.0	...	...
S43000	430	0.12	16.0–18.0	...	...
S43020	430F	0.12	16.0–18.0	0.6	0.06 P; 0.15 min S
S43023	430FSe	0.12	16.0–18.0	...	0.15 min Se
S43400	434	0.12	16.0–18.0	0.75–1.25	...
S43600	436	0.12	16.0–18.0	0.75–1.25	Nb + Ta = 5 × %C min
S44200	442	0.20	18.0–23.0	...	...
S44600	446	0.20	23.0–27.0	...	...

(a) Single values are maximum values unless otherwise indicated. Source: Ref 1

Table 3 Chemical compositions of selected wrought duplex stainless steels

UNS No.	Common designation	C	Mn	S	P	Si	Cr	Ni	Mo	Cu	W	N₂
S31500	3RE60	0.03	12.0–2.0	0.03	0.03	1.4–2.0	18.0–19.0	4.25–5.25	2.5–3.0	...	...	0.05–0.10
S31260	DP3	0.03	1.00	0.030	0.030	0.75	24.0–26.0	5.5–7.5	2.5–3.5	0.20–0.80	0.10–0.50	0.10–0.30
S31803	2205	0.03	2.00	0.02	0.03	1.00	21.0–23.0	4.5–6.5	2.5–3.5	...	...	0.08–0.20
S32550	255	0.03	1.5	0.03	0.04	1.00	24.0–27.0	4.5–6.5	2.9–3.9	1.5–2.5	...	0.10–0.25
S32900	10RE51	0.06	1.00	0.03	0.04	0.75	23.0–28.0	2.5–5.0	1.0–2.0	...	...	...
S32950	7-Mo Plus	0.03	2.00	0.01	0.035	0.60	26.0–29.0	3.5–5.20	1.0–0.5	...	...	0.15–0.35

(a) Single values are maximum. Source: Ref 1

ments. Therefore, the only heat treatments that are used for these grades are full annealing by rapid cooling from elevated temperatures, stress relieving, and surface hardening by nitriding.

The 200 and 300 grades cannot be quench hardened like alloy steels. As shown in Table 1, all of these grades have very low carbon contents, but relatively high nickel and/or manganese contents, both of which are strong austenite (gamma) formers. Type 301 can have a combined nickel and manganese content of about 8.0% (Table 1), but for most grades shown in Table 1, the nickel and manganese (principally nickel) content is much higher (around 24% for type 310). However, some of these grades—in sections up to 3.2 mm (⅛ in.) thick—can be hardened up to 35 HRC by cold rolling, after annealing, with a corresponding improvement in mechanical properties.

Now referring to Fig. 1, the vertical axis as a line only is essentially the constitution diagram for pure iron (see Chapter 2, "Fundamentals of the Heat Treating of Steel," for a complete explanation). Now, for the moment, instead of adding carbon we add chromium in increasing amounts (horizontal axis). Now, the examination must be restricted to the inner loop (vertical lines identified on Fig. 1 as "gamma loop" of pure Fe-Cr alloys). Note that chromium has a marked effect on the characteristics of pure iron. As mentioned previously, quench hardening can be accomplished only when a gamma (austenite) to alpha (ferrite) transformation takes place. Now, without considering the effects of either nickel or carbon, it is evident that an alloy with approximately 12.0% chromium, when heated above about 900 °C (1650 °F), transforms to austenite, thus establishing the basis for a hardenable alloy.

Effects of Nickel and/or Manganese. When nickel or manganese (nickel is more frequently used) are added to an iron-chromium alloy, the "gamma loop" (Fig. 1) expands to the right and downward so that, with the amount of alloying elements used in the 200 and 300 stainless steels, the entire area down to and below room temperature in Fig. 1 is austenite.

Table 4 Chemical compositions of standard AISI martensitic stainless steels

UNS No.	Type/designation	C	Mn	Si	Cr	Ni	P	S	Other
S40300	403	0.15	1.00	0.50	11.5–13.0	...	0.04	0.03	...
S41000	410	0.15	1.00	1.00	11.5–13.5	...	0.04	0.03	...
S41400	414	0.15	1.00	1.00	11.5–13.5	1.25–2.50	0.04	0.03	...
S41600	416	0.15	1.25	1.00	12.0–14.0	...	0.06	0.15 min	0.6 Mo(b)
S41623	416Se	0.15	1.25	1.00	12.0–14.0	...	0.06	0.06	0.15 min Se
S42000	420	0.15 min	1.00	1.00	12.0–14.0	...	0.04	0.03	...
S42020	420F	0.15 min	1.25	1.00	12.0–14.0	...	0.06	0.15 min	0.6 Mo(b)
S42200	422	0.20–0.25	1.00	0.75	11.5–13.5	0.5–1.0	0.04	0.03	0.75–1.25 Mo; 0.75–1.25 W; 0.15–0.3 V
S43100	431	0.20	1.00	1.00	15.0–17.0	1.25–2.50	0.04	0.03	...
S44002	440A	0.60–0.75	1.00	1.00	16.0–18.0	...	0.04	0.03	0.75 Mo
S44003	440B	0.75–0.95	1.00	1.00	16.0–18.0	...	0.04	0.03	0.75 Mo
S44004	440C	0.95–1.20	1.00	1.00	16.0–18.0	...	0.04	0.03	0.75 Mo

(a) Single values are maximum values unless otherwise indicated. (b) Optional. Source: Ref 1

Chapter 10: Heat Treating of Stainless Steels / 179

Table 5 Chemical compositions of precipitation-hardening stainless steels

UNS No.	AISI	Name(a)	C	Cr	Ni	Mn	Si	Ti	Al	Mo	N	Other
Martensitic grades											
S17600	635	Stainless W	0.08	15.0–17.5	6.0–7.5	1.0	1.0	0.4–1.2	0.40	...	...	...
S17400	630	17-4PH	0.07	15.5–17.5	3.0–5.0	1.0	1.0	...	...	...	...	0.15–0.45 Nb; 3.0–5.0 Cu
S15500	...	15-5 PH (XM-12)	0.07	14.0–15.5	3.5–5.5	1.0	1.0	...	...	...	...	0.15–0.45 Nb; 2.5–4.5 Cu
S16600	...	Croloy 16-6 PH	0.045	15.0–16.0	7.0–8.0	0.70–0.90	0.5	0.3–0.5	0.25–0.40	...	...	...
S45000	...	Custom 450 (XM-25)	0.05	14.0–16.0	5.0–7.0	1.0	1.0	...	...	0.5–1.0	...	1.25–1.75 Cu; Nb = 8 × C min
S45500	...	Custom 455 (XM-16)	0.05	11.0–12.5	7.5–9.5	0.50	0.50	...	...	0.50	...	0.10–0.50 Nb; 1.5–2.5 Cu
S13800	...	PH 13-8 Mo (XM-13)	0.05	12.25–13.25	7.5–8.5	0.20	0.10	...	0.90–1.35Al	2.0–2.5	...	...
S36200	...	Almar 362 (XM-9)	0.05	14.0–14.5	6.25–7.0	0.50	0.30	0.55–0.9	...	...	0.01	...
Semiaustenitic grades											
S17700	631	17-7 PH	0.09	16.0–18.0	6.5–7.75	1.0	1.0	...	0.75–1.5	...	...	...
S15700	632	PH 15-7 Mo	0.09	14.0–16.0	6.5–7.25	1.0	1.0	...	0.75–1.5	2.0–3.0	...	...
S35000	633	AM-350	0.07–0.11	16.0–17.0	4.0–5.0	0.50–1.25	0.50	...	...	2.5–3.25	0.07–0.13	...
S35000	634	AM-355	0.10–0.15	15.0–16.0	4.0–5.0	0.50–1.25	0.50	...	...	2.5–3.25	0.07–0.13	...
S14800	...	PH 14-8 Mo (XM-24)	0.05	13.75–15.0	7.5–8.75	1.0	1.0	...	0.75–1.5	2.0–3.0	...	...
Austenitic grades											
S66286	660	A-286	0.08	13.5–16.0	24.0–27.0	2.0	1.0	1.9–2.35	0.35	1.0–1.5	...	0.10–0.50 V; 0.001–0.01 B

(a) Designations in parentheses are ASTM designations. (b) Single values are maximum values unless otherwise indicated. Source: Ref 1

Under these conditions, no phase change takes place during heating and cooling; thus, none of these alloys can be quench hardened. They can, however, be annealed as required for fabrication because some grades (notably type 301) are highly susceptible to work hardening. It is now evident why the 200 and 300 series stainless steels are not really "steels" when compared with carbon and alloy steels. The procedure used for annealing austenitic stainless steels is similar to that used for hardening carbon and alloy grades; that is, austenitic stainless steels are annealed by heating to an elevated temperature, followed by rapid cooling. This results in a simple microstructure of equiaxed (uniform structure) grains, as shown in Fig. 2.

Heat Treating of Austenitic Grades

Recommended annealing temperatures for the grades shown in Table 1 are given in Table 6. The precise heating temperature is not extremely critical, although with one exception (type 321) the grades shown in Table 1 should be heated to at least 1010 °C (1850 °F) to ensure complete solid

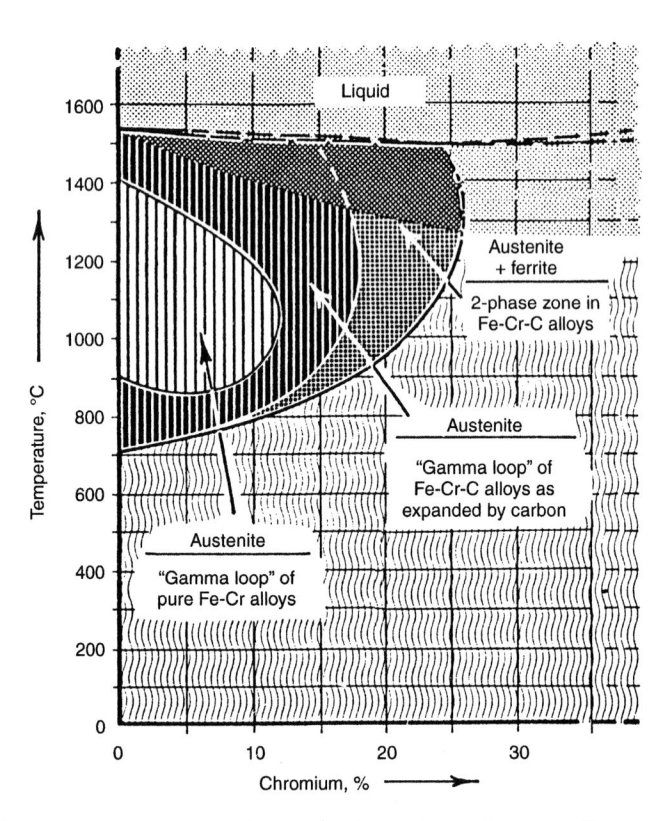

Fig. 1 Effect of carbon on the "gamma loop" of iron-chromium alloys. Source: Ref 3

solution of the carbon that could be tied up with chromium as chromium carbides.

Cooling practice is the critical operation in the annealing of these steels. It is essential that all of these steels be cooled from the annealing tem-

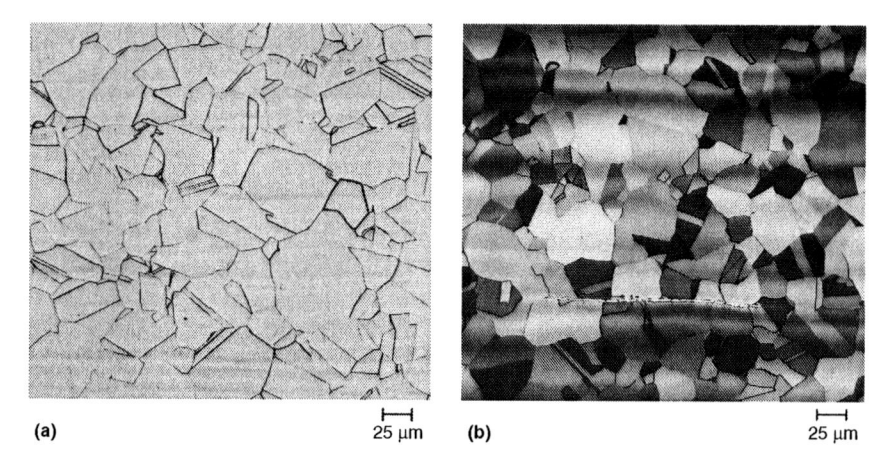

Fig. 2 Typical equiaxed grain structure in a type 316L austenitic stainless steel that was solution annealed at 955 °C (1750 °F) and etched with (a) waterless Kalling's and (b) Beraha's tint etch. Source: Ref 4

Table 6 Full annealing temperatures for austenitic stainless steels

Type		Annealing temperature(a)	
AISI	UNS No.	°C	°F
201	S20100	1010–1120	1850–2050
202	S20200	1010–1120	1850–2050
205	S20500	1065	1950
301	S30100	1010–1120	1850–2050
302	S30200	1010–1120	1850–2050
302B	S30215	1010–1120	1850–2050
303	S30300	1010–1120	1850–2050
303Se	S30323	1010–1120	1850–2050
304	S30400	1010–1120	1850–2050
304L	S30403	1010–1120	1850–2050
...	S30430	1010–1120	1850–2050
304N	S30451	1010–1120	1850–2050
305	S30500	1010–1120	1850–2050
308	S30800	1010–1120	1850–2050
309	S30900	1040–1120	1900–2050
309S	S30908	1040–1120	1900–2050
310	S31000	1040–1150	1900–2100
310S	S31008	1040–1150	1900–2100
314	S31400	1150	2100
316	S31600	1010–1120	1850–2050
316L	S31603	1010–1120	1850–2050
316F	S31620	1095	2000
316N	S31651	1010–1120	1850–2050
317	S31700	1010–1120	1850–2050
317L	S31703	1040–1095	1900–2000
321	S32100	955–1120	1750–2050
330	N08330	1065–1175	1950–2150
347	S34700	1010–1120	1850–2050
348	S34800	1010–1120	1850–2050
384	S34800	1040–1150	1900–2100

(a) In all cases, cooling must be rapid, usually by water quenching, depending on section thickness. Source: Ref 5

perature to below 450 °C (850 °F) in a very short time (1 or 2 min). Allowing these steels to "linger" within the temperature range of about 450 to 800 °C (850 to 1475 °F) results in precipitation of chromium carbides at the grain boundaries (known as sensitization), which embrittles the alloy and can allow selective corrosion to occur in the grain boundaries where the chromium has been depleted and lead to the deterioration of the steel. Exceptions are types 321 and 347, which contain small additions of titanium, or niobium and tantalum, respectively. These are referred to as stabilized grades and are not susceptible to grain-boundary carbide precipitation. They are used in applications where annealing is not practical, or for applications involving service temperatures within the critical temperature range stated previously. Also, grades with extra-low-carbon are less susceptible to sensitization, simply because there is very little carbon to form chromium carbides.

Cooling from the annealing temperature may be by water or oil quenching. For very thin sections, air cooling is satisfactory—the objective is to cool rapidly through the critical range. Heavier sections require water quenching.

Stress relieving (if desired) following cold working operations can be accomplished for the austenitic grades by heating up to 400 °C (750 °F); the cooling rate from this temperature is not critical.

Corrosion Resistance of Austenitic Grades

As a group, the austenitic grades (AISI 200 and 300 series) have greater corrosion resistance than the ferritic, martensitic, and precipitation-hardening stainless steels, in terms of the number of different corrosive environments in which they remain passive (corrosion resistant). However, there are large differences in corrosion-resistant properties among the grades listed in Table 1. For example, grades alloyed with molybdenum (types 316 and 317) exhibit greater resistance to pitting corrosion than do lesser-alloyed grades (for example, types 301 and 304). Higher chromium and nickel contents (types 309 and 310) result in improved corrosion resistance in high-temperature environments. Although chromium is the principal contributing factor to corrosion resistance, nickel is a "powerful helpmate."

Ferritic Grades

The ferritic stainless steels are identified as the AISI 400 series. In corrosion resistance, these steels generally rank higher than the martensitic grades, but substantially lower than most of the austenitic grades.

The ferritic grades are so named because their structure is ferritic at all temperatures, thus, like the austenitic grades, they cannot be hardened by heating and quenching. Annealing to lower hardnesses than that developed after cold working is the only heat treatment applied to the ferritic grades.

Heat Treating of Ferritic Grades

Figure 1 also can be used to explain why the ferritic grades cannot be hardened by austenitizing and quenching. It is similar to the explanation as to why austenitic grades cannot be quench hardened, except the condition is reversed. Austenitic grades cannot be hardened because the alloy does not transform from gamma to alpha, while the ferritic grades cannot be quench hardened because there is no alpha-to-gamma transformation.

As shown in Fig. 1, if a line is drawn vertically from the horizontal at 10% chromium, it intersects the original "gamma loop" (disregard the shaded extensions of the loop). This indicates that a 10% chromium alloy can be austenitized. However, if a line is drawn vertically from the horizontal at the 15% chromium level, the vertical line does not intersect any portion of the "gamma loop," and therefore, that material cannot be austenitized (all of the area below and to the right of the loop is ferrite, or alpha phase).

Thus far, discussion has been limited to the iron-chromium system and carbon has not been considered. As shown in Fig. 1, the expanded loop areas show the effects of increased carbon, which has a profound effect in enlarging the loop in the pattern as shown. For high-carbon contents, the gamma phase (austenite), under conditions of high temperature, can still be formed with a chromium content of around 25%. Actually, the practical upper limit is usually about 20%.

Therefore, whether an iron-chromium-carbon alloy can be quench hardened depends greatly on the balance of carbon and chromium. This is a complex subject; consequently, a detailed discussion is not included. For more information, see Ref 6 and 7.

Type 430 (Table 2) contains a maximum of 0.12% carbon and chromium in the range of 16.0 to 18.0%—a typical ferritic steel that is widely used. Normally, this alloy cannot be hardened by quenching. However, if the carbon content is increased to 0.40%, or the chromium content is decreased, the material readily becomes a quench-hardening alloy.

On a practical basis, however, when the carbon is on the high side of the allowable range and the chromium is on the low side, a borderline condition can be created whereby some austenite is formed on heating. This results in some hardening from the phase change mechanism.

Annealing of the ferritic steels can be achieved readily by the heating and cooling practice presented in Table 7. The annealing temperatures in this case are substantially lower than those used for annealing the austenitic grades. Further, the cooling rate is less critical compared with cooling rates required for annealing the austenitic grades. To avoid embrittlement, however, cooling should be as rapid as possible, as indicated by the recommended practice shown in Table 7.

Duplex Grades

Wrought duplex stainless steels are two-phase alloys based on the iron-chromium-nickel system. These materials typically comprise approximately equal amounts of ferrite and austenite in their microstructure and are characterized by their low-carbon content (<0.03 wt%) and additions of molybdenum, nitrogen, copper, and/or tungsten. Typical chromium and nickel contents are about 20 to 30% and 4 to 8%, respectively (Table 3). The duplex grades are highly resistant to chloride stress-corrosion cracking.

Heat Treating of Duplex Grades

Duplex stainless steels are not hardenable by heat treatment. Annealing is accomplished by rapid cooling from the recommended annealing temperature, which ranges from about 1020 to 1120 °C (1870 to 2050 °F), depending on the alloy in question. Table 8 lists recommended annealing temperatures for selected duplex grades.

Table 7 Annealing treatments of ferritic stainless steels

Type	Temperature(a) °C	°F	Cooling method(b)
405	735–815	1345–1500	AC or WQ
409	885–900	1625–1650	AC
429	780–845	1450–1555	AC or WQ
430	760–815	1400–1500	AC or WQ
430F	675–760	1250–1400	AC or WQ
430FSe	675–760	1250–1400	AC or WQ
434	790–845	1455–1555	AC or WQ
436	790–845	1455–1555	AC or WQ
442	735–815	1355–1505	AC or WQ
446	790–870	1455–1600	AC or WQ

(a) Time at temperature depends on section thickness, but is usually 1 to 2 h, except for sheet, which may be soaked 3 to 5 min per 2.5 mm (0.10 in.) of thickness. (b) AC, air cool; WQ, water quench. Source: Ref 8

Table 8 Recommended annealing temperatures for selected duplex stainless steels

UNS No.	Designation	Annealing temperature(a) °C (°F)
S32900	329	925–955 (1700–1750)
S32950	7 Mo Plus	995–1025 (1825–1875)
S31500	3RE60	975–1025 (1785–1875)
S31803	SAF 2205	1020–1100 (1870–2010)
S31260	DP-3	1065–1175 (1950–2150)
S32550	Ferralium 255	1065–1175 (1950–2150)

(a) Cooling from the annealing temperature must be rapid, but it also must be consistent with limitations of distortion. Source: Ref 2

Martensitic Grades

These alloys are capable of changing their crystal structure on heating and cooling, thus responding to heat treatment much the same as carbon and alloy steels. Compositions of martenstic stainless steels are listed in Table 4. Steels of this group also carry the AISI 400-series designation. Nickel is specified in only three grades shown, and the maximum amount is 2.5%. Further, chromium content is generally lower than for the austenitic grades. Consequently, corrosion resistance of the martensitic grades is far lower compared with that of the austenitic or duplex grades and, in most instances, somewhat lower than that of the ferritic grades.

The relative corrosion resistance of the different martensitic grades is not necessarily equal, and corrosion resistance also does not necessarily increase with increasing chromium content. As indicated earlier, carbon has a high affinity for chromium, and chromium carbide contributes little if any to the total corrosion resistance of the alloy. In fact, when massive chromium carbides form (as they may in type 440C), these carbides may be corrosion nuclei. Therefore, type 410 (nominal composition of 12.5% chromium) often has better corrosion resistance than does 440C with 17.0%.

Hardenability

All of the martensitic grades have extremely high hardenability, to the extent that they can be fully hardened by quenching in still air from their austenitizing temperatures, as shown in Fig. 3. Figure 3(a) is a time-temperature-transformation (TTT) curve for type 410, which shows that the "nose" of the curve is shifted far to the right, thus indicating a very low critical cooling rate compared with carbon and alloy steels (Fig. 3b).

Hardenability is shown as a straight line; the end that was drastically quenched in water is no harder than the opposite end of the specimen that was cooled in still air (see Chapter 3, "Hardness and Hardenability").

Heat Treating of Martensitic Grades

The maximum hardness that can be achieved by austenitizing martensitic stainless steels is governed by the carbon content. The martensitic microstructures of stainless steels in the as-quenched or quenched and tempered condition are similar in appearance to those of alloy steels.

Annealing practice for the grades of martensitic alloys is given in Table 9. Full and subcritical annealing processes are shown. The subcritical approach is used whenever possible because the cooling rate from the

186 / Practical Heat Treating: Second Edition

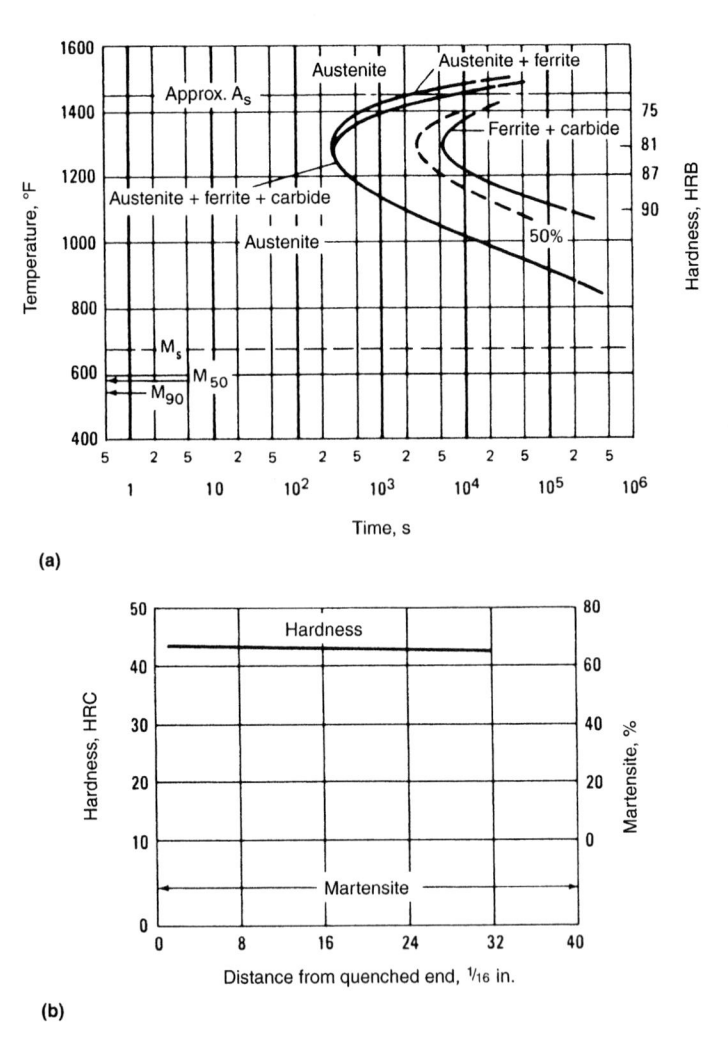

Fig. 3 Type 410 stainless steel. Composition: 0.11 C, 0.44 Mn, 0.37 Si, 0.16 Ni, 12.18 Cr. Austenitized at 980 °C (1800 °F). Grain size 6 to 7. (a) Time-temperature-transformation (TTT) curve. (b) End-quench hardenability. Source: Ref 3

Table 9 Temperatures for heat treating several martensitic stainless steels

Type AISI	UNS No.	Annealing temperature °C	Annealing temperature °F	Subcritical annealing temperature °C	Subcritical annealing temperature °F	Hardening temperature °C	Hardening temperature °F	Tempering temperature °C	Tempering temperature °F
410	S41000	815–900	1500–1650	650–760	1200–1400	925–1010	1700–1850	205–760	400–1400(a)
414	S41400	...	...	650–705	1200–1300	930–1040	1800–1900	205–705	400–1300(b)
416	S41600	815–900	1500–1650	650–760	1200–1400	925–1010	1700–1850	205–760	400–1400(a)
420	S42000	845–900	1550–1650	730–790	1350–1450	930–1040	1800–1900	150–370	300–700
440A	S44002	845–900	1550–1650	730–790	1350–1450	...	1850–1950	150–370	300–700
440B	S44003	845–900	1550–1650	730–790	1350–1450	...	1850–1950	150–370	300–700
440C	S44004	845–900	1550–1650	730–790	1350–1450	...	1850–1950	150–370	300–700

(a) Tempering in the range of 370 to 565 °C (700 to 1050 °F) results in decreased impact strength and corrosion resistance. (b) Tempering in the range of 400 to 600 °C (750 to 1100 °F) results in decreased impact strength and corrosion resistance. Source: Ref 3

subcritical range is not critical. Cooling from the full annealing temperature requires a cooling rate no faster than about 17 °C (30 °F) per hour.

Austenitizing and Quenching. Austenitizing (hardening) temperatures for the martensitic grades are given in Table 9. Temperatures in the middle of those shown usually are recommended. Air cooling frequently is used, although oil quenching may be used. For the higher-carbon grades, common practice is to oil quench to "black" then air cool; this procedure minimizes scale formation.

Especially with higher-carbon grades, quenched parts should be tempered immediately, and preferably just before they reach room temperature. Ideally, when parts are "warm to the touch," the tempering operation should be started. Failure to follow this practice often results in cracking.

Of this group of steels, 440C is capable of the highest hardness—60 to 62 HRC.

Tempering. All quenched parts of martensitic stainless must be tempered. These steels are highly alloyed and do not behave like carbon and alloy steels in tempering. For most alloy steels, as tempering temperature increases, the hardness decreases. This is not the case with the martensitic stainless steels.

For these high-alloy steels, transformation of the austenite is not complete upon quenching, which allows a great deal of austenite to remain as retained austenite—a metastable constituent. Some of the retained austenite transforms to martensite during tempering, which actually may result in high hardness after tempering at some elevated temperature such as 480 °C (900 °F). After this hardness peak, the decrease of hardness with an increase of tempering temperature is generally rapid, as shown in Fig. 4 for type 440C. This hardness "hump," which occurs at about 480 °C (900 °F), is the result of a phenomenon known as secondary hardening.

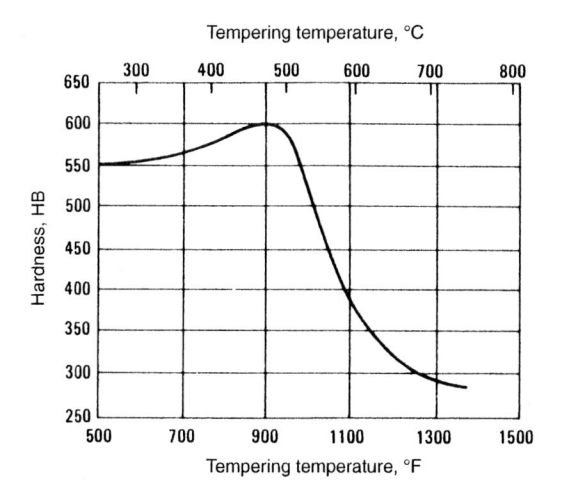

Fig. 4 Hardness vs. tempering temperature for type 440C stainless steel after oil quenching from 1040 °C (1900 °F). Source: Ref 9

Despite the fact that higher hardnesses may be attained by hitting the peak of the secondary hardening, a marked decrease in corrosion resistance results when the tempering temperature exceeds about 370 °C (700 °F). Therefore, 370 °C (700 °F) is usually the maximum tempering temperature. The minimum is 150 °C (300 °F), as indicated in Table 9.

For all martensitic stainless steels, but more especially for the higher-carbon grades, double tempering should be used. Steels should be heated to the preestablished tempering temperature, cooled to room temperature, and then tempered again at the same temperature.

Precipitation-Hardening Grades

Compositions of precipitation-hardening (PH) steels are listed in Table 5; these steels are further subdivided as to the type of structure they develop. While the AISI numbering system (600 series) exists for some of these steels, they are more frequently described by their UNS designation or proprietary alloy name.

In corrosion resistance, these steels may vary considerably among the different grades within the group, but generally, their corrosion resistance approaches that of some of the austenitic grades.

Most of the PH grades can be hardened to at least 42 HRC and higher, but not by conventional quench-hardening techniques used for martensitic grades. Hardening techniques for the PH grades are similar to those used for nonferrous metals. General practice is to solution treat by heating to an elevated temperature, followed by rapid cooling, then age hardening by heating to an intermediate temperature. There is, however, considerable difference in techniques used for various grades.

The metallurgy of precipitation hardening stainless steels is complex and cannot be completely covered herein. For more detailed information, see Ref 2, 6, and 7. However, successful precipitation hardening depends on the presence of one or more elements that are readily soluble in the alloy at elevated temperatures but are less soluble or insoluble at ambient temperatures.

For example, in AISI type 630 (17-4PH), copper is the principal element responsible for precipitation hardening. Copper, in this case, is assisted by small amounts of niobium and tantalum (Table 5).

Recommended Heat Treating Practice for PH Grades

Continuing with 17-4PH as an example, largely because it is one of the most widely used of the PH grades, the recommended heat treating procedure is as follows. For this grade and for most other PH grades, the final properties differ somewhat with variations in aging temperature. However, the following procedure is most commonly used:

1. Solution treat by heating at 1025 to 1050 °C (1875 to 1925 °F).
2. Oil quench to room temperature.
3. Age at 480 °C (900 °F) for 1 h and air cool (known as condition H900).

Following this heat treatment, a hardness of approximately 42 to 44 HRC can be expected.

Regardless of the specific alloy, the material is generally as "soft" as it can be made following solution treating. Because the alloy is in a metastable (unstable) condition, some hardening (aging) occurs even at room temperature over a long period of time. However, full hardness can be accomplished in an hour or so by heating to 480 °C (900 °F), as prescribed previously.

For the recommended heat treatment for any of the other PH stainless steels not detailed in this chapter, see Ref 2.

Suitable Heat Treatment Conditions

Because the surface of stainless steel must remain unaltered in composition in order to retain the proper corrosion protection, it is very important that the proper temperature control and proper atmosphere be used during the processing of stainless steels. This becomes somewhat less important if all of the exposed surfaces have adequate stock removal by machining after heat treatment.

Ideally, stainless steels are best suited for heat treatment in vacuum furnaces, although many stainless products are successfully heat treated in other types of equipment. Generally, direct-fired furnaces are not suitable for high-temperature heat treatments of stainless steels because of poor temperature uniformity and/or scaling or oxidizing conditions.

Radiant tube-heated or electrically heated furnaces using either metallic or silicon carbide resistors are satisfactory for most stainless steel heat processing operations. In some instances, liquid salt baths are suitable for some treatments.

Exothermic atmospheres with a gas ratio of 6.5 to 1 or 7 to 1 are usually satisfactory for austenitizing or annealing stainless steels. Disassociated ammonia can be used for "bright annealing" of stainless steels. Endothermic atmospheres can be used provided the atmosphere carbon level is matched to the surface carbon of the stainless material being treated. Carburization (adding carbon to the surface) of the stainless can cause the material to be ruined for the use intended.

Tempering or aging operations generally are satisfactorily performed in recirculating air furnaces. However, some applications may require vacuum tempering or aging.

REFERENCES

1. J.R. Davis, Ed., *Alloy Digest Source Book: Stainless Steels,* ASM International, 2000
2. J.R. Davis, Ed., *ASM Specialty Handbook: Stainless Steels,* ASM International, 1994
3. H.E. Boyer, Chapter 7, *Practical Heat Treating,* 1st ed., American Society for Metals, 1984, p 136–148
4. G.F. Vander Voort and E.P. Manilova, Metallography and Microstructures of Stainless Steels and Maraging Steels, *Metallography and Microstructures,* Vol 9, *ASM Handbook,* ASM International, 2004, p 670–700
5. H. Chandler, Ed., Austenitic Stainless Steels, *Heat Treater's Guide: Practices and Procedures for Irons and Steels,* 2nd ed., ASM International, 1995, p 724–725
6. J. Beddoes and J.G. Parr, *Introduction to Stainless Steels,* 3rd ed., ASM International, 1999
7. D. Peckner and I.M. Berstein, Ed., *Handbook of Stainless Steels,* McGraw-Hill, 1977
8. H. Chandler, Ed., Ferritic Stainless Steels, *Heat Treater's Guide: Practices and Procedures for Irons and Steels,* 2nd ed., ASM International, 1995, p 759
9. H. Chandler, Ed., 440C, *Heat Treater's Guide: Practices and Procedures for Irons and Steels,* 2nd ed., ASM International, 1995, p 785

CHAPTER **11**

Heat Treating of Tool Steels

IT WOULD SEEM LOGICAL that the term "tool steels" would refer to steels used for making tools, but this is not entirely true, depending largely on individual interpretation of what tools are and what they are not. To the layman, tools are items such as hammers, chisels, and saws commonly found in a hardware store. While these are tools, it is rare that any of these hardware store items are made from materials that conform to a manufacturer's understanding of tool steels. Tool steels represent a small, but very important, segment of the total production of steel. Their principal use is for tools and dies that are used in the manufacture of commodities.

Besides being used in dies, these steels, particularly hot-work die steels, have also been used in structural and aerospace applications, but not as widely as they once were, primarily because of the development of several other steels at essentially the same cost but with substantially greater fracture toughness at equivalent strength.

Classification of Tool Steels

Over 100 different compositions of tool steels are produced. They vary in composition from simple plain carbon steels that contain iron and as much as 1.2% C with no significant amounts of alloying elements, up to and including very highly alloyed grades in which the total alloy content approaches 50%. In between these extremes, practically every combination of the principal alloying elements including manganese, silicon, chromium, nickel, molybdenum, tungsten, vanadium, and cobalt has been employed. The great diversity among tool steels has posed a major problem in classifying tool steels.

Historically, tool steels were known and marketed by trade name. Unfortunately, this practice still persists in many areas. In reviewing the compositions of tool steels, it is evident immediately that many are identical in composition to carbon and alloy steels that are produced in large tonnages. Then, why pay the higher cost for a tool steel? The only answer is to ask another question: what quality level is required? Many tooling applications can be fulfilled satisfactorily by using lower-priced carbon or alloy grades; conversely, as many nontooling applications require the high level of quality provided by tool steels. It should be clearly understood that tool steels are made and processed in small quantities to extremely high levels of quality control.

Tool steels do not lend themselves to the type of classification used by the Society of Automotive Engineers (SAE) and the American Iron and Steel Institute (AISI) for carbon and alloy steels, in which an entire series of steels is defined numerically and based on a variation of carbon content alone. While some carbon tool steels and low-alloy tool steels are made in a wide range of carbon contents and permit such a classification, most of the higher-alloyed types of tool steels have a comparatively narrow carbon range, and such a classification would be meaningless. Tool steels cannot be classified on the basis of the predominant alloying element, for in many of the more complex compositions, one alloying element can be partially or wholly substituted for another with little change in the mechanical properties of the steel. Classification by application appears possible for certain types of tool steels, but quite impractical for others. For example, tool steels used for hot extrusion tools can be closely classified. It would be impossible, however, to combine in one class all of the tool steels that are used to make a tap. Steels suitable for taps range from carbon tool steels through the ultrahigh-speed steels.

There are certain groups of tool steels that, while varying considerably in composition, have mechanical properties and other characteristics that are so similar they naturally fall within a common group.

One logical approach to classification is to use a mixed classification in which some steels are grouped by use, others by composition or by certain mechanical properties, and still others by the method of heat treatment (precisely by the quenching technique). Actually, through the mediums of shop language and the published literature, tool steels have become somewhat automatically classified on such a basis. For instance, the terms high-speed steels; water-hardening steels; hot-work steels; and high-carbon, high-chromium steels represent just such a mixed classification.

High-speed steels are grouped together because they have certain common properties; the water-hardening steels because they are hardened in a common manner; the hot work steels because they have certain common properties; and the high-carbon, high-chromium steels are grouped together because of their similar compositions and applications.

There are over 70 different compositions designated in the AISI system. Table 1 lists 26 of the most widely used of these compositions, which are representative of the total group. The Unified Numbering System (UNS) identifications are also given in Table 1. All elements are given in nominal amounts, which may vary somewhat for different tool steel producers. When heat treating shops receive tools for heat treatment that are identified only by proprietary name, every effort should be made to obtain the AISI identification before performing any heat treating operation.

The grouping of tool steels published by AISI has proved workable; the ten main groups and their corresponding symbols are:

Name	Identifying symbol
Water-hardening tool steels	W
Shock-resisting tool steels	S
Oil-hardening cold-work tool steels	O
Air-hardening, medium-alloy cold-work tool steels	A
High-carbon, high-chromium cold-work tool steels	D
Low-alloy, special-purpose tool steels	L
Low-carbon mold steels	P
Hot-work tool steels, chromium, tungsten, and molybdenum	H
Tungsten high-speed tool steels	T
Molybdenum high-speed tool steels	M

For three groups (W, O, and A), the quenching medium serves as the identifying symbol. In other instances, the use is indicated by the letter symbol. In most instances, each group bears a common letter symbol as indicated in the table. High-speed steels are the exception because of the first group, which employs tungsten as the principal alloying element, and the succeeding groups wherein molybdenum is the principal alloying element.

Heat Treating Processes for Tool Steels—General

For the most part, the processes used for heat treating carbon and alloy steels are also used for heat treating tool steels—annealing, austenitizing, tempering, and so forth. However, normalizing is used only to a limited extent almost exclusively for the W grades. Normalizing, for most steels, is used to refine the coarse grain size that results from hot working. However, most of the highly alloyed tool steels are annealed. Normalizing of the highly alloyed grades after forging would likely cause cracking and would accomplish nothing. The P grades shown in Table 1 (generally very low in carbon) are most often hardened by case hardening—carburizing or nitriding (see Chapter 8, "Case Hardening of Steel").

Because the surface of most tool steels must remain unaltered in composition in order to retain the inherent advantages of the material, it is very important that there is accurate temperature control and that proper atmosphere conditions are used during processing. Ideally, most tool steels are best suited for heat treatment in vacuum furnaces, although many tool steels are successfully heat treated in other types of equipment, including salt baths and atmosphere furnaces (refer to Chapters 4, "Furnaces and Related Equipment for Heat Treating," and 5, "Instrumentation and Control of Heat Treating Processes").

Table 1 Classification and approximate compositions of some principal types of tool steels

AISI	UNS No.	C	Mn	Si	Cr	V	W	Mo	Co	Ni
Water-hardening tool steels										
W1	T72301	0.60–1.40(a)	...	...	...	...	...	...	...	...
Shock-resisting tool steels										
S1	T41901	0.50	...	...	1.50	...	2.50	...	...	...
S5	T41905	0.55	0.80	2.00	...	...	...	0.40	...	...
Oil-hardening cold-work tool steels										
O1	T31501	0.90	1.00	...	0.50	...	0.50	...	...	...
O2	T31502	0.90	1.60	...	...	...	...	...	...	...
Air-hardening medium-alloy cold-work steels										
A2	T30102	1.00	...	...	5.00	...	...	1.00	...	...
A3	T30103	1.25	...	...	5.00	1.00	...	1.00	...	...
High-carbon, high-chromium cold-work steels										
D2	T30402	1.50	...	...	12.00	1.00	...	1.00	...	...
D5	T30405	1.50	...	...	12.00	...	...	1.00	3.00	...
Low-alloy, special-purpose tool steels										
L2	T61202	0.50–1.10(a)	1.00	0.20	...	...	...	...	...	...
L6	T61206	0.70	0.75	...	...	...	0.25(b)	...	...	1.50
Mold steels										
P2	T51602	0.07	...	...	2.00	...	...	0.20	...	0.50
P4	T51604	0.07	...	...	5.00	...	...	0.75	...	...
Chromium hot-work tool steels										
H11	T20811	0.35	...	...	5.00	0.40	...	1.50	...	...
H12	T20812	0.35	...	...	5.00	0.40	1.50	1.50	...	...
H13	T20813	0.35	...	...	5.00	1.00	...	1.50	...	...
Tungsten hot-work tool steels										
H21	T20821	0.35	...	...	3.50	...	9.00	...	...	...
H26	T20826	0.50	...	...	4.00	1.00	18.00	...	...	...
High-speed tool steels										
T1	T12001	0.75(a)	...	...	4.00	1.00	18.00	...	...	...
T8	T12008	0.75	...	...	4.00	2.00	14.00	...	5.00	...
T15	T12015	1.50	...	...	4.00	5.00	12.00	...	5.00	...
M1	T11301	0.80(a)	...	...	4.00	1.00	1.50	8.00	...	...
M2	T11302	0.85–1.00(a)	...	...	4.00	2.00	6.00	5.00	...	...
M10	T11310	0.85–1.00(a)	...	...	4.00	2.00	...	8.00	...	...
M36	T11336	0.80	...	...	4.00	2.00	6.00	5.00	8.00	...
M44	T11344	1.15	...	...	4.25	2.00	5.25	6.25	12.00	...

(a) Available with different carbon contents as specified. (b) Optional. Source: Ref 1

Apparent Hardness versus Actual Hardness

At this time, before discussing the heat treating of tool steels, the reader should clearly understand the difference between "apparent" and "actual" hardness.

The general principles of hardness and hardness measurement are discussed in Chapter 3 ("Hardness and Hardenability"), where it is indicated that microhardness testing is often required to reveal true conditions.

For most carbon and alloy steels, the apparent, or measured, hardness as indicated by test instruments is generally accurate and represents nearly true conditions because the structures of carbon and alloy steels are generally homogeneous.

For highly alloyed steels, such as high-carbon stainless and many tool steels, the structures are not necessarily homogeneous. For example, the structure of hardened W1 tool steel (Fig. 1) shows a high degree of homogeneity and measures 64 HRC, thus registering nearly true conditions.

Now let us examine the hardness of a highly alloyed tool steel such as D2. Figure 2 shows the structure of D2 in the quenched and tempered condition. The dark matrix is tempered martensite, not unlike the matrix of W1 shown in Fig. 1. Likewise, the registered hardness of the two is essentially the same—62 HRC or about 760 on the Knoop scale. However, an entirely different condition exists in Fig. 2—a substantial addition of undissolved complex alloy carbides (white constituent). Measurements of the hardness of these individual carbides as determined by microhardness testing show readings of 2000 HK or higher. In testing with instruments such as the Rockwell tester, the carbide particles are simply "pushed" away and have little effect on the reading. Therefore, apparent or measured

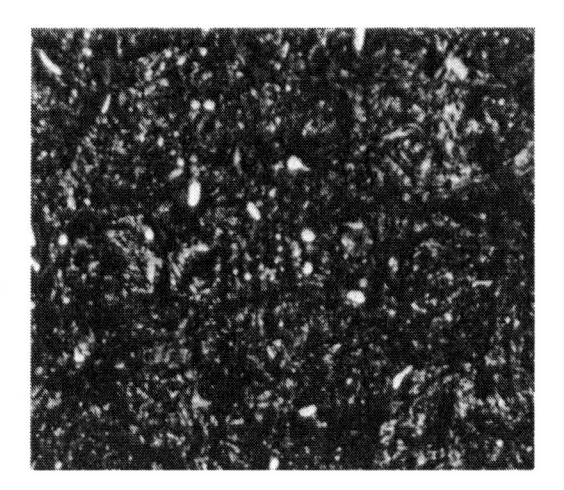

Fig. 1 Microstructure of W1 tool steel after brine quenching from 790 °C (1450 °F) and tempering at 175 °C (350 °F). Dark matrix is tempered martensite. A few undissolved particles of carbide are visible (white constituent). Hardness is 64 HRC. Source: Ref 1

hardness is the hardness value provided by testing with conventional hardness testers, whereby the actual or true hardness takes into account the hardness of the individual constituents.

The practical difference between the two microstructures in Fig. 1 and 2 is their varying resistance to abrasion or wear. For example, sheet metal forming dies made of D2 have been known to outlast W1 counterparts by a factor of 5 to 1, simply because of the difference in resistance to wear of the two structures.

Heat Treating Practice for Specific Groups

Water-Hardening Tool Steels (W Grades). The AISI lists a total of three W steel compositions, but W1 (Table 1) is by far the most widely used. Recommended practice for annealing and hardening W1 is given in Table 2. Except for extremely thin sections (generally under 4.8 mm, or $3/16$ in.), the W steels must be quenched in water or brine to attain full hardness. All W tool steels can be hardened to about 65 HRC at the surface, but their hardenability is very low, so that this high-hardened zone is generally very shallow. However, a hard exterior and a softer core are desirable for many applications, such as punches.

Even though all of the W steels have relatively low hardenability, these grades usually are available as shallow, medium, or deep hardening; this property is controlled by the manufacturer.

Shock-Resisting Tool Steels (S Grades). Five steels are listed by AISI under the symbol S, although only the two most widely used grades are shown in Table 1 (S1 and S5).

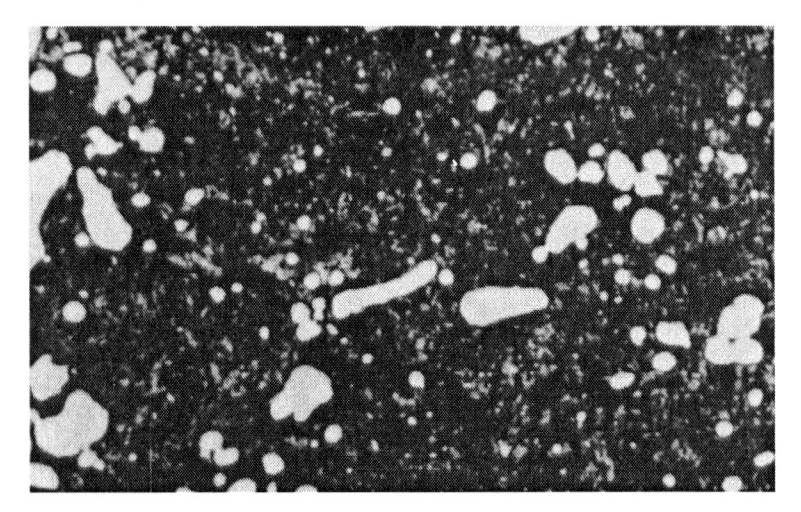

Fig. 2 Microstructure of D2 tool steel after air cooling from 980 °C (1800 °F) and tempering at 540 °C (1000 °F). Hardness is approximately 62 HRC. Dark matrix is tempered martensite with a dispersion of very hard carbide particles (white). See text for discussion. Source: Ref 1

There is a considerable difference between the compositions of S1 and S5, as shown in Table 1. However, they are both intended for similar applications—applications that require extreme toughness such as punches, shear knives, and air-operated chisels.

These steels have sufficient hardenability so that full hardness can be attained by oil quenching. They must be tempered to at least 175 to 205 °C (350 or 400 °F) and preferably higher if some hardness can be sacrificed (see general section on tempering later in this chapter).

Details of recommended practice for annealing and hardening SI and S5 are given in Table 2.

Oil-Hardening Cold-Work Tool Steels (O Grades). Four O grades are listed by AISI, but only the two most widely used grades (O1 and O2) are listed in Table 1. Of these, O1 is used more frequently. A portion of the carbon in O6 is in the form of graphite, which allows better machinability, a factor in making intricate dies. In addition, the graphite particles in the microstructure provide a built-in lubricant.

Heat treating practices for both grades are given in Table 3. As indicated, the high hardenability of O1 is derived from the relatively high manganese content (1.0%) and from small additions of chromium and tungsten. Grade O2 depends on manganese content for hardenability. Both of these steels have far greater hardenability than W1 and thus are used where deep hardening by oil quenching is desired.

Air-Hardening, Medium-Alloy Cold-Work Tool Steels (A Grades). The cold-work tool steels listed under the letter A cover a wide range of carbon and alloy contents, but all have high hardenability and exhibit a high degree of dimensional stability in heat treatment. A total of 8 different compositions are listed by AISI, although A2 and A3 only are listed in Table 1. These are, by far, the two most widely used for the A group. As shown in Table 1, A3 has a higher carbon content than does A2 and also contains 1.0% V, thus some vanadium carbides are formed, which pro-

Table 2 Heat treating practice for the W- and S-grade tool steels

Treatment	Type W1 (T72301)	Type S1 (T41901)	Type S5 (T41905)
Annealing			
Temperature, °C (°F)	740–790 (1360–1450)	790–815 (1450–1500)	775–800 (1425–1475)
Rate of cooling, °C (°F) max per hour	22 (40)	22 (40)	14 (25)
Typical annealed hardness, HB	156–201	183–229	192–229
Hardening			
Rate of heating	Slowly	Slowly	Slowly
Preheat temperature, °C (°F)	(a)	650 (1200)	760 (1400)
Hardening temperature, °C (°F)	760–845 (1400–1550)	900–955 (1650–1750)	870–925 (1600–1700)
Time at temperature, min	10–30	15–45	5–20
Quenching medium	Brine or water	Oil	Oil
Tempering			
Tempering temperature, °C (°F)	175–345 (350–650)	205–650 (400–1200)	175–425 (350–800)
Approximate tempered hardness, HRC	64–50	58–40	60–50

(a) For large tools and tools having intricate sections, preheating at 565–650 °C (1050–1200 °F) is recommended. Source: Ref 1

vides A3 with superior wear resistance. Each of these grades contains 5.0% Cr so that some free chromium carbide characterizes the microstructures of these grades as shown for D2 in Fig. 2. Also, both A2 and A3 are sufficiently high in alloy content to develop some secondary hardening (see the general section on tempering later in this chapter). Type A7 (not listed in Table 1), which has high carbon (2.00 to 2.85%) and vanadium (3.90 to 5.15) contents, exhibits maximum abrasion resistance but should be restricted to applications where toughness is not a prime consideration.

Specific heat treating procedures are given for A2 and A3 in Table 3. As indicated, the recommended heat treating procedures are almost the same for these two grades.

High-Carbon, High-Chromium Cold-Work Tool Steels (D Grades). The AISI lists five different compositions in the D group. These steels are all characterized by high carbon content (1.5 to 2.5%) and contain a nominal 12% Cr content. All grades of this group have extremely high resistance to abrasive wear, which increases as the carbon and vanadium increase. Compositions for the two most popular grades are given in Table 1. Note that D2, which is by far the most widely used D-type steel, also contains 1.0% V and 1.0% Mo; thus, it can become fully hardened by air cooling from the austenitizing temperature, forming a martensitic matrix with large amounts of excess carbide as shown in Fig. 2.

As a rule, D2 tool steel is frequently used for dies employed for cold-working operations, although it does have some resistance to softening from elevated temperatures.

Grade D5 is similar to D2 in composition, except that it contains 3.0% Co instead of vanadium. This compositional difference lowers its abrasion resistance, but the cobalt addition increases its resistance to softening from heat. Consequently, D5 often is used for hot forming tools. Heat treating procedures for both D2 and D5 are presented in Table 4.

Table 3 Heat treating practice for the O- and A-grade tool steels

Treatment	Type O1 (T31501)	Type O2 (T31502)	Type A2 (T30102)	Type A3 (T30103)
Annealing				
Temperature, °C (°F)	760–790 (1400–1450)	745–775 (1375–1425)	845–870 (1550–1600)	845–870 (1550–1600)
Rate of cooling, °C (°F) maximum per hour	22 (40)	22 (40)	22 (40)	22 (40)
Typical annealed hardness, HB	183–212	183–217	201–235	207–229
Hardening				
Rate of heating	Slowly	Slowly	Slowly	Slowly
Preheat temperature, °C (°F)	650 (1200)	650 (1200)	790 (1450)	790 (1450)
Hardening temperature, °C (°F)	790–815 (1450–1500)	760–800 (1400–1475)	925–980 (1700–1800)	955–1010 (1750–1850)
Time at temperature, min	10–30	5–20	20–45	25–60
Quenching medium	Oil	Oil	Air	Air
Tempering				
Tempering temperature, °C (°F)	175–260 (350–500)	175–260 (350–500)	175–540 (350–1000)	175–540 (350–1000)
Approximate tempered hardness, HRC	62–57	62–57	62–57	62–57

Source: Ref 1

Low-Alloy, Special-Purpose Tool Steels (L Grades). The tool steels listed under the symbol L cover a wide range of alloy content and mechanical properties. They are used for die components and machinery parts. Both L6 and the lower carbon versions of L2 are often used for applications requiring extreme toughness including punches and heading tools. Recommended temperatures for heat treating L2 and L6 are:

- Annealing: 760 to 790 °C (1400 to 1450 °F)
- Hardening: 790 to 845 °C (1450 to 1600 °F)
- Tempering: 175 to 540 °C (350 to 1000 °F)

Reference 2 offers more detailed information on heat treating L-group steels as well as other tool steels not discussed in this chapter.

Mold Steels (P Grades). Of the seven mold steel compositions listed by AISI, compositions of two of these grades are listed in Table 1. Heat treatments commonly used for each are given in Table 5. Both of these

Table 4 Heat treating practice for grades D2 and D5 tool steels

Treatment	Type D2 (T30402)	Type D5 (T30405)
Annealing		
Temperature, °C (°F)	870–900 (1600–1650)	870–900 (1600–1650)
Rate of cooling, °C (°F) maximum per hour	22 (40)	22 (40)
Typical annealed hardness, HB	217–255	223–255
Hardening		
Rate of heating	Very slowly	Very slowly
Preheat temperature, °C (°F)	815 (1500)	815 (1500)
Hardening temperature, °C (°F)	980–1025 (1800–1875)	980–1025 (1800–1875)
Time at temperature, min	15–45	15–45
Quenching medium	Air	Air
Tempering		
Tempering temperature, °C (°F)	205–540 (400–1000)	205–540 (400–1000)
Approximate tempered hardness, HRC	61–54	61–54

Source: Ref 1

Table 5 Heat treating practice for grades P2 and P4 mold steels

Treatment	Type P2 (T51602)	Type P4 (T51604)
Annealing		
Temperature, °C (°F)	730–815 (1350–1500)	870–900 (1600–1650)
Rate of cooling, °C (°F) maximum per hour	22 (40)	14 (25)
Typical annealed hardness, HB	103–123	116–128
Hardening		
Carburizing temperature, °C (°F)	900–925 (1650–1700)	970–995 (1775–1825)
Hardening temperature, °C (°F)	830–845 (1525–1550)	970–995 (1775–1825)
Time at temperature, min	15	15
Quenching medium	Oil	Air
Tempering		
Tempering temperature, °C (°F)	175–260 (350–500)	175–480 (350–900)
Approximate tempered hardness, HRC	64–58	64–58

Source: Ref 1

steels are very low in carbon content, which is intentional to permit forming of plastic molds by hobbing. As shown in Table 5, case hardening (carburizing, see Chapter 8) is the most commonly used heat treatment. Grade P2 is widely used for injection and compression molds for plastics. With the addition of 5.0% Cr, P4 has greater resistance to deterioration from heat so that it usually is used for molding plastics that require higher temperatures. Grade P4 also is used for dies for die casting of zinc.

Hot-Work Tool Steels (H Grades). The total AISI list comprises of 12 different grades; compositions for the most widely used grades are given in Table 1. Note that they are all characterized by medium carbon content and that they are all highly alloyed, although their total alloy contents are by no means equal.

The chromium types (grades H11 and H13) are the most widely used. Principle tooling applications include forging dies and die inserts, blades for hot shearing, and dies for die casting of aluminum alloys. These grades have also been used for structural and aerospace applications.

The tungsten grades (H21 to H26) are capable of withstanding higher temperatures than are the chromium types. H21 and H26 are used extensively for hot extrusion tools and dies for die casting of copper alloys.

Heat treating procedures for each of the grades shown in Table 1 are given in Table 6.

High-Speed Steels (T, M, and Ultrahigh-Speed Grades). The total AISI listing of high-speed steels comprises a family of 29 grades. Compositions of eight of the most important types are given in Table 1.

High-speed steels are so named because of their resistance to heat and abrasion. Also, one of their principal uses is for machining other metals at high speeds.

The high-speed steels are divided into three groups: (a) those bearing the symbol T where tungsten is the major alloying element; (b) those bearing the symbol M, indicating that molybdenum is the principal alloying element; and (c) a group of more highly alloyed steels that are capable of attaining unusually high hardness values.

T1 was one of the original high-speed steels, although all tungsten grades are used to a limited extent because of the cost and questionable availability of tungsten. Of the T steels, general purpose T1 and high-vanadium-cobalt T15 are most commonly used. T15 is used for cutting tools that are exposed to extremely rigorous heat or abrasion in service.

The M tool steels generally are considered to have molybdenum as the principal alloying element, although several contain an equal or a slightly greater amount of such elements as tungsten or cobalt. Types with higher carbon and vanadium contents offer improved abrasion resistance, but machinability and grindability may be adversely affected. The series beginning with M41 is characterized by the capability of attaining exceptionally high hardness in heat treatment, reaching hardnesses as high as 70 HRC. In addition to being used for cutting tools, some of the

Table 6 Heat treating practice for hot-work tool steels

Treatment	Type H11 (T20811)	Type H12 (T20812)	Type H13 (T20813)	Type H21 (T20821)	Type H26 (T20826)
Annealing					
Temperature, °C (°F)	845–900 (1550–1650)	845–900 (1550–1650)	845–900 (1550–1650)	870–900 (1600–1650)	870–900 (1600–1650)
Rate of cooling, °C (°F) maximum per hour	22 (40)	22 (40)	22 (40)	22 (40)	22 (40)
Typical annealed hardness, HB	192–235	192–235	192–229	207–235	217–241
Hardening					
Rate of heating	Moderately from preheat	Moderately from preheat	Moderately from preheat	Rapidly from preheat	Rapidly from preheat
Preheat temperature, °C (°F)	815 (1500)	815 (1500)	815 (1500)	815 (1500)	870 (1600)
Hardening temperature, °C (°F)	995–1025 (1825–1875)	995–1025 (1825–1875)	995–1040 (1825–1900)	1095–1205 (2000–2200)	1175–1260 (2150–2300)
Time at temperature, min	15–40	15–40	15–40	2–5	2–5
Quenching medium	Air	Air	Air	Air, oil	Salt, air, oil
Tempering					
Tempering temperature, °C (°F)	540–650 (1000–1200)	540–650 (1000–1200)	540–650 (1000–1200)	595–675 (1100–1250)	565–675 (1050–1250)
Approximate tempered hardness, HRC	54–38	55–38	53–38	54–36	58–43

Source: Ref 1

M high-speed steels are successfully used for such cold-work applications as cold header die inserts, thread rolling dies, and blanking dies. For such applications, the high-speed steels are hardened from a lower temperature than that used for cutting tools to increase toughness.

Heat treating practice for the eight compositions of high-speed steels shown in Table 1 are given in Tables 7 and 8. High austenitizing temperatures are required to dissolve sufficient amounts of the very difficult-to-dissolve complex carbides of tungsten, molybdenum, vanadium, and chromium. The quenching medium may be either oil or air. Even though all of the high-speed steels have hardenability that is sufficient to attain full hardness by air cooling, oil quenching may be used to minimize scaling. One approach that is often used for heat treating tools made from high-speed steels is to oil quench "to black" from the austenitizing temperature, then to cool to near room temperature in still air. If the tools have been austenitized in a molten salt bath, they should be given a "quick quench," which is actually a rinse at approximately 595 °C (1100 °F) in a water-soluble salt and then air cooled. If a high-temperature salt is allowed to solidify on the tool, it is almost impossible to remove.

High-speed steels usually are tempered at relatively high temperatures, and double or multiple tempering is essential.

Tempering of Tool Steels

All austenitized and quenched tool steels should be tempered to at least 150 °C (300 °F) for a length of time that ensures that tools have been heated throughout the section thickness involved (usually a minimum time of 1 h per inch of maximum section). From here on, tempering time and temperature depend greatly on the grade of steel and/or its intended end use.

Table 7 Heat treating practice for tungsten high-speed steels

Treatment	Type T1 (T12001)	Type T8 (T12008)	Type T15 (T12015)
Annealing			
Temperature, °C (°F)	870–900 (1600–1650)	870–900 (1600–1650)	870–900 (1600–1650)
Rate of cooling, °C (°F) maximum per hour	22 (40)	22 (40)	22 (40)
Typical annealed hardness, HB	217–255	229–255	241–277
Hardening			
Rate of heating	Rapidly from preheat	Rapidly from preheat	Rapidly from preheat
Preheat temperature, °C (°F)	815–870 (1500–1600)	815–870 (1500–1600)	815–870 (1500–1600)
Hardening temperature, °C (°F)	1260–1300 (2300–2375)	1260–1300 (2300–2375)	1205–1260 (2200–2300)
Time at temperature, min	2–5	2–5	2–5
Quenching medium	Oil, air, salt	Oil, air, salt	Oil, air, salt
Tempering			
Tempering temperature, °C (°F)	540– 595 (1000–1100)	540– 595 (1000–1100)	540– 650 (1000–1200)
Approximate tempered hardness, HRC	65–60	65–60	68–63

Source: Ref 1

Table 8 Heat treating practice for molybdenum high-speed steels

Treatment	Type M1 (T11301)	Type M2 (T11302)	Type M10 (T11310)	Type M36 (T11336)	Type M44 (T11344)
Annealing					
Temperature, °C (°F)	815–870 (1500–1600)	870–900 (1600–1650)	815–870 (1500–1600)	870–900 (1600–1650)	870–900 (1600–1650)
Rate of cooling, °C (°F) maximum per hour	22 (40)	22 (40)	22 (40)	22 (40)	22 (40)
Typical annealed hardness, HB	207–235	212–241	207–255	235–269	248–285
Hardening					
Rate of heating	Rapidly from preheat	Rapidly from preheat	Rapidly from preheat	Rapidly from preheat	Rapidly from preheat
Preheat temperature, °C (°F)	730–845 (1350–1550)	730–845 (1350–1550)	730–845 (1350–1550)	730–845 (1350–1550)	730–845 (1350–1550)
Hardening temperature, °C (°F)	1175–1220 (2150–2225)	1190–1230 (2175–2250)	1175–1220 (2150–2225)	1220–1245 (2225–2275)	1200–1225 (2190–2240)
Time at temperature, min	2–5	2–5	2–5	2–5	2–5
Quenching medium	Oil, air, salt	Oil, air, salt	Oil, air, salt	Oil, air, salt	Oil, air, salt
Tempering					
Tempering temperature, °C (°F)	540–595 (1000–1100)	540–595 (1000–1100)	540–595 (1000–1100)	540–595 (1000–1100)	540–625 (1000–1160)
Approximate tempered hardness, HRC	65–60	65–60	65–60	65–60	70–62

Source: Ref 1

Three important rules to establish tempering techniques for tool steels are: (a) use a temperature that is as high as can be tolerated in terms of loss of hardness; (b) always temper the steel at a temperature that is at least as high, and preferably slightly higher, than the maximum temperature to which it will be subjected in service; and (c) never allow the quenched part to quite reach room temperature before beginning the tempering operation. Careful adherence to the third rule becomes more important as carbon and/or alloy content increases. While the workpiece must be allowed to complete its transformation, when the workpiece is still "just a little hot" to the touch (about 50 °C, or 125 °F) is generally the ideal time to place it in the tempering furnace.

Tempering Temperature versus Hardness. For the W and O grades of tool steels, hardness after tempering can be expected to decrease gradually as tempering temperature increases (Fig. 3). To a great extent, this also holds true for the S grades.

For the more highly alloyed A grades, the conditions change, and secondary hardening (humps in the curves) becomes evident. The mechanism of secondary hardening is explained in Chapter 10, "Heat Treating of Stainless Steels," in connection with heat treating of high-carbon stainless steels. Specific curves for A2 after austenitizing are given in Fig. 4.

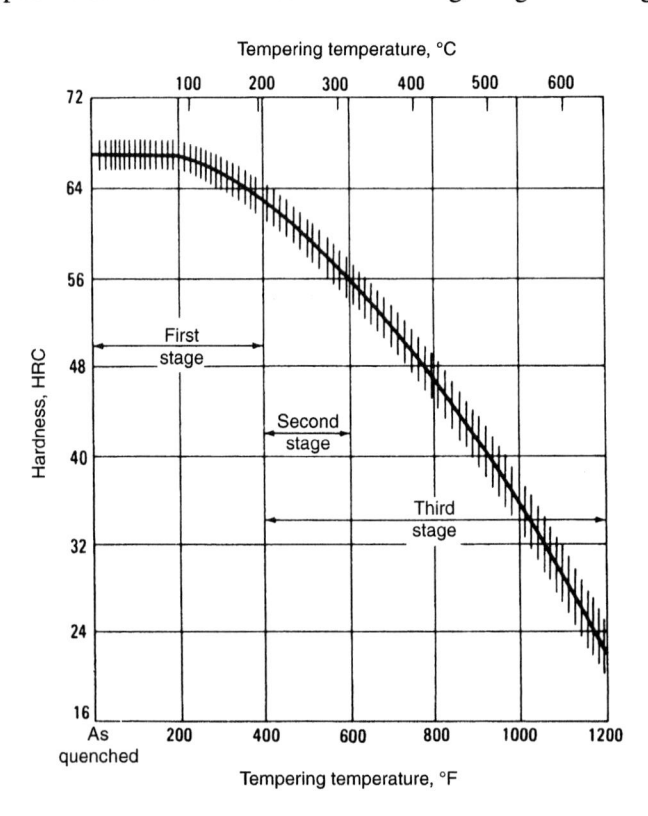

Fig. 3 Hardness vs. tempering temperature for W1 tool steel after brine quenching from 790 °C (1450 °F). Source: Ref 1

For the D grades, the tempering temperature versus hardness curves are similar to Fig. 4, except that secondary hardening is more pronounced for the more highly alloyed grades. Often, the hardness at the peak of secondary hardening equals the as-quenched hardness.

Tempering curves for chromium-type hot-work grades (H11, H12, and H13) generally are similar to those shown in Fig. 4. Because of their lower carbon content, however, initial hardness (as-quenched) is lower. The tungsten-type hot-work steels (H21 and H26) behave more like the high-speed steels during tempering.

Hardness versus tempering temperature curves for the most widely used grade of high speed tool steel (M2) are presented in Fig. 5. As shown, hardness drops at first; that is, after tempering at about 290 °C (550 °F), hardness is lower than the initial hardness by approximately five points HRC. Secondary hardening then begins, and tempering between 480 and 595 °C (900 and 1100 °F) produces hardnesses that are approximately the same as the initial hardness. Similar tempering temperature versus hardness patterns prevail for all of the high-speed steels. Consequently, a tempering temperature of 565 °C (1050 °F) is almost universally used for hardware items made from high-speed steels.

Double and Multiple Tempering. While double tempering—tempering for a given time at an established temperature, cooling to near room temperature, and then tempering again at the established temperature—is

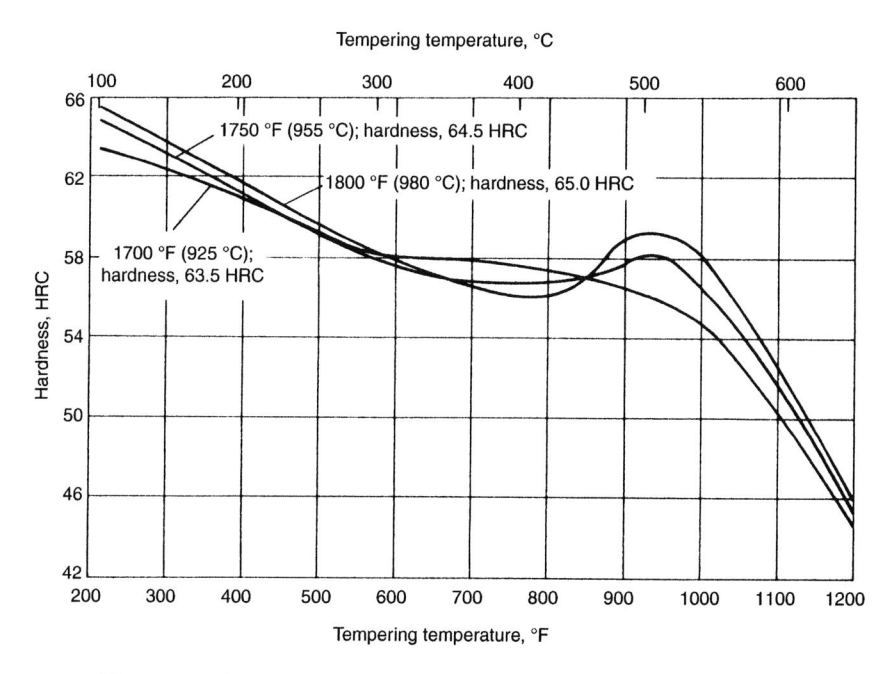

Fig. 4 Hardness vs. tempering temperature for A2 tool steel, austenitized as shown and air cooled. Source: Ref 3

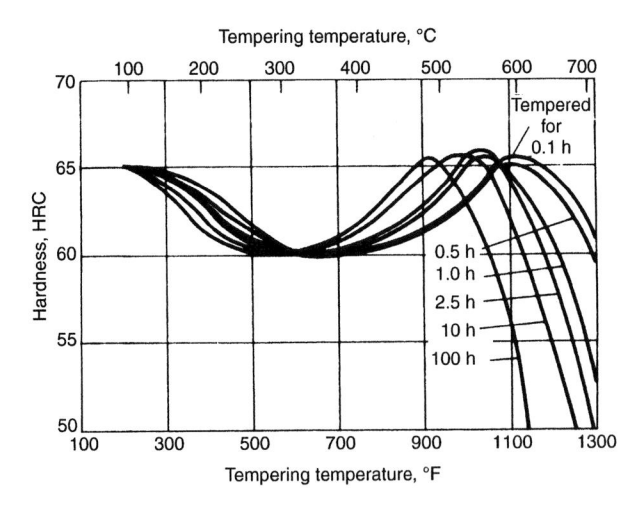

Fig. 5 Hardness vs. tempering temperature for M2 high speed tool steel after austenitizing at 1220 °C (2225 °F) and tempering for time periods and temperatures as shown. Source: Ref 4

never wrong, it is usually considered a waste of time and energy for steels that exhibit no sign of secondary hardening (Fig. 3).

Steels that do exhibit secondary hardening (Fig. 4 and 5) should at least be double tempered. Triple tempering (three complete cycles) is frequently used for high-cobalt grades such as M44. In fact, some heat treaters employ as many as five cycles for the highly alloyed grades in an attempt to complete the transformation of austenite to tempered martensite.

REFERENCES

1. H.E. Boyer, Chapter 8, *Practical Heat Treating,* 1st ed., American Society for Metals, 1984, p 149–164
2. H. Chandler, Ed., *Heat Treater's Guide: Practices and Procedures for Irons and Steels,* 2nd ed., ASM International, 1995
3. H. Chandler, Ed., Medium-Alloy, Air-Hardening Cold Work Tool Steels (A Series), *Heat Treater's Guide: Practices and Procedures for Irons and Steels,* 2nd ed., ASM International, 1995, p 544–559
4. H. Chandler, Molybdenum High-Speed Tool Steels (M Series), *Heat Treater's Guide: Practices and Procedures for Irons and Steels,* 2nd ed., ASM International, 1995, p 639–669

SELECTED REFERENCES

- B.A. Becherer, T.J. Witheford, and L. Ryan, Heat Treating of Tool Steels, *Heat Treating,* Vol 4, *ASM Handbook,* ASM International, 1991, p 711–766
- G. Roberts, G. Krauss, and R. Kennedy, *Tool Steels,* 5th ed., ASM International, 1998

CHAPTER **12**

Heat Treating of Cast Irons

THE TERM *CAST IRON* as covered in this chapter includes gray iron, white iron, ductile iron, and malleable irons. The use of malleable irons has been gradually decreasing in favor of ductile irons. Because the heat treatment of malleable iron is done largely by the manufacturer, heat treating procedures described herein will be confined to the higher tonnage irons—gray and ductile. References 1 and 2 provide more complete information on all grades of cast irons including the compacted graphite and the special-purpose high-alloy grades.

Essentially, cast iron is an alloy of iron, carbon, and silicon with a total carbon content much higher than that found in steel (see the right side of the iron-carbon equilibrium diagram shown in Fig. 1). Silicon is an im-

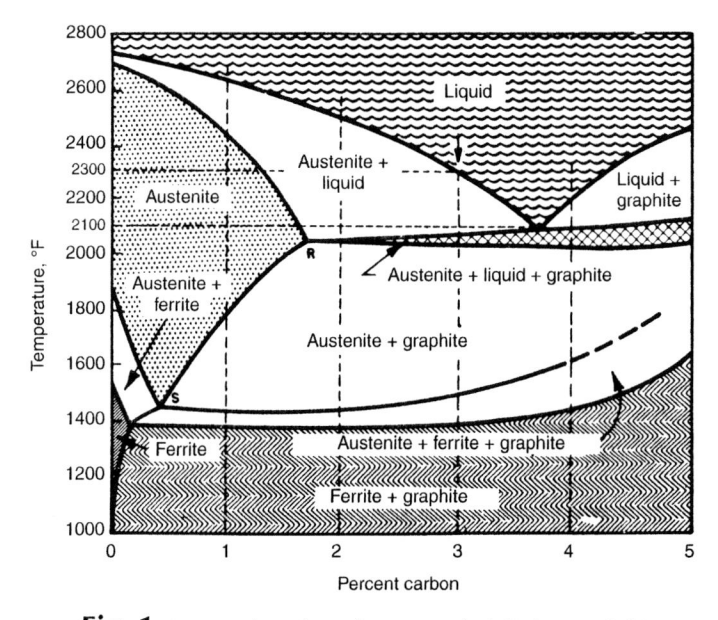

Fig. 1 Iron-graphite phase diagram at 2½% Si. Source: Ref 3

portant control element in cast iron and, therefore, must be given full consideration. The iron-graphite phase diagram shown in Fig. 1 represents a modification of the iron-cementite diagram presented in Chapter 2, "Fundamentals of the Heat Treating of Steel." One principal difference between the two diagrams is that the composition shown in Fig. 1 of this chapter contains 2.5% Si (a typical amount in gray and ductile irons). To interpret Fig. 1, use the same technique as described for use of the iron-cementite diagram shown in Chapter 2—the intersection of any line drawn from the vertical and horizontal axes shows the phase or phases present in that area, as a function of temperature and carbon content.

Further, in cast irons, carbon is present in excess of the amount that is retained in solid solution in austenite at the eutectic temperature. This excess carbon is in the form of graphite, so that cast irons are not only alloys, they are also considered composites or mixtures. As a rule, the range of total carbon in cast irons is approximately 1.75 to 4.0%, with varying amounts of silicon up to about 2.8%. Some special-purpose grades of cast iron contain up to about 12.0% Si. In addition, commercial grades commonly contain manganese in the range of approximately 0.40 to 0.90% and sometimes higher. Sulfur content is generally less than 0.15%, and phosphorus is usually within the range of 0.02 to 0.90%. Alloying elements such as nickel, chromium, molybdenum, vanadium, tin, and copper—singly or in combination—may be added to develop specific properties, much the same as for steels.

The ductility of gray irons in the as-cast condition is very low, ranging from virtually zero up to 2% in terms of elongation. Ductile cast irons, however, possess ductile properties that approach those of steels with similar microstructures. Heat treated malleable irons also possess considerable ductility.

Differences among Types of Cast Irons

The principal basis for distinguishing among various types of cast irons is to determine the form in which carbon exists. The two vital factors in determining the carbon behavior are chemical composition and cooling rate. The composition determines whether carbon will be initially stable as a metal carbide (white cast iron) or would prefer to exist as graphite in flake form (gray cast iron), in spheroidal form (ductile cast iron), as a mixture of spheroidal and vermicular graphite shapes (compacted graphite iron), or as metal carbide temporarily and later forming nodules of graphite (malleable cast iron) when annealed at 815 to 925 °C (1500 to 1700 °F). Table 1 lists typical compositions for several classes of cast irons. For further information on irons other than conventional gray irons and ductile irons, see Ref 1 and 2.

Table 1 Typical compositions of major classes of cast irons

Irons	Total C(b)	Si	Mn	P	S	Other
Gray cast iron						
Ordinary grade	3.4	2.0	0.6	0.20	0.10	...
High-strength grade	3.0	1.5	0.8	0.20	0.10	...
White cast iron	2.0–3.5 (all combined)	0.5–1.5	0.5	0.2–0.4	0.10	...
Malleable cast iron	2.3	1.0	0.4	0.20	0.10	...
Ductile cast iron						
Ferrite	3.5	2.5	**0.2**	0.05	0.01	**0.04 Mg**
Pearlitic	3.5	2.5	**0.4**	0.05	0.01	**0.04 Mg**

(a) Important differences appear in boldfaced type. (b) Total carbon content (includes graphite plus combined carbon). Source: Ref 3

For cast irons with chemical compositions that place them between white and gray cast irons, cooling rate is the dominant variable. Rapid cooling favors the formation of white cast irons (chilled cast iron), whereas slow cooling favors gray cast irons. Intermediate cooling rates can result in the simultaneous formation of metal carbide and flake graphite (mottled cast irons). Mottled cast irons are simply a mixture of white and gray cast iron. The number of different irons that can exist is consequently almost infinite.

For the two general types of iron that are emphasized in this chapter—gray and ductile—the form in which the graphite particles exists has a distinct affect on mechanical properties in their as-cast as well as their heat treated condition. The composition of a ductile iron may be almost identical to that of a plain gray iron, and yet their mechanical properties are widely different because of the difference in graphite shape (Fig. 2).

Carbon Equivalent

Before discussing heat treating techniques for cast irons, the reader should understand the concept of carbon equivalent.

Silicon and phosphorus considerably decrease the amount of carbon necessary to form eutectic composition with iron. This is important because as eutectic composition is approached, the melting point decreases, fluidity increases, and strength of the iron decreases. This effect is measured by carbon equivalent (CE) and is of great importance in the heat treating response and mechanical properties of cast iron. The most commonly used formula for calculating CE is:

$$CE = \%C + \%Si/3 + \%P/3$$

Gray iron foundries often keep phosphorus very low, so they delete it from the CE equation and simply state:

$$CE = \%C + \%Si/3$$

As indicated in the sections that follow, cast irons having relatively low carbon equivalents respond better and more consistently to quench hardening because the combined carbon is higher.

Measuring Hardness of Cast Irons

Conventional hardness measurements on cast irons always indicate lower values than the true hardness of the matrix because the graphite particles have essentially no hardness; thus, an indenter such as a Brinell ball or a Rockwell brale cannot show true conditions because the area occupied by the graphite offers no resistance to an indenter. The amount of error depends not only on the total amount of carbon in the iron, but depends more on what percentage of the total carbon is in the form of graphite. Thus, if the precise composition of an iron is known (including

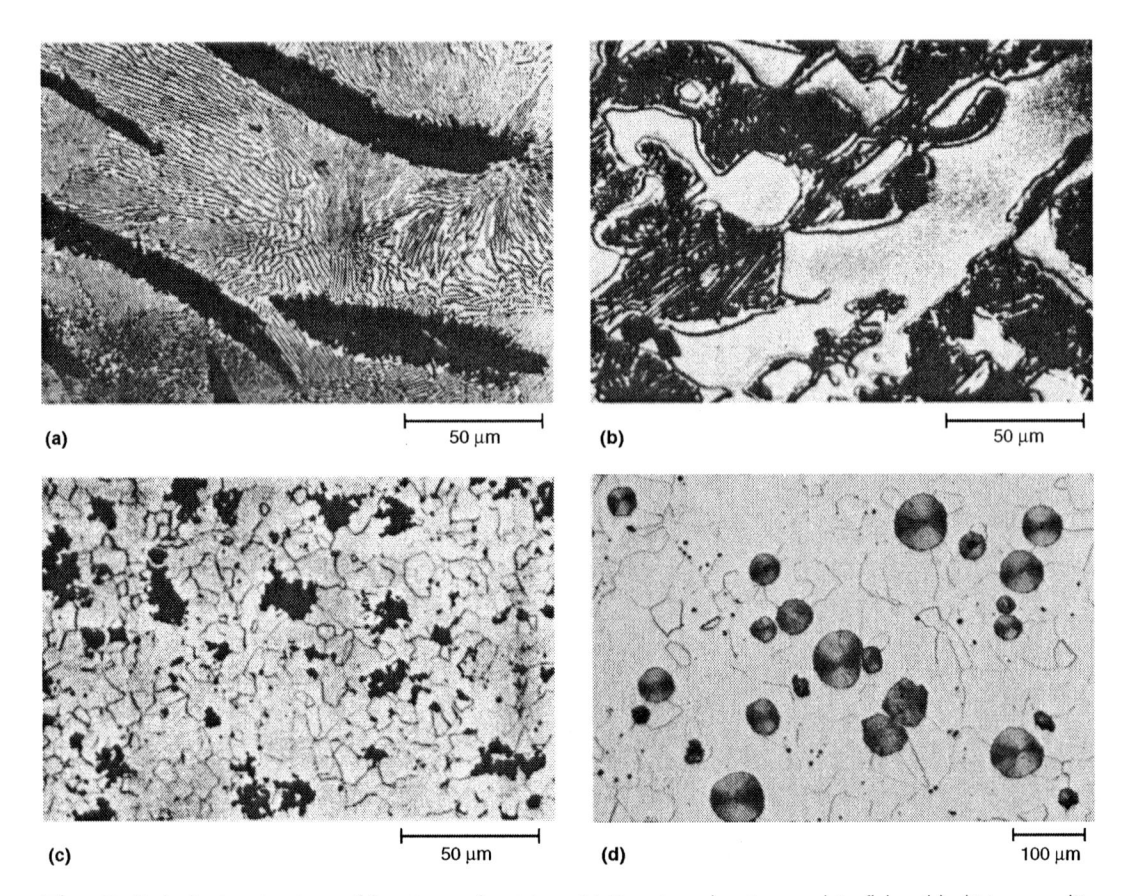

Fig. 2 Typical microstructures of four types of cast iron. (a) Gray iron showing graphite flakes (black) in a pearlite matrix. 380×. (b) White cast iron showing massive carbides (white) and pearlite. 380×. Malleable iron showing graphite nodules in a ferrite matrix. 380×. (d) Ductile iron showing spheroidal graphite in an annealed ferrite matrix. 100×. Source: Ref 3

the form in which the graphite exists), tables and charts are available that can provide a reasonably accurate conversion.

The situation is generally the same as discussed for tool steels in Chapter 11, "Heat Treating of Tool Steels," except that extremely hard microconstituents were involved; for cast irons, soft constituents cause inaccuracies in hardness testing. In either case, microhardness testing is the only approach for determining true hardness.

As an example, comparative hardness readings obtained on ten quenched and tempered ductile irons are given in Table 2. Observed HRC readings range from 3.8 to 8.3 points lower than those converted from microhardness readings.

Figure 3 shows the relation between observed HRC readings and those converted from microhardness values for five gray irons of different car-

Table 2 Comparative hardness values for quenched and tempered ductile irons

Iron	HB(a)	HRC converted from HB(b)	Observed HRC(c)	Microhardness HV(d)	HRC converted from HV(b)	HRC converted from HV minus observed HRC
1	415	44.5	44.4	527	50.9	6.5
2	444	47.2	45.0	521	50.6	5.6
3	444	47.2	45.7	530	51.1	5.4
4	444	47.2	47.6	593	54.9	7.3
5	461	48.8	46.7	595	55.0	8.3
6	461	48.8	48.3	560	53.0	4.7
7	461	48.8	49.1	581	54.2	5.1
8	477	50.3	49.6	572	53.7	4.1
9	477	50.3	50.1	618	56.2	6.1
10	555	55.6	53.4	637	57.2	3.8

(a) Average of three readings for each iron. (b) Values based on SAE-ASM-ASTM hardness conversions for steel. (c) Average of five readings for each iron. (d) Average of a minimum of five readings for each iron; 100 kg (220 lb) load. Source: Ref 3

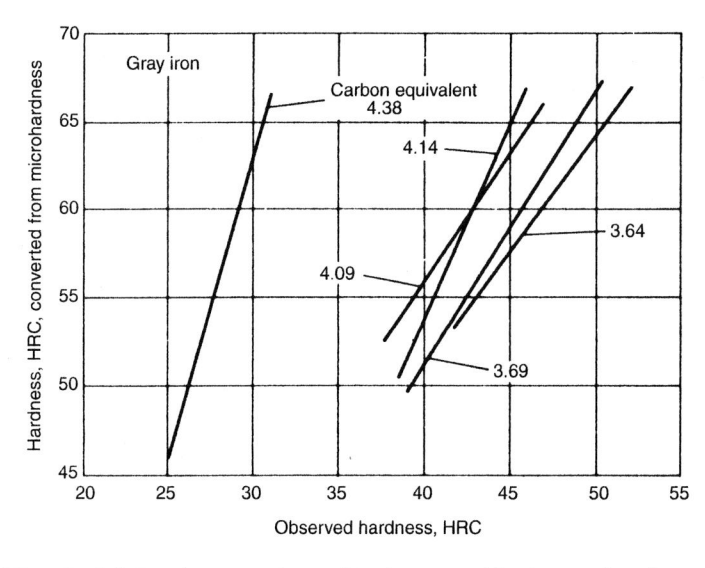

Fig. 3 Relations between observed and converted hardness values for gray iron. Source: Ref 4

bon equivalents. Hardness measurements were taken at two laboratories after quenching and after tempering of each iron. The data in Fig. 3 show why the observed values obtained by conventional hardness testing may be misleading and help to explain the good wear resistance of gray irons with apparently low hardness. Note that there is a correlation with carbon equivalent for all five irons tested and that the discrepancy between observed and converted hardness values diminishes at the lower level of carbon equivalent.

Hardenability of Gray and Ductile Irons

The meaning of and method of determining hardenability for gray and ductile irons (carbon and alloy) are essentially the same as for steel (see Chapter 3, "Hardness and Hardenability"). It must be emphasized that hardenability is that property of the iron that controls the depth of hardness that occurs when quenched from above the transformation temperature and should not be confused with hardness. Maximum attainable hardness is governed principally by the content of combined carbon. However, cast irons contain two forms of carbon—combined and graphitic. At the austenitizing temperature, some of the graphite (carbon) dissolves in the ferrite and can influence hardness. This depends upon the initial structure and the time-temperature relationship; even a 100% pearlitic matrix increases in carbon content at the austenitizing temperature if silicon content is low. Thus, it becomes a matter of controlling the time at temperature.

The methods of measuring hardenability are the same as those used for steel, which are described in Chapter 3.

Heat Treatments for Gray Irons

The types of heat treatments used for gray irons include stress relieving, annealing, normalizing, heating for hardening by quenching, tempering, austempering, martempering, and flame and induction hardening (see Chapters 6, 7, and 9 for discussions on these special treatments).

In general, the principles involved for heat treating of gray irons do not differ greatly from those used for carbon and alloy steels. One significant difference is the care that must be taken to avoid cracking of gray iron castings. They are inherently brittle and are susceptible to cracking from thermal shock.

The cardinal rule for heating gray iron, regardless of whether it is to be normalized, annealed, or austenitized for quenching, is to heat slowly. Any gray iron castings should be heated slowly through the lower temperature range. Above a range of 595 to 650 °C (1100 to 1200 °F), heating may be as rapid as desired. In fact, time may be saved by heating the casting slowly to about 650 °C (1200 °F) in one furnace and then transferring it to a second furnace and bringing it rapidly up to the austenitizing temperature.

Stress Relieving of Gray Irons

Gray iron castings may contain high-magnitude residual stresses because cooling (and therefore contraction) proceeds at different rates throughout various sections of a casting. The resultant residual stresses cause distortion. In extreme cases, they even result in failure or cracking when they approach the ultimate strength of the material. The magnitudes of these stresses depend on the shape and dimensions of the casting, on the casting technique employed, on the composition and properties of the material cast, and on whether the casting has been stress relieved.

As a rule, a reasonably complete removal of residual stresses can be achieved by heating at 540 to 565 °C (1000 to 1050 °F). This temperature seldom decreases the hardness or strength of the as-cast condition. The results of stress relieving four different compositions of gray iron are shown in Fig. 4.

The rate of heating for stress relief depends on the shape of the part. Except for parts having thick and thin sections, rate of heating is not especially critical. When a batch-type furnace is employed, it is of the utmost importance that furnace temperature not exceed 95 °C (200 °F) at the time of loading. After the furnace is loaded, the heating rate may be fairly rapid. For example, it is common practice to heat to 565 °C (1050 °F) in about 3 h, hold at temperature for 1 h per inch of section, and cool

Iron	TC(a)	CC(b)	Si	Cr	Ni	Mo
A	3.20	0.80	2.43	0.13	0.05	0.17
B	3.29	0.79	2.58	0.24	0.10	0.55
C	3.23	0.70	2.55	0.58	0.06	0.12
D	3.02	0.75	2.38	0.40	0.07	0.43

(a) Total carbon. (b) Combined carbon

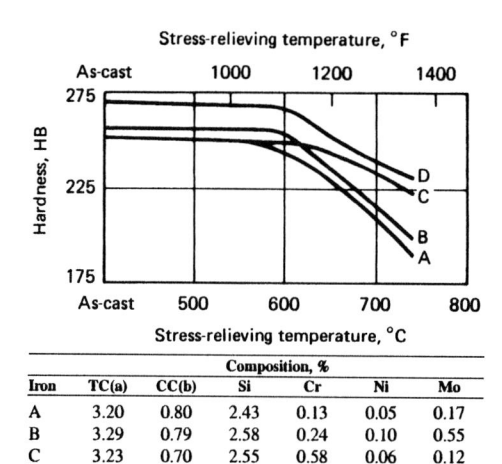

Fig. 4 Effect of stress relieving temperature on hardness of gray irons. Bar specimens 30 mm (1.2 in.) in diam where held for 1 h at the indicated temperatures and then air cooled. Source: Ref 5, 6

to 315 °C (600 °F) in about 4 h before removing castings from the furnace and permitting them to cool in air. These conditions apply also to continuous furnaces in which the various zones can be controlled to avoid introducing additional thermal stress in the castings. For very precise machining tolerances, slower heating and cooling rates are generally specified.

Annealing of Gray Irons

With the possible exception of stress relieving, annealing is the heat treatment most frequently applied to gray irons. Annealing of gray irons consists of heating it to a temperature that is high enough to soften it, thereby improving its machinability. This is frequently the main reason for annealing.

Up to approximately 595 °C (1100 °F), the effect of temperature on the structure and hardness of gray irons is insignificant. As the temperature increases above 595 °C (1100 °F), the rate at which iron carbide decomposes to ferrite plus graphite increases markedly, reaching a maximum at the lower transformation temperature (about 760 °C, or 1400 °F, for unalloyed or low-alloy iron).

Gray irons usually are subjected to one of three annealing treatments, each of which involves heating to a different range of temperature. These treatments are the ferritizing anneal, the medium (or full) anneal, and the graphitizing anneal.

Ferritizing Anneal. This process may be used for unalloyed or low-alloy gray iron of normal composition, when the only result desired is the conversion of pearlitic carbide to ferrite and graphite for improved machinability. For most gray irons, a ferritizing annealing temperature between 700 and 760 °C (1290 and 1400 °F) is recommended. It is not generally necessary to heat the casting above 760 °C (1400 °F). Precise temperatures within this range depend on the exact composition of the iron. When machining properties are of primary importance, it is advisable to anneal a number of samples at various temperatures between 705 and 760 °C (1300 and 1400 °F) to determine the temperature that yields the desired final hardness. At temperatures between 705 and 760 °C (1300 and 1400 °F), holding time varies with chemical composition and may be as short as 10 min for thin sections of unalloyed cast irons but is generally one hour per inch of section thickness. If an unusually slow rate of cooling is used, the time at temperature may be reduced further.

In ferritizing annealing, the cooling rate is seldom of great importance. If the stress relief that automatically occurs during annealing is to be retained, however, a maximum cooling rate of 110 °C/h (200 °F/h) is recommended.

While a ferritizing anneal greatly increases machinability of a gray iron, a considerable loss in hardness and strength results.

Full annealing is also called medium annealing because the results lie somewhere between the ferritizing and graphitizing anneals. Full annealing usually is performed by heating in the range of 790 to 900 °C (1455 to 1650°F).

This treatment is used when a ferritizing anneal would be ineffective because of the high-alloy or high combined carbon content of a particular iron. However, it is recommended that the effect of temperature at or below 760 °C (1400 °F) be tested before a higher annealing temperature is adopted as part of a standard procedure.

In full annealing, holding times comparable to those of the ferritizing anneal are usually employed. However, when the high temperatures of full annealing are used, the casting must be cooled slowly through the transformation range from about 790 to 675 °C (1450 to 1250 °F).

In general, cast irons that have been subjected to full annealing are still lower in hardness and/or strength compared with irons that were annealed by ferritizing. Full annealing is not practiced extensively because of the loss of strength, which can be regained only by further heat treatments that are expensive.

As a rule, the correlation of casting design and composition should be such that full annealing should not be required. This refers specifically to section thickness. With some grades of iron—namely those with a low carbon equivalent and/or those containing chromium or molybdenum—the minimum section thickness becomes extremely critical. When section thicknesses are thin, there is a danger of white iron forming (Fig. 2b). This constituent is nearly impossible to machine; subsequent annealing is required to decompose the white iron into its constituents—ferrite and graphite. It can thus be concluded that, if the castings are to be machined, white iron should not be allowed to form; breaking down white iron with heat simply degrades the properties of the casting.

The graphitizing anneal represents the ultimate in annealing of gray iron—not only in decomposition of chilled iron or massive carbides, but it also generally reduces strength to a minimum. If the microstructure of a gray iron contains massive carbides or chilled areas, higher annealing temperatures may be needed to meet acceptable machinability requirements. The graphitizing anneal may simply convert massive carbide to pearlite and graphite. In some applications, it may be desired to carry decomposition all the way to a ferrite-graphite structure for maximum machinability. Here again, producing irons that require drastic annealing is usually accidental and should be avoided. However, in practice, such conditions often occur, and the ultimate in annealing is then required.

To break down the massive carbide with reasonable speed, temperatures of at least 870 °C (1600 °F) are required. With each additional 55 °C (100 °F) increment in holding temperature, the rate of carbide decomposition doubles; consequently, it is general practice to employ holding temperatures of 900 to 955 °C (1650 to 1750 °F) and above.

The holding time at temperature may vary from a few minutes to several hours. The chill (white iron) in some high-silicon, high-carbon irons can be eliminated in as little as 15 min at 940 °C (1720 °F). In all applications, unless a controlled-atmosphere furnace is used, the time at temperature should be as short as possible. This is because at these high temperatures, gray irons are susceptible to scaling.

The cooling rate chosen depends on the final use of the iron. If the principal object of the treatment is to break down carbides, and if the retention of maximum strength and wear resistance is desired, the casting should be air cooled from the annealing temperature to about 540 °C (1000 °F) to promote the formation of a pearlitic structure. If maximum machinability is the object, the casting should be furnace cooled to 540 °C (1000 °F), and special care should be taken to cool slowly through the transformation range. In both instances, cooling from 540 °C (1000 °F) to about 290 °C (550 °F) at not more than 110 °C/ h (200 °F/ h) is recommended to avoid formation of residual stresses.

Normalizing of Gray Irons

Gray irons are normalized by being heated to a temperature above the transformation range, held at this temperature for a period of about 1 h for each 25 mm (1 in.) of maximum section thickness, and cooled in still air to room temperature. Normalizing may be used to enhance mechanical properties such as hardness and tensile strength. Also, it can restore as-cast properties that have been modified by another heating process such as graphitizing or the preheating and post-heating associated with repair welding.

The temperature range for normalizing gray iron is approximately 885 to 925 °C (1625 to 1695 °F). Heating temperature has a marked effect on microstructure and on mechanical properties such as hardness and tensile strength. Because temperatures of 815 °C (1500 °F) and higher are above the transformation temperature, the matrix of the as-cast iron is converted to austenite on heating and is transformed during air cooling to ferrite-carbide aggregates (pearlite). These vary in fineness depending on the maximum temperature (the normalizing temperature) and the alloy content. For the alloyed irons, the higher the normalizing temperature, the harder and stronger is the normalized structure. For the unalloyed iron, all normalizing temperatures produce the same hardness and strength, and all structures produced are generally softer than the as-cast material. Thus, normalizing is a hardening process for alloy irons and a softening process of unalloyed gray irons.

Some control of hardness can be exercised during normalizing by allowing castings to cool in the furnace to a temperature below the normalizing temperature. Figure 5 shows the results obtained with gray iron rings that were heated to 955 °C (1750 °F), then furnace cooled to different temperatures before being removed from the furnace and cooled in air. These data also indicate that annealing can be accomplished by cooling castings in the furnace to 650 °C (1200 °F) and then air cooling. However, if stress-free castings are desired, they should be cooled in the furnace to 290 °C (550 °F) before removal.

It can be concluded that normalizing of gray irons serves to restore as-cast properties. Further, if the carbon equivalent is sufficiently low, normalizing causes as-cast properties to be exceeded.

Hardening and Tempering of Gray Irons

Gray irons can be hardened and tempered to improve their mechanical properties and their resistance to wear by using techniques similar to those used for steel. Hardened and tempered gray iron has approximately five times more resistance to wear than pearlitic gray iron.

Austenitizing. In hardening gray irons, the casting generally is preheated to approximately 540 °C (1000 °F) and then is heated to a high-enough temperature to form austenite, held at that temperature for a sufficient length of time to affect solution of the desired amount of carbon, and then quenched at a rate suitable for that particular iron's composition. Heating for austenitizing can be accomplished in a salt bath or in an electrically heated, gas-fired, or oil-fired furnace.

The temperature to which the casting must be heated is determined by the transformation range of the particular gray iron of which it is made. A formula for determining the approximate transformation temperature of unalloyed gray iron is:

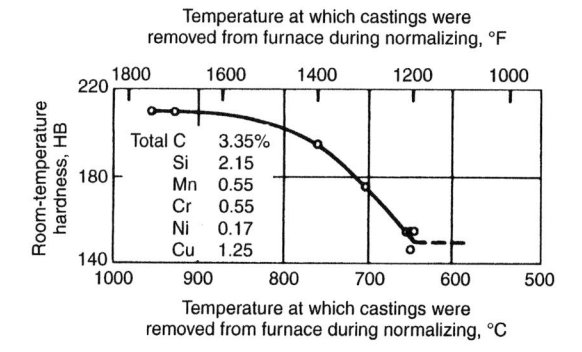

Fig. 5 Room-temperature hardness of gray iron normalizing. Effect of temperature at start of air cooling on hardness of normalized gray iron rings 120 mm (4¾ in.) in outside diam 95 mm (3¾ in.) in inside diam and 38 mm (1¾ in.) in length. Source: Ref 5, 6

$$°C = 730 + 28.0\ (\%Si) - 25.0\ (\%Mn)$$
$$°F = 1345 + 50.4\ (\%Si) - 45.0\ (\%Mn)$$

Chromium raises the transformation range of gray iron. In high-nickel, high-silicon irons, for example, each percent of chromium raises the transformation range by about 40 °C (72 °F). Nickel, on the other hand, lowers the transformation range.

Provided recommended limits are not exceeded, the higher the casting is heated above the transformation range, the greater will be the amount of carbon dissolved in the austenite and the higher will be the hardness of the casting after quenching. In practice, temperatures higher than the calculated transformation temperature are used to ensure full austenitizing. However, excessively high temperatures should be avoided. Quenching from such high temperatures increases the danger of distortion and cracking and promotes retention of austenite, particularly in alloyed irons.

Quenching. Cast irons are far more susceptible to cracking in quenching than most steels, which must be given full consideration in quenching. Molten salt or oil are the most frequently used quenchants. Oil at 80 to 105 °C (180 to 220 °F) can be used to minimize severity of quench when quenching castings of contrasting section size. Generally, water is not a satisfactory quenching medium for furnace-heated gray irons because it extracts heat so rapidly that distortion and cracking are likely in all but small parts of simple design. Water-soluble polymer quenchants are an option to provide cooling rates between those of water and oil to minimize thermal shock.

During quenching, agitation is desirable because it ensures an even temperature distribution in the bath and improves quenching efficiency. Because as-quenched castings at room temperature are extremely sensitive to cracking, they should be removed from the quench bath as soon as their temperature falls to about 150 °C (300 °F) and tempered immediately.

Tempering. After quenching, castings are tempered at a temperature well below the transformation temperature. The tempering temperature depends mainly on the desired final hardness. Tempering time is generally 1 h per inch of section. As the quenched casting is tempered, the hardness decreases as tempering temperature is increased (Fig. 6). Also, as indicated in Fig. 6, the rate of hardness decrease is influenced by the composition of the iron.

Austempering and Martempering of Gray Irons

The maximum hardness obtainable by austempering is usually less than that obtainable by martempering, although this difference may be largely offset during the tempering treatment that is usually necessary following martempering. Both austempering and martempering can result in less distortion and growth than conventional oil quenching and tempering.

Austempering. In this treatment, gray iron is quenched from a temperature above the transformation range in a hot quenching bath and held at constant temperature until the austempering transformation is complete. Quenching is usually in salt, oil, or lead baths at temperatures in the range of 230 to 425 °C (445 to 795 °F). When high hardness and resistance to wear are specified, quenchant temperature usually is held between 230 and 290 °C (445 and 555 °F). Fig. 7 shows the effect of quenching temperature on hardness for an unalloyed and nickel alloyed gray iron.

Holding time for maximum transformation is determined by quenching bath temperature and composition of the iron. The effect of the latter on

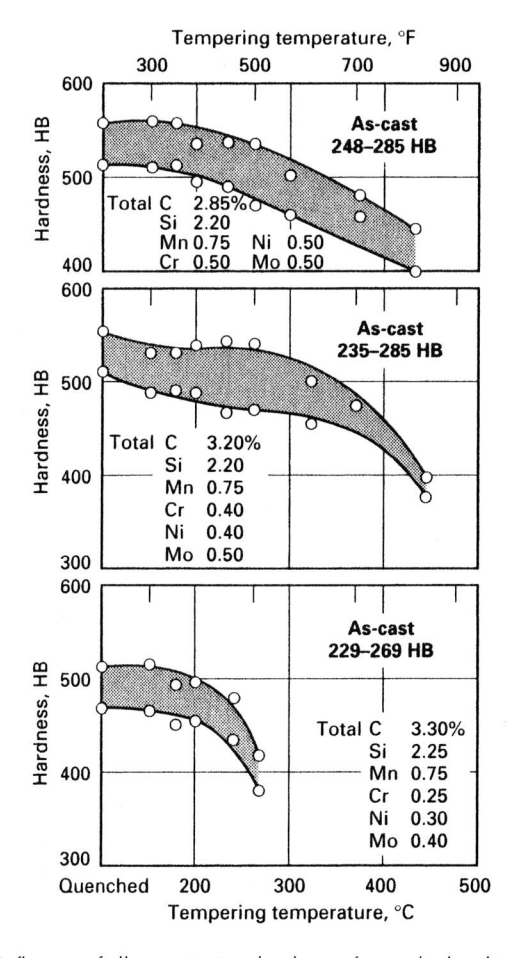

Fig. 6 Influence of alloy content on hardness of quenched and tempered gray iron test castings. Castings were normalized to the same hardness range before being austenitized for hardening and were oil quenched from 850 °C (1560 °F). Source: Ref 5, 6

holding time may be considerable. Alloy additions such as nickel, chromium, and molybdenum increase the time required for transformation.

Martempering is used to produce martensite without developing the high stresses that usually accompany its formation. It is similar to conventional hardening except that distortion is minimized. Nevertheless, the characteristic brittleness of the martensite remains in a gray iron casting after martempering, and martempered castings are almost always tempered. The casting is quenched from above the transformation range in a salt, oil, or lead bath; held in the bath at a temperature slightly above the range at which martensite forms (200 to 260 °C, or 400 to 500 °F, for unalloyed irons) only until the casting has reached the bath temperature; and then cooled to room temperature.

If a wholly martensitic structure is desired, the casting must be held in the hot quench bath only long enough to permit it to reach the temperature of the bath. Thus, the size and shape of the casting dictate the duration of martempering.

Flame and Induction Hardening of Gray Irons

Flame Hardening. Flame hardening is the method of surface hardening most commonly applied to gray iron. The mechanics of the process are dealt with in detail in Ref 7.

After flame hardening, a gray iron casting consists of a hard, wear-resistant outer layer of martensite and a core of softer gray iron, which during treatment does not reach the transformation temperature (in fact, the unhardened metal immediately below the hardened case, which has been heated by the flame to some extent, may even be partially annealed during flame hardening if it is unalloyed).

Both unalloyed and alloyed gray irons can be successfully flame hardened. However, some compositions yield much better results than do oth-

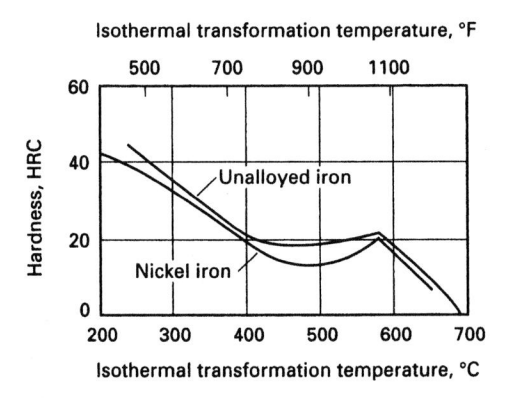

Fig. 7 Effect of isothermal transformation temperature on hardness of austempered gray irons. Holding times were sufficient to complete transformation. Source: Ref 5, 6

ers. One of the most important aspects of composition is the combined carbon content, which should be in the range of 0.50 to 0.70%, although irons with as little as 0.40% combined carbon can be flame hardened. In general, flame hardening is not recommended for irons that contain more than 0.80% combined carbon because such irons (mottled or white irons) may crack in surface hardening.

Whenever practicable or economically feasible, flame-hardened castings should be stress relieved at 150 to 200 °C (300 to 400 °F) in a furnace, in hot oil, or by passing a flame over the hardened surface. Such a treatment will minimize distortion or cracking and will increase the toughness of the hardened layer.

Stress relieving at 150 °C (300 °F) for 7 h was found to remove 25 to 40% of the residual stresses in a flame-hardened casting, while reducing the hardness of the surface by only 2 to 5 points on the HRA scale. Although stress relieving is desirable, it can often be safely omitted.

Induction Hardening. Gray iron castings can be surface hardened by the induction method when the number of castings to be processed is large enough to warrant the relatively high equipment cost and the need for special induction coils. Considerable variation in the hardness of the cast irons may be expected because of a variation in the combined carbon content.

A minimum combined carbon content of 0.40 to 0.50% C (as pearlite) is recommended for cast iron to be hardened by induction, with the short heating cycles that are characteristic of this process. Heating castings with lower combined carbon content to high hardening temperatures for relatively long periods of time may dissolve some free graphite, but such a procedure is likely to coarsen the grain structure at the surface and will result in undesirably large amounts of retained austenite in the surface layers. The recommended minimum induction-hardening temperature for gray iron is 870 to 925 °C (1600 to 1700 °F).

The surface hardness attained from the induction hardening of gray iron is influenced by the carbon equivalent (%C + 1/3% Si) when this hardness is measured by conventional Rockwell tests. The more graphite that is present in the microstructure, the lower the surface hardness will appear to be after hardening. Table 3 shows the surface hardness of induction-hardened gray iron castings of various carbon equivalents from 3.63 to 4.23. The microstructure of these castings, which were cast in the same manner and cooled at similar rates, contained more and larger graphite flakes as the carbon equivalent increased. This resulted in lower apparent surface hardness after hardening, yet the hardened matrix was consistently 57 to 61 HRC (converted from microhardness).

Stress relieving after induction hardening is recommended—150 °C (300 °F) minimum or at a temperature high enough to meet the desired hardness range. Heating in a recirculating air tempering furnace is generally recommended.

Heat Treating of Ductile Irons

Ductile irons are heat treated primarily to create matrix microstructures and associated mechanical properties not readily obtained in the as-cast condition. In general, the same types of heat treatments used for gray irons can be successfully applied to ductile irons. Ductile irons are somewhat less susceptible to cracking during heat treating compared with gray iron.

Stress Relieving of Ductile Irons

When not otherwise heat treated, complex engineering castings are stress relieved at 510 to 675 °C (950 to 1245 °F). Temperatures at the lower end of the range are adequate in many applications. Those at the higher end eliminate virtually all residual stresses, but at the cost of some reduction in hardness and tensile strength.

Recommended stress-relieving temperatures for different types of ductile iron are:

- *Unalloyed:* 510 to 565 °C (950 to 1050 °F)
- *Low alloy:* 565 to 595 °C (1050 to 1105 °F)
- *High alloy:* 595 to 650 °C (1105 to 1200 °F)
- *Austenitic:* 620 to 675 °C (1150 to 1245 °F)

Cooling should be uniform to avoid the reintroduction of stresses. Practice is to furnace cool to 290 °C (555 °F), then air cool, if desired. Austenitic iron can be uniformly air cooled from the stress-relieving temperature.

Annealing of Ductile Irons

Castings usually are given a full ferritizing anneal for good machinability when high strength is not required. The treatment produces a com-

Table 3 Effect of carbon equivalents on surface hardness of induction-hardened gray irons

Composition, %(a)			Hardness HRC, converted from		
C	Si	Carbon equivalent(b)	As read	Rockwell 30-N	Microhardness
3.13	1.50	3.63	50	50	61
3.14	1.68	3.70	49	50	57
3.19	1.64	3.74	48	50	61
3.34	1.59	3.87	47	49	58
3.42	1.80	4.02	46	47	61
3.46	2.00	4.13	43	45	59
3.52	2.14	4.23	36	38	61

(a) Each iron also contained 0.50 to 0.90 Mn, 0.35 to 0.55 Ni, 0.08 to 0.15 Cr, and 0.15 to 0.30 Mo. (b) Carbon equivalent = %C + ⅓% Si. Source: Ref 5, 6

pletely ferritic grade, the microstructure of which is converted to ferrite and spheroidal graphite. Manganese and phosphorus, and alloying elements such as chromium, nickel, copper, and molybdenum, should be as low as possible, if the best machinability is desired, because these elements retard the annealing process.

Recommended practice for castings with differing alloy contents is:

- *Full anneal for unalloyed 2 to 3% Si iron that does not contain eutectic carbide:* Heat and hold at 870 to 900 °C (1600 to 1650 °F) for 1 h per inch of section. Furnace cool to 345 °C (655 °F) at 55 °C (100 °F) per h; air cool
- *Full anneal with carbides present:* Heat and hold at 900 to 925 °C (1650 to 1695 °F) for 2 h minimum—longer for heavier sections. Furnace cool to 700 °C (1290 °F) at 55 °C (100 °F) per h. Hold 2 h at 700 °C (1290 °F). Furnace cool to 345 °C (655 °F) at 55 °C (100 °F) per h; air cool
- *Subcritical anneal to convert pearlite to ferrite:* Heat and hold at 705 to 720 °C (1300 to 1330 °F), 1 h per inch of section. Furnace cool to 345 °C (655 °F) at 55 °C (100 °F) per h; air cool

Normalizing of Ductile Irons

Normalizing can result in a considerable improvement in tensile properties but with a large decrease in ductility. The microstructure obtained by normalizing depends on the composition of the casting and its cooling rate. The cooling rate depends on the mass of the casting. However, it also may be influenced by the temperature and movement of the surrounding air during cooling. Generally, normalizing produces a structure of fine pearlite, with graphite spheroids, if the metal is not too high in silicon and has at least a moderate manganese content. Heavier castings should contain alloying elements such as nickel, copper, molybdenum, and additional manganese for satisfactory normalizing. Lighter castings made of alloy iron may be martensitic after normalizing.

Normalizing should be followed by tempering to provide uniform hardness and resulting mechanical properties and to relieve residual stresses that develop when various section thicknesses of a casting are cooled in air at different rates. The effect of tempering on the hardness and tensile properties depends on the composition of the iron and on the hardness level that was obtained in normalizing. In general, pearlitic structures such as those resulting from normalizing soften less than the harder martensitic structures that are obtained by quenching.

Quenching and Tempering of Ductile Irons

A temperature of 845 to 925 °C (1550 to 1700 °F) normally is used for austenitizing commercial ductile iron castings and produces the highest as-quenched hardness. To minimize stresses, oil is preferred as a quenching medium, but water or brine may be used for simple shapes.

The influence of the austenitizing temperature on the hardness of water quenched 13 mm (½ in.) cubes of ductile iron is shown in Fig. 8. These data show that the highest range of hardness (55 to 57 HRC) was obtained with austenitizing temperatures between 845 and 870 °C (1550 and 1600 °F). Specimens quenched from 925 °C (1700 °F) contained enough retained austenite to lower the hardness to 47 HRC.

Time at austenitizing temperature also is important for obtaining full hardness. This was determined by first heating as-cast specimens in molten salt at 870 °C (1600 °F) and then quenching in water. Specimens that were heated for 2 min contained 30 to 35% ferrite in the microstructure and developed a hardness of 32 to 45 HRC. When similar specimens were held for 4 min, the retained ferrite decreased to 12 to 15% and hardness increased to 44 to 51 HRC. Further, when specimens were held for 10 min at 870 °C (1600 °F), the ferrite had disappeared and hardness was fully developed (53 to 57 HRC).

After quenching, ductile iron castings are usually tempered for 1 h plus 1 h per inch of maximum section thickness. The results in terms of hardness versus tempering temperature can be held to closer ranges than those in gray iron, although hardness values after tempering are generally similar to those shown for gray iron in Fig. 6 for similar compositions.

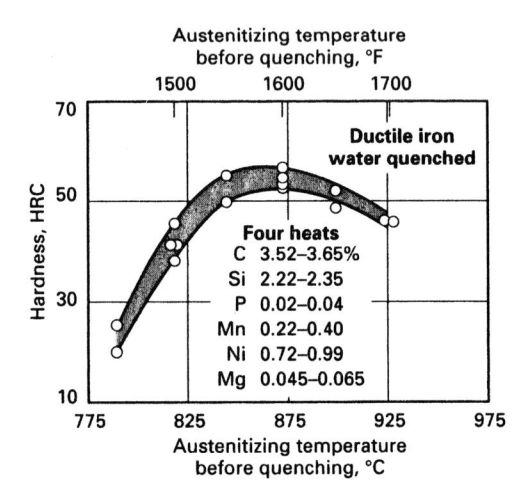

Fig. 8 Influence of austenitizing temperature on hardness of ductile iron. Each value represents the average of three hardness readings. Specimens (13 mm, or ½ in., cubes) were heated in air for 1 h and water quenched. Source: Ref 8, 9

The influence of tempering temperature on the mechanical properties of 4 heats of ductile iron is shown in Fig. 9. The samples were all quenched from 870 °C (1600 °F) and tempered for 2 h. With increasing temperature, the tensile and yield strengths drop, the hardness decreases, and the elongation has a marked increase for temperatures over about 530 °C (985 °F).

Austempering of Ductile Irons

An austempered structure of austenite and ferrite provides optimum strength and ductility. The austempered matrix is responsible for a significantly better tensile strength-to-ductility ratio than is possible with any other grade of ductile iron as shown in Fig. 10. Production of these properties calls for careful attention to section size and to time/temperature exposure during austenitizing and austempering.

As section size increases, the rate of temperature change between the austenitizing and austempering temperature decreases. Quenching and austempering techniques include:

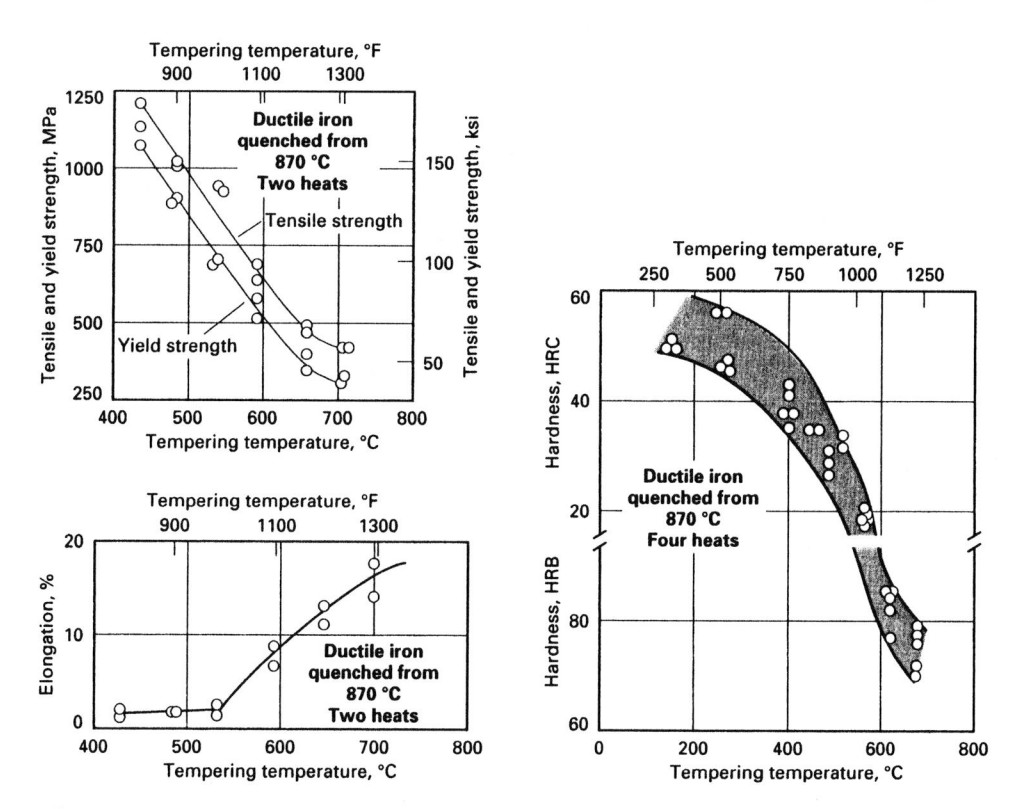

Fig. 9 Influence of tempering temperature on mechanical properties of ductile iron quenched from 870 °C (1600 °F) and tempered 2 h. Data represent irons from four heats with composition ranges of: 3.52 to 3.68% C, 2.28 to 2.35% Si, 0.02 to 0.04% P, 0.22 to 0.41% Mn, 0.69 to 0.99% Ni, and 0.045 to 0.065% Mg. Data for tensile strength, tensile yield strength, and elongation are for irons (from two of these heats) that contained 0.91 and 0.99% Ni. Source: Ref 8, 9

- Hot oil quenching (≤240 °C, or 460 °F, only)
- Nitrate/nitrite salt quenches
- Fluidized bed quenching (only for small, thin parts)
- Lead baths (for tool-type applications)

Austenitizing temperatures between 845 and 925 °C (1555 and 1695 °F) are normal, and times of about 2 h have been shown to recarburize the matrix fully.

Properties obtained in austempering vary with temperature and time. Effects of temperatures ranging from 240 to 400 °C (460 to 770 °F) on yield strength, tensile strength, and impact strength are shown in Fig. 11.

Reaching maximum ductility at a given temperature is a time-sensitive function. After maximum ductility is attained, further austempering reduces ductility. Typical austempering times vary from 1 to 4 h.

Surface Hardening of Ductile Irons

Ductile iron responds readily to surface hardening by flame, induction, nitriding (conventional, salt bath, or plasma processes) and laser or plasma torch heating. Because of the short heating cycle in these processes, the pearlitic types of ductile iron (ASTM 80-60-03 and 100-70-03) are preferred. Irons without free ferrite in their microstructure respond almost instantly to flame or induction heating and require very little holding time at the austenitizing temperature in order to be fully hardened.

With a moderate amount of free ferrite, the response may be satisfactory, but an entirely ferritic matrix, typical of the grades with high ductility

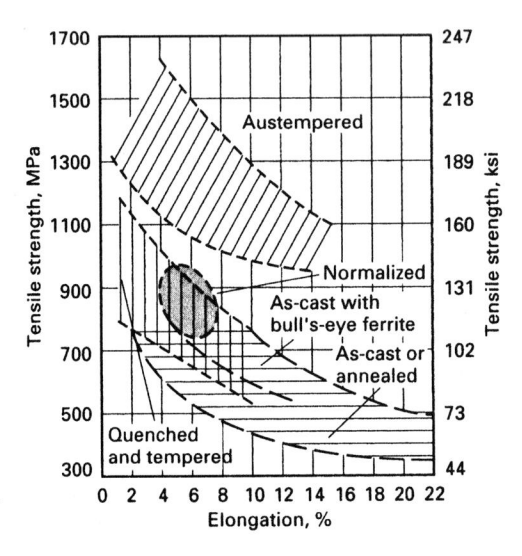

Fig. 10 Strength and ductility ranges of as-cast and heat treated ductile irons. Source: Ref 8, 9

(ASTM 60-40-18 or 60-45-12), requires several minutes at 870 °C (1600 °F) to be fully hardened by subsequent cooling. A matrix microstructure of fine pearlite, readily obtained by normalizing, has a rapid response to surface hardening and provides excellent core support for the hardened case.

The response of ductile iron to induction hardening depends on the amount of pearlite in the matrix in the as-cast, normalized, and normalized and tempered conditions. Percentages of pearlite or ferrite required in each condition are:

- *As-cast condition:* A minimum of 50% pearlite is considered necessary for satisfactory hardening with heating cycles of 3.5 s and longer at hardening temperatures of 955 to 980 °C (1750 to 1795 °F). Castings can be hardened at higher temperatures when structures contain less pearlite, but at the risk of damaging surfaces. With pearlite above 50%, hardening temperatures may be reduced within the range of 900 to 925 °C (1650 to 1695 °F).
- *Normalized condition:* For quenching cycles of 3.5 s and longer, at temperatures of 955 to 980 °C (1750 to 1795 °F), 50% pearlite in a prior structure is a minimum.
- *Quenched and tempered:* The response is excellent over a wide range of microstructures containing up to 95% ferrite. Quenching and tempering, as a prior treatment, permits a lower prior hardness, but there is a risk of distortion and quench cracking.

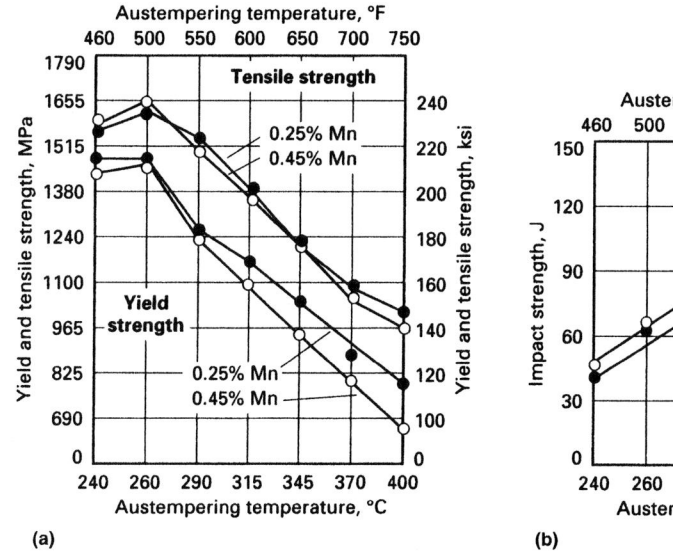

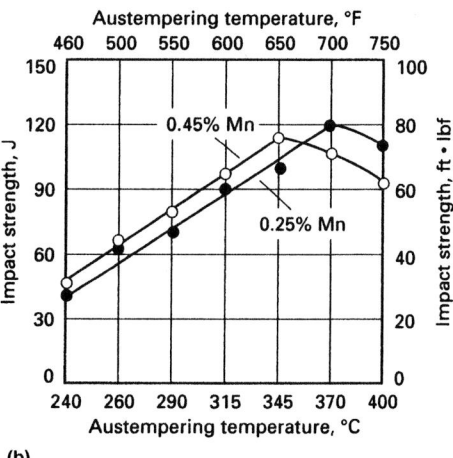

Fig. 11 Effect of austempering temperature on properties of ductile iron. (a) Yield strength and tensile strength vs. austempering temperature. (b) Impact strength vs. austempering temperature. Source: Ref 8, 9

Surface Hardness. With proper surface hardening technique and the control of temperature between 845 and 900 °C (1550 and 1650 °F), the ranges of surface hardness for ductile iron with different matrices expected in commercial production are:

- Fully annealed (ferritic), water quenched behind the flame or induction coil, 35 to 45 HRC
- Predominantly ferritic (partly pearlitic), stress relieved prior to heating, self quenched, 40 to 45 HRC
- Predominantly ferritic (partly pearlitic), stress relieved prior to heating, water quenched, 50 to 55 HRC
- Mostly pearlitic, stress relieved before heating, water quenched, 58 to 62 HRC

Heating time and temperature, amount of dissolved carbon, section size, and rate of quench help to determine final hardness values. Often soluble-oil or polymer quench media are used to minimize quench cracking where the casting section changes.

Nitriding. Case hardening can be carried out with conventional nitriding, or in liquid salt baths based on cyanide salts, or in a plasma (ion nitriding process).

In the conventional nitriding process, nitrogen diffuses into surfaces at temperatures ranging from approximately 550 to 600 °C (1020 to 1110 °F). Ammonia usually is the source of nitrogen. A surface layer, typically white and featureless in an etched microstructure, of approximately 0.1 mm (0.004 in.) deep is produced. Surface hardness is close to 1100 HV. Additions of alloying elements (0.5 to 1% Al, Ni, or Mo) can increase hardness. Advantages of nitriding, in addition to very high hardnesses, are better resistance to wear and scuffing, plus improved fatigue life and resistance to corrosion.

In the salt bath nitriding treatment, temperatures are lower, but case depth is less than that obtained with the conventional process.

Plasma (ion) nitriding is a more recently developed heat treatment used to improve resistance to wear, fatigue, and corrosion of ductile iron. For more information on the process, see Chapter 8, "Case Hardening of Steel," or Ref 10.

Laser/Plasma Torch Processes. This technology makes it possible to produce tiny, remelted areas on selected surfaces of castings. The heated area rapidly solidifies because of the self-quenching effect of the mass of the casting. The result is a white iron structure that is substantially free of graphite and has a combination of high hardness and resistance to wear.

REFERENCES

1. Cast Irons, *Properties and Selection: Irons, Steels, and High-Performance Alloys,* Vol 1, *ASM Handbook,* ASM International, 1990, p 1–104

2. J.R. Davis, Ed., *ASM Specialty Handbook: Cast Irons,* ASM International, 1996
3. H.N. Oppenheimer, Heat Treatment of Cast Irons, Course 42, Lesson 4, *Practical Heat Treating,* Materials Engineering Institute, ASM International, 1995
4. C.F. Walton, Introduction to Heat Treating of Cast Irons, *Heat Treating,* Vol 4, *ASM Handbook,* ASM International, 1991, p 667–669
5. B. Kovacs, Heat Treating of Gray Irons, *Heat Treating,* Vol 4, *ASM Handbook,* ASM International, 1991, p 670–681
6. H. Chandler, Ed., Heat Treating of Gray Iron, *Heat Treater's Guide: Practices and Procedures for Irons and Steels,* 2nd ed., ASM International, 1995, p 816–823
7. T. Ruglic, Flame Hardening, *Heat Treating,* Vol 4, *ASM Handbook,* ASM International, 1991, p 268–285
8. K.B. Rundman, Heat Treating of Ductile Irons, *Heat Treating,* Vol 4, *ASM Handbook,* ASM International, 1991, p 682–692
9. H. Chandler, Ed., Heat Treating of Ductile Iron, *Heat Treater's Guide: Practices and Procedures for Irons and Steels,* 2nd ed., ASM International, 1995, p 824–834
10. J.M. O'Brien and D. Goodman, Plasma (Ion) Nitriding, *Heat Treating,* Vol 4, *ASM Handbook,* ASM International, 1991, p 420–424

CHAPTER **13**

Heat Treating of Nonferrous Alloys

THE SCOPE OF NONFERROUS METALS and metal alloy systems that are routinely heat treated is quite broad and would include aluminum alloys, copper alloys, magnesium alloys, nickel and nickel alloys, titanium and titanium alloys, tin-rich alloys, lead and lead alloys, uranium and uranium alloys, and precious metals. Therefore, this chapter is intended to present only an overview of the heat treating of nonferrous alloys. First, a brief discussion of the effects of cold work and annealing on nonferrous alloys is presented. This is followed by a discussion of the mechanisms involved in the more commonly used heat treating procedures for hardening or strengthening—solution treating and aging. No attempt is made to cover the more complex procedures required for duplex structures that characterize some nonferrous alloys, notably titanium and copper alloys. References 1 to 3 provide more complete information on the heat treating of nonferrous alloys and the properties that may be obtained.

Work Hardening

Most metals and metal alloys (ferrous and nonferrous) respond to hardening from cold work. There is, however, a wide variation among the many metal and metal alloy combinations in their rate of work hardening by successive cold working operations.

Basically, work hardening occurs by converting the original grain shape into deformed, elongated grains by rolling or other cold working procedures. This grain shape change is illustrated schematically in Fig. 1. In this case, a copper-zinc alloy is subjected to 60% reduction by cold rolling. The change in grain shape is evident. In practice, this much reduction in one pass would be rare, but the end result is essentially the same regardless of the number of passes.

In this example, hardness of the alloy was originally 78 HRB, but changed to 131 HRB after cold reduction. This amount of hardness increase is typical of many copper alloys, most of which are very high in response to work hardening.

Annealing of Nonferrous Metals and Alloys

Most nonferrous alloys that have become hardened by cold work can be essentially restored to their original grain structure by annealing, during which process recrystallization occurs. This process is closely related to process annealing of low-carbon steels, which includes a certain amount of resulting grain growth that can be controlled to a great extent by annealing time and temperature.

A typical example of the effects of annealing temperature on hardness (diamond pyramid hardness, DPH, or Vickers microhardness, HV) of pure copper and a copper-zinc alloy is presented in Fig. 2. Several facts are evident:

- The copper-zinc alloy work hardens to a higher degree than pure copper.
- The effects of annealing are evident at a much lower temperature for pure copper compared with its companion alloy.
- Time at temperature has far less effect than temperature.
- Hardness of the pure metal and the alloy are nearly equal when near full annealing is accomplished at approximately 600 °C (1110 °F).

Figure 2 is presented as a simple example of recrystallization annealing. Each of the many nonferrous metals and metal alloy combinations has

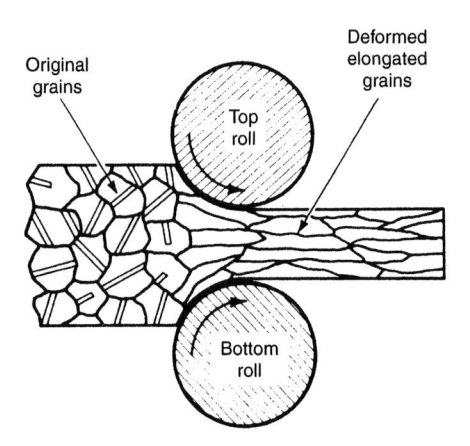

Fig. 1 Schematic presentation of cold rolling a copper-zinc alloy (60% reduction). Hardness was 78 HRB before reduction, which increased to 131 HRB after reduction. Source: Ref 2

unique annealing characteristics that produce a unique graph of data. In many instances, adjustment may be required in the annealing time and/or temperature to attain desired grain size. Although there are many sources for data that apply to specific alloys, Ref 1 to 3 are highly recommended.

Basic Requirements for Hardening by Heat Treatment

As stated previously, practically all nonferrous alloys respond readily to annealing treatments, but only a relatively small portion of the total number respond significantly to hardening by heat treatment. With the exception of titanium, common high-use aluminum, copper, and magnesium alloys are not allotropic; thus, they do not respond in the same way as steels when subjected to heating and cooling treatments (see Chapter 2, "Fundamentals of the Heat Treating of Steel").

At one time, technologists thought that ancient civilizations had developed a method for hardening copper alloys that had become a lost art. Copper articles removed from tombs showed a relatively high degree of hardness. However, it was proved that the impurities in these copper articles were responsible for their hardness. This hardness had evidently developed over a period of centuries by the precipitation mechanism rather than intentionally induced by the people who made them.

Solubility and Insolubility. In many instances, two or more metals, alloyed above their melting temperature, are completely soluble in each

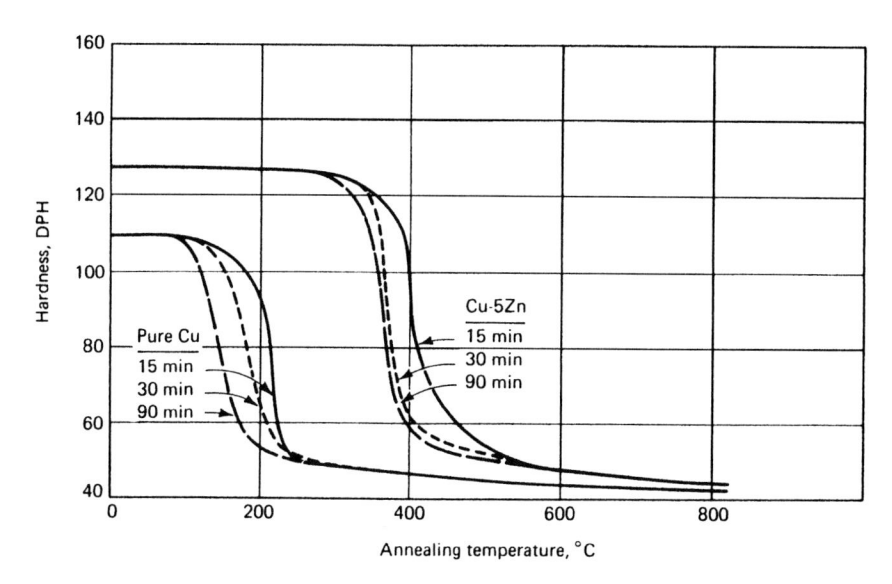

Fig. 2 Hardness as a function of annealing temperature for pure copper and a Cu-5Zn alloy, using three different annealing times. Both materials were originally cold rolled at 25 °C (75 °F) to a 60% reduction in thickness. Source: Ref 2

other, which continues through the solid solution range. For example, copper and tin alloy readily to form an entire "family" of bronzes. For these alloys, the solid solutions that form at elevated temperature remain completely stable at room temperature or below. Such an alloy, therefore, can be hardened only by cold working.

On the other hand, many alloys contain phases or constituents that are readily soluble at elevated temperature but are far less soluble or insoluble at room temperature, which is the basic requirement for hardening by heat treatment.

General Heat Treating Procedures

Alloys with the aforementioned characteristics are hardenable at least to some degree. The metallurgical principles responsible for this phenomenon are the same (or at least closely related) for all of the nonferrous alloys. Likewise, methods of processing are basically the same. However, temperature and time cycles cover a wide range, depending not only on alloy composition but also on whether the alloy is in the wrought or cast condition.

As a rule, the highest possible increase in mechanical properties for a given alloy is accomplished in two distinct heating operations. First, the alloy is heated to a temperature just below (say, at least 28 °C, or 50 °F) its solidus (beginning of melting) temperature. Holding time at temperature depends on the solubility of the various phases and whether the alloy is wrought or cast. It must be considered that the primary purpose of this operation is to affect a solid homogeneous solution of all phases. Consequently, this part of the operation is commonly known as solution treating or homogenizing.

Because the structures of castings are relatively heterogeneous, a much longer time (at least twice as long) is required at the solution treating temperature compared with a wrought product of similar composition.

Cooling from the Solutionizing Temperature. After solid solution has been attained, the alloy is cooled as quickly as possible to room temperature (most often by water or polymer quenching) to arrest the structure obtained by heating. The required cooling rate is more critical for some alloys than for others. In addition, section thickness markedly affects cooling rate; still air is sometimes adequate.

At this stage, the alloy is in the solution-treated condition and, in most cases, is fully annealed (usually as soft as it can be made). However, the alloy is in an unstable condition because the phase or phases that possess a low degree of solubility at room temperature are virtually all dissolved.

Aging (Precipitation Hardening). As soon as room temperature is reached, these phases start to precipitate out of solid solution in the form of fine particles within the crystals and at the grain boundaries. In most alloys, precipitation occurs very slowly at room temperature and is even more sluggish at lower temperatures. In fact, this action in most alloys can be completely arrested by lowering to subzero temperatures. However, if the temperature of the solution-treated alloy is raised, the action is greatly increased and can be completed in a relatively short time.

Aging Cycles. The precipitation or dispersion of fine particles tends to key the crystals together at their slip planes so that they are more resistant to deformation; thus, the alloy is stronger, harder, and less ductile. For each alloy, there is an optimum time and temperature. For the majority of nonferrous alloys, this is from about 165 to 345 °C (325 to 650 °F), although some alloys require higher aging temperatures.

As is true for most heat treating operations, precipitation is a function of time and temperature. For example, a specific alloy might assume its maximum strength by aging for 5 h at 165 °C (325 °F). Overaging by either increasing the time or increasing the temperature is highly detrimental and decreases strength of the alloy.

Specific Alloy Systems

When precipitation hardening mechanisms are discussed, one is usually referring to nonferrous metals, although there are some iron-base alloys that are hardened by the precipitation mechanisms (see Chapter 10, "Heat Treating of Stainless Steels"). Compositions and typical uses of some commonly used alloys of aluminum, copper, magnesium, and nickel that respond readily to precipitation hardening are given in Table 1.

The discussion has, thus far, been quite general. We shall now take two commercially important alloys—one from the aluminum-copper system

Table 1 Examples of some precipitation-hardenable alloys

Alloy type	Principal alloying elements, %	Typical uses
Aluminum-base alloys		
2014	4.4 Cu, 0.8 Si, 0.8 Mn, 0.4 Mg	Forged aircraft fittings, aluminum structures
2024	4.5 Cu, 1.5 Mg, 0.6 Mn	High-strength forgings, rivets, and structures
6061	1 Mg, 0.6 Si, 0.25 Cu, 0.25 Cr	Furniture, vacuum cleaners, canoes, and corrosion-resistant applications (used in ASM's geodesic dome)
7075	5.5 Zn, 2.5 Mg, 1.5 Cu, 0.3 Cr	Strong aluminum alloy; used in aircraft structures
Copper-base alloys		
Beryllium bronze (beryllium copper)	1.9 Be, 0.2 Co or Ni	Surgical instruments, electrical contacts, nonsparking tools, springs, nuts, gears, and other heavy duty applications
Aluminum bronze	10 Al, 1 Fe	Applications requiring resistance to corrosion, stress-corrosion, and wear.
Magnesium-base alloys		
AM100A	10 Al, 0.1 Mn	Though, leakproof sand castings
AZ80A	8.5 Al, 0.5 Zn, 0.15 Mn min	Extruded products and press forgings
Nickel-base alloys		
René 41	0.1 C, 19 Cr, 10 Co, 10 Mo, 3 Ti, 1.5 Al, 1.5 Fe, 0.005 B	High-temperature applications where strength is very important (up to about 980 °C, or 1800 °F)
Inconel 718	0.35 C, 19 Cr, 5 Nb, 3 Mo, 1 Ti, 0.4 Al, 0.006 B, 0.3 Cu, bal Fe	Applications requiring high tensile, yield, creep, and rupture strength up to about 700 °C (1300 °F)
Waspaloy	0.07 C, 19.5 Cr, 13.5 Co, 4.3 Mo, 3 Ti, 1.4 Al, 2 Fe, 0.006 B, 0.09 Zr	Applications requiring outstanding tensile and stress-rupture strength up to about 750 °C (1400 °F) or oxidation resistance up to 870 °C (1600 °F)

and one from the copper-beryllium system—and demonstrate more clearly the mechanism involved in solution treating and aging. As previously stated, time and temperature cycles vary greatly for different alloys, but the principles are basically the same for all precipitation hardening.

To accomplish this, it is necessary to use phase diagrams. For more information on the use of phase diagrams, see Chapters 2, "Fundamentals of the Heat Treating of Steel," and 12, "Heat Treating of Cast Irons," and Ref 4.

Aluminum-Copper Alloys

The first example that demonstrates the conditions that must exist for precipitation hardening is the partial aluminum-copper phase diagram presented in Fig. 3. In this phase diagram, percent copper is plotted on the horizontal axis, and temperature is plotted on the vertical axis. The left side of the diagram (up to about 4.5% Cu) is the significant part of the diagram for showing the mechanisms of solution and aging.

First, note that there is terminal solid solution (denoted as Al) on the left of the diagram (shaded area). The curved line *ABC* shows the maximum solid solubility of copper in aluminum for the temperature range of 0 to 650 °C (32 to 1200 °F). The horizontal line at 547 °C (1018 °F) is the eutectic temperature. If the copper content is greater than that shown by line *AB*, the solid aluminum is saturated with copper and a new phase, $CuAl_2$, appears.

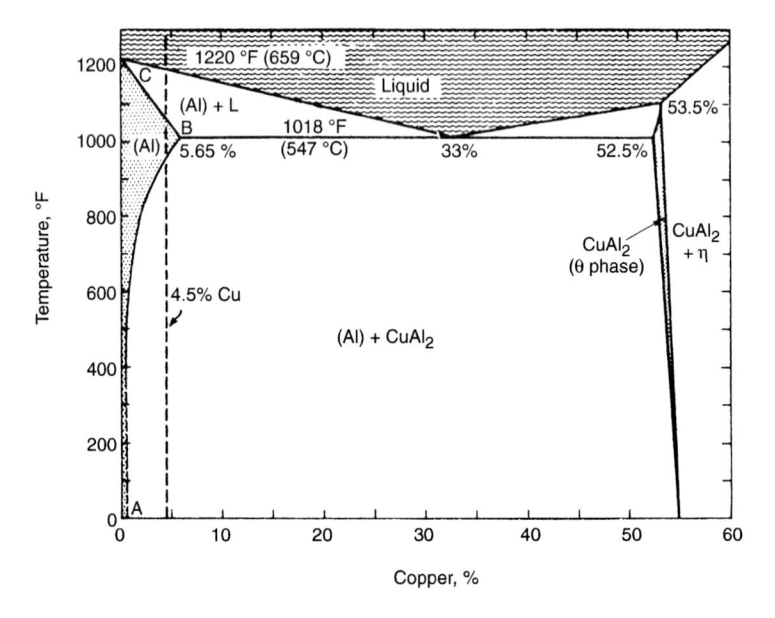

Fig. 3 Part of the aluminum-copper phase diagram. The kappa phase, bounded by ABC, is a solid solution of copper in aluminum; $CuAl_2$ precipitates from this phase on slow cooling or on aging after solution treatment. Source: Ref 5

Therefore, when aluminum containing about 4.5% Cu is heated to around 540 °C (1000 °F), it becomes a single-phase stable alloy at that temperature. Because of the similarity in size of the copper and aluminum atoms, the alloying is hardening by substitution (trading atoms from their lattice position) rather than by interstitial alloying, as explained in Chapter 2 for the alloying of iron and carbon. Thus, at various points in the aluminum lattice, copper atoms are substituted for aluminum atoms. When the alloy is quenched in water (or otherwise quickly cooled), there is insufficient time for the copper to precipitate as $CuAl_2$. According to Fig. 3, copper at room temperature is nearly insoluble in aluminum (shown as less than 1.0%). At this point, the alloy is in an unstable condition.

Aging. Under these very unstable conditions, some precipitation of $CuAl_2$ occurs at room temperature. This can be checked by first measuring the hardness in the as-quenched condition (essentially as soft as the alloy can be made), then continuing to take hardness measurements at regular intervals. In a matter of hours, the alloy will register an increase in hardness caused by precipitation of $CuAl_2$ particles. The rate of hardness at room temperature, however, gradually diminishes, and after a time (say a few days), action practically ceases. However, if the as-quenched alloy is heated to about 170 °C (340 °F) for several hours, the copper atoms regain enough mobility to effect the optimum amount of precipitation. Therefore, maximum attainable strength for the specific alloy is developed quickly by this artificial aging (purposely heating above room temperature).

Overaging. There is an optimum time and temperature for aging of any specific alloy. When overaging takes place—in terms of either time or temperature, or both—there is a rapid drop in hardness. This is clearly shown in Fig. 4. It is also obvious from Fig. 4 that a fair-sized area exists

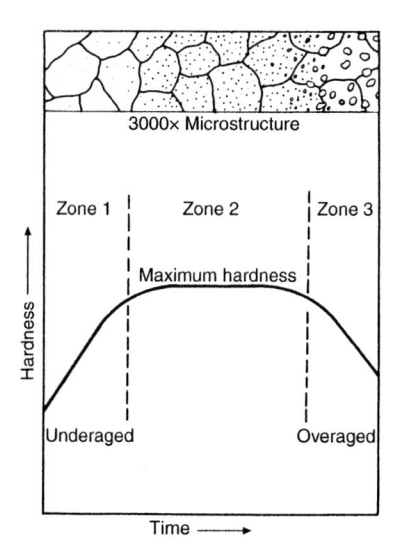

Fig. 4 Schematic aging curve and microstructure. At a given aging temperature, the hardness of aluminum-copper alloys increases to a maximum, then drops off. Source: Ref 4

that shows "maximum hardness" (zone 2), which indicates that there is some variation in the time-temperature cycles.

Changes in microstructure also are included in Fig. 4. Fine particles are precipitated in zone 2, whereas coarse particles (overprecipitation) are shown within the grains in zone 3, where a definite decrease in hardness exists.

Copper-Beryllium Alloys

Copper-beryllium alloys hold a unique position among all other engineering alloys that are in use today. Copper by itself is used extensively throughout the electrical industry because it conducts electric current better than any other metal (excluding silver). Its resistance to the flow of current is low. By itself, however, copper does not have a great deal of strength. It is too soft to use in most structural applications. To impart some strength to copper, other elements are alloyed with it. One of the more potent of these is beryllium. The alloys of copper containing beryllium are among the strongest known of the copper alloy family. This high strength combined with good electrical conductivity makes copper-beryllium alloys (also known as beryllium bronze or beryllium copper) particularly suited for electrical applications, as in contacts and relays that must open and close a great number of times. Often these switches must open and close many times each minute, and during years of continuous service, they may be stressed a billion times or more. Copper-beryllium alloys are highly suited for applications of this type. The same part may also serve as a structural member in the actual relay to hold it together and to support other parts. Heat treated copper-beryllium alloys also have been used extensively in aircraft and aerospace applications.

Phase Diagrams. A partial copper-beryllium phase diagram is presented in Fig. 5. This diagram can be interpreted using the guidelines described in Chapter 2, and in much the same manner as the aluminum-copper system in Fig. 3. Figure 5 shows beryllium contents up to only 10%. The remainder of the phase diagram is not included because there are no commercial copper-base alloys in use with beryllium content above $-2\frac{1}{2}\%$. Alloys with about 2% Be are used where high strength is required. A ½% beryllium alloy is used where high electrical conductivity is the main objective. However, to get good aging response, about 2% Ni or Co must be added to such an alloy.

This phase diagram for the copper-beryllium system is slightly more complex than those presented previously, and for that reason we will not discuss it here in any detail. The important feature to note is the all-important decreasing of solubility of beryllium in the alpha (α) phase with decreasing temperature (large shaded area on the left of Fig. 5).

Chapter 13: Heat Treating of Nonferrous Alloys / 239

The precipitation process in the copper-beryllium alloy system is complicated by the beta (β) phase changing into gamma (γ). However, for the alloy under discussion the reader is interested only in the left portion of the phase diagram. Note that the primary requirement for precipitation hardening has been met; that is, beryllium has a very low solubility in copper at low temperatures and a relatively high solubility at elevated temperatures.

Solution Treating and Aging. In considering heat treatment, the shape of the hardness versus aging time curves for alloys with varying amounts of beryllium should be considered. Figure 6 shows the aging curves for several copper-beryllium alloys containing from 0.77 to 4.0% Be. These alloys were all solution treated at 800 °C (1470 °F) and aged at 350 °C (660 °F) for various lengths of time, as shown in Fig. 6.

Note that an alloy containing as little as 0.77% Be does not harden as a result of aging. However, when the beryllium content is raised to 1.32%, a significant hardness increase takes place after 16 h. By increasing the beryllium content to 1.82%, a great deal of hardening takes place after 4 h, and the hardness stays at essentially the same level for the duration of the aging period. A beryllium content of 2.39% raises the peak hardness attained for this alloy. By increasing the beryllium content to 4%, still higher maximum hardnesses can be attained (Fig. 6).

Several interesting observations can be made from examining Fig. 6. First, note that at zero aging time (the solution treated condition), alloys with higher beryllium content are harder. This is known as solid solution hardening. The atoms of the parent lattice are displaced in making an

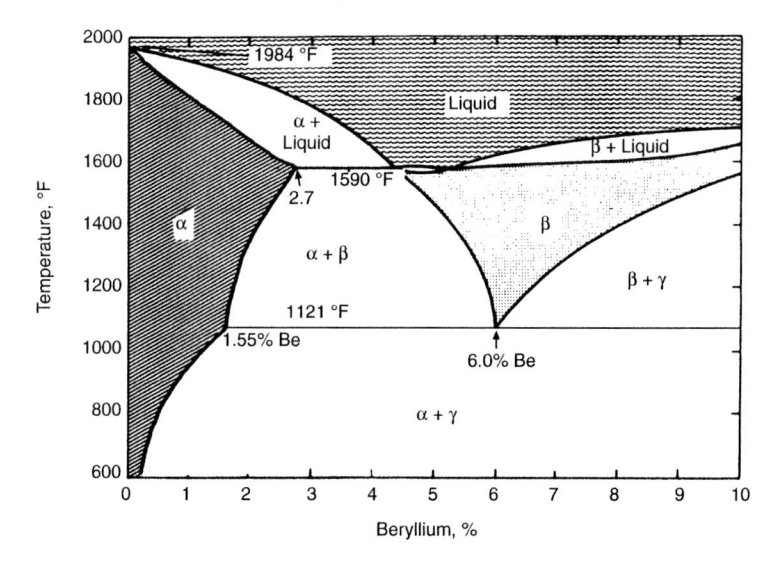

Fig. 5 Beryllium-copper phase diagram. The alpha phase holds about 1.55% Be at 605 °C (1121 °F) and about 2.7% at 865 °C (1590 °F). Decreasing solubility of the beryllium causes precipitation of a hard beryllium-copper phase on slow cooling or on aging after solution treatment. Source: Ref 4

adjustment. As a result of the shifting of position, the hardness of the alloy increases. A similar but less intensely stressed condition results in solid solution hardening. When beryllium atoms are present in the copper lattice, the copper atoms are forced to move slightly out of their normal position to accommodate the somewhat different size of the beryllium atom. This accommodation or slight shifting in position gives rise to a moderate increase in hardness. When more beryllium atoms are present, more copper atoms have to shift. The result is a higher hardness as the number of beryllium atoms increases.

Attaining Maximum Hardness. As the beryllium content is increased up to 2.39%, the time required to reach the maximum hardness decreases (Fig. 6). Thus, aging for 36 h does not allow time enough to produce maximum hardness in a 1.32% beryllium alloy. With 1.82% Be, however, the maximum hardness occurs after approximately 5 h. With 2.39% Be or more, the maximum hardness occurs after only 1 h. It can also be seen from Fig. 6 that by increasing the beryllium content above 2.39%, the rate of aging is not significantly increased. The 3.31 and 4% beryllium alloys reach a maximum hardness at about the same time as the 2.39% alloy. By looking at the phase diagram of the beryllium-copper alloy system, the reason for this can be seen. The maximum solid solubility of beryllium in copper is just about 2.7% (at a temperature of 865 °C, or 1590 °F). Thus, the 3.31% alloy contains more than the maximum amount of beryllium that can be dissolved in copper. This means that the increased hardness of the 3.31 and 4% alloys is not due to any additional aging

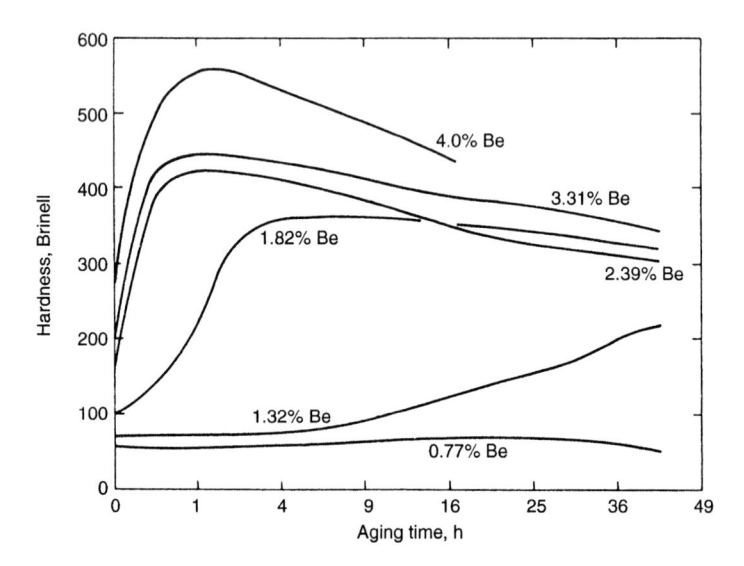

Fig. 6 Precipitation-hardening curves of beryllium-copper binary alloys. As the percentage of beryllium increases, the aging time required to reach maximum hardness is shortened, and the maximum hardness is increased. These alloys were quenched form 800 °C (1470 °F) and aged at 350 °C (660 °F) for the time shown. Source: Ref 4

effects but is simply caused by the presence of undissolved beta phase, as may be predicted from the phase diagram. This increase in hardness due to undissolved precipitate is known as *dispersion hardening*. The undissolved precipitate is harder than the matrix surrounding it, and the final hardness of the mixture depends on how much undissolved precipitate is present, as well as its manner of distribution.

REFERENCES

1. *Heat Treating,* Vol 4, *ASM Handbook,* ASM International, 1991
2. C.R. Brooks, *Heat Treatment, Structure and Properties of Nonferrous Alloys,* American Society for Metals, 1982
3. *Heat Treater's Guide: Practices and Procedures for Nonferrous Alloys,* ASM International, 1996
4. H. Baker, Introduction to Alloy Phase Diagrams, *Alloy Phase Diagrams,* Vol 3, *ASM Handbook,* ASM International, 1992, p 1-1 to 1-29
5. H.E. Boyer, Chapter 12, *Practical Heat Treating,* 1st ed., American Society for Metals, 1984, p 209–220

CHAPTER **14**

Assuring the Quality of Heat Treated Product

A SUCCESSFUL HEAT TREATING OPERATION is determined by the ability to satisfy the customer's quality requirements consistently and economically. Quality requirements may be defined by such characteristics as hardness, dimensions, surface condition, uniformity, properties, microstructure, and so forth. This chapter reviews the steps that are important to produce quality parts in heat treating with a brief practical explanation of each. References 1 to 3 provide more comprehensive discussions of quality concepts.

Final inspection is necessary to verify that specifications have been met, but testing of as-received material and in-process testing are important to determine whether a problem occurred during processing and at what stage it occurred. It is certainly uneconomical to complete an expensive heat treatment only to find a problem that could have been caught in an early stage of processing. From strictly an economic point of view, the part being heat treated generally has a value of 10 to 20 times the actual cost of the heat treatment; therefore, if a heat treating operation produces a scrap rate of only 1 to 2%, any profit from that operation is lost.

Components of a Quality Assurance Program

The components of a program to ensure quality in heat treating should include:

- Selection of proper material and design of the part being treated
- Use of the proper equipment and control of its operation (control of heat treating processes is discussed in Chapter 5, "Instrumentation and Control of Heat Treating Processes")
- A determination whether the process is capable of heat treatment to the required specification

- Use of statistical process control (SPC), control charting, and in-process inspection and testing
- Statistical quality control (SQC) and final testing (sampling) to verify the results

If these factors are incorporated into the heat treating operation, including the proper methods of statistical analysis, quality can be assured. Quality cannot be inspected into a part, but proper attention to each step of the process can identify sources of problems and prevent their occurrence.

As demand for increased quality and documentation has been experienced by heat treaters, the subject of automatic collection and use of process information in a SPC/SQC format has become mandatory. Data acquisition and documentation a few years ago meant a chart recorder for temperature and a log sheet for the operator's dew point readings. Today, it more than likely means a computer system tied into key points of the heat treating equipment and process with the objective of logging important information for later review or perhaps being taken into account in real time.

Material Considerations

Before applying statistical control techniques to monitor process or product uniformity, it is important to know how the base material uniformity is controlled prior to heat treat processing. Having the incoming materials identified and kept separate by heat numbers in the case of steel or by batch number in the case of cast materials can greatly influence the uniformity of heat treatment response between pieces and lots being processed.

Cast irons are probably the best example of a material where hardness test results can be a function of the hardness testing scale used. This sensitivity of hardness value to the testing method and the hardness scale used is due to the fact that the different phases present in the workpiece vary significantly in hardness (hard matrix versus soft graphite). This same effect exists in other materials that are heat treated.

Another type of problem that can influence testing results that are not the direct result of heat treat processing is "banding." Many steels, particularly resulfurized ones such as SAE-AISI 1100 or 1200 series, exhibit banding or microalloy segregation. The bands exist prior to heat treatment and the ferrite-rich and pearlite-rich areas run in bands across the longitudinal rolling direction of the bar stock from which parts are made (Fig. 1). It has been found that this condition can result in a 4 to 10 point of Rockwell C hardness variation between these bands of different chemical composition after hardening. This problem is greatest when the bands are widest and the heat treatment times are very short, such as for the induction hardening or laser processes.

Decarburization or surface carbon reduction to a greater or lesser degree exists on most steels having more than 0.30% C (see Appendix B, "Decarburization of Steels"). This defect results from basic steel processing and, if not removed in the part manufacturing process prior to heat treatment, can influence the surface hardness of parts after induction, flame, or direct hardening processes that may not be capable of correcting the condition. However, it should be recognized that many heat treating processes can also cause this same condition. It is thus important for one to have characterized the incoming product to be processed so that the controllable incoming material variability can be isolated and corrected independently from the product variations due to the process.

Statistical Control

Understanding SPC and SQC techniques requires some definitions. The principles of SPC rely on the science of statistics: the collection and classification of facts from which conclusions can be drawn with a given degree of certainty. Statistics may be used to analyze such data as obtained in coin tossing, throwing dice, measuring dimensions, or the hardness of parts after heat treating.

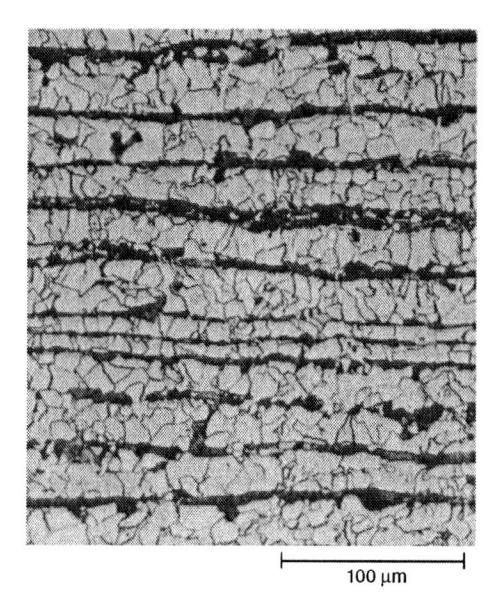

Fig. 1 Microstructure of hot-rolled AISI 1022 steel showing severe banding. Bands of pearlite (dark) and ferrite were caused by segragation of carbon and other elements during solidification and later decomposition of austenite. Etched in natal. 250×. Source: Ref 1

In applying statistical analysis, the assumption is implicit that the data analyzed are random; that is, the data tend to scatter around some average value. The pattern of variation or frequency distribution typically found when data are taken from some natural or artificial process is called a normal distribution. When plotted on a graph of frequency of occurrence of a given value versus property value, a curve such as that shown in Fig. 2 results. This curve is called a normal, or bell-shaped, curve.

Another important property is called the standard deviation and is designated as sigma (σ). It is defined as the squares of the difference between each measured value and the average value summed over all data points and divided by the number of data points. The square root of the result is the standard deviation. It can be shown that in a normal statistical distribution, 68.3% of all data lie between $\pm \sigma$ (a range of 2 σ), 95.5% between $\pm 2\sigma$, and 99.7% between $\pm 3\sigma$ (Fig. 2).

A final quantity used in SPC is the range, R. This is the difference between the maximum and minimum values measured.

As an example, suppose HRC tests are made on 20 pieces from each of two different batches. The results are recorded in Table 1. For each batch, individual hardness values are listed (x), the average ($\bar{x}$ or x-bar) calculated, the difference between each value and the average, and the square of this number. It is seen that the average for each batch is 50.3, but the standard deviations are 3.79 and 1.30, respectively. Also, the range for batch A is 16 and 5 for batch B. When these data are plotted (Fig. 3), bell-shaped curves are obtained in both cases, but the values in batch B are grouped much closer to the average than in batch A. Clearly, the processes used to harden the two batches differed considerably in the control exercised.

The calculations in Table 1 are quite tedious to carry out by hand (and most processes analyzed by statistical methods would consist of many more data values). Fortunately, computer programs are available to calculate all important quantities by merely entering data points.

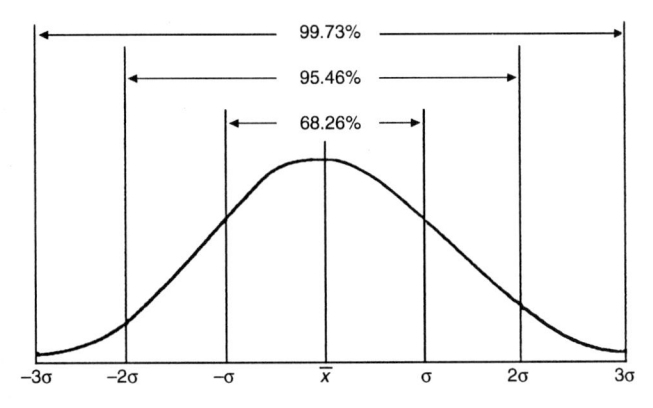

Fig. 2 Normal distribution curve. Source: Ref 2

Process Capability

Before statistical sampling techniques can be applied to a process, a process capability study should be conducted that demonstrates that the process can hold the prescribed specification or specifications. Capability

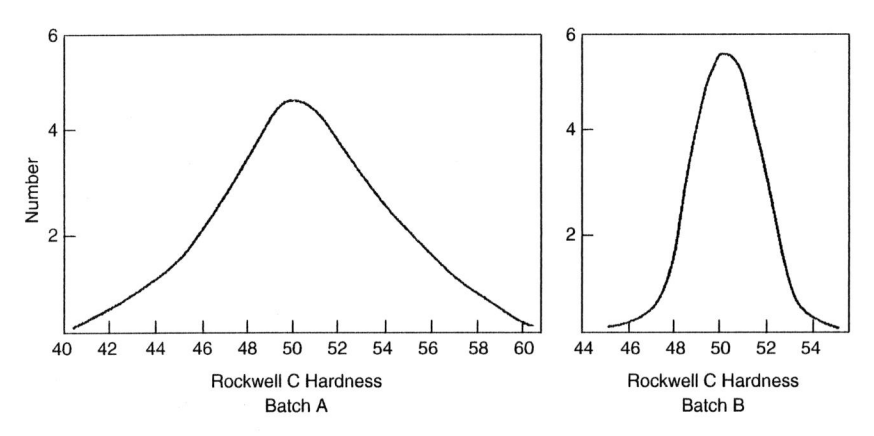

Fig. 3 Distribution curves for two sets of hypothetical data. See text and Table 1. Source: Ref 2

Table 1 Hypothetical hardness values for two batches of parts

	Batch A				Batch B		
Test No.	(x_i), HRC	$x_i - \bar{x}$	$(x_2 - \bar{x})^2$	Test No.	(x_i), HRC	$x_i - \bar{x}$	$(x_2 - \bar{x})^2$
1	46	4.3	18.49	1	52	1.7	2.89
2	50	0.3	0.09	2	50	0.3	0.09
3	53	2.7	7.29	3	51	0.7	0.49
4	48	2.3	5.29	4	49	1.3	1.69
5	42	8.3	68.89	5	51	0.7	0.49
6	50	0.3	0.09	6	53	2.7	7.29
7	56	5.7	32.49	7	50	0.3	0.09
8	52	1.7	2.89	8	50	0.3	0.09
9	52	1.7	2.89	9	51	0.7	0.49
10	50	0.3	0.09	10	49	1.3	1.69
11	58	7.7	59.29	11	52	1.7	2.89
12	52	1.7	2.89	12	50	0.3	0.09
13	46	4.3	18.49	13	49	1.3	1.69
14	50	0.3	0.09	14	48	2.3	5.29
15	48	2.3	5.29	15	50	0.3	0.09
16	53	2.7	7.29	16	51	0.7	0.49
17	50	0.3	0.09	17	49	1.3	1.69
18	48	2.3	5.29	18	52	1.7	2.89
19	56	5.7	32.49	19	49	1.3	1.69
20	52	1.7	2.89	20	50	0.3	0.09

Batch A:
$\Sigma x_i = 1006$
$\bar{x} = \dfrac{\Sigma x}{n} = 50.3$
$\Sigma(x_i - \bar{x})^2 = 272.75$
$\sigma = \sqrt{\dfrac{272.75}{19}} = 3.79$
Range = 58 − 42 = 16

Batch B:
$\Sigma x_i = 1006$
$\bar{x} = \dfrac{\Sigma x_i}{n} = 50.3$
$\Sigma(x_i - \bar{x})^2 = 32.2$
$\sigma = \sqrt{\dfrac{32.2}{19}} = 1.30$
Range = 53 − 48 = 5

Source: Ref 2

studies are conducted on many other types of manufacturing processes to determine the statistical variation of a product with respect to a measured characteristic. For heat treating processes, characteristics frequently measured are hardness and case depth. Because these metallurgical characteristics are sometimes difficult to define, specifications may initially need to be clarified with regard to the exact test scales or test methods to be used and the critical locations where these tests are to be made before a capability study is conducted. Process results for many metallurgical and heat treating processes are dependent on material-related characteristics such as hardenability, material chemistry, and/or part geometry that also make the process test results sensitive to those variables.

After the metallurgical requirements are clearly established, a basic process capability study may be conducted. A minimum of 31 samples must be tested for each variable. Care should be taken so that the parts tested are from the loading locations representing the extremes in process variability. A good guideline for test sample locations is to use the nine loading locations prescribed for temperature uniformity surveys—each corner and the center of a rectangular furnace load.

For continuous processes, it is important to collect the samples over a sufficiently long period of time in order to reflect process fluctuations or other process abnormalities that could be time dependent.

The use of frequency distribution and normal probability paper or a suitable computer program for data representation and plotting is highly recommended. If the data do not plot as a straight line indicating a normal distribution, a metallurgical or process-related reason for this "skewness" should be apparent or be determined. An example of a capability study of an atmosphere harden and temper operation for automotive seat belt parts made from SAE 4037 steel is shown in Fig. 4 and 5.

As can be seen in Fig. 6, the overall process capabilities results may be the result of many contributing factors:

- *Base materials contributions:* Unique materials characteristics, materials defects, and hardenability differences. These can vary from lot to lot and also between materials.
- *Part-related contributions:* Part geometry and section size variations
- *Process-related contributions:* Temperature uniformity as affected by process control and mass effects, time control, atmosphere control, and cooling method (as determined by uniformity and average severity)
- *Evaluation method contribution:* Standards accuracy and testing method accuracy

Thus, to successfully use the process capability study as a dynamic tool to reduce and narrow process variability, the following three steps should be used in conjunction with process capability studies:

Step No. 1:

- Identify critical control variables and their relative contributions to process attribute variations (this can be done by process modeling techniques).
- Measure process inputs with corresponding process output results.
- Document process control procedures.

Step No. 2:

- Modify control procedures, manufacturing procedures, or equipment in order to reduce process variability.

Step No. 3:

- Remeasure process capability (as in step 1) to ascertain the effectiveness of the changes.

MIDLAND METAL TREATING, INC.
VARIABLE DISTRIBUTION ANALYSIS

CUSTOMER	ABC COMPANY	INSPECTOR: J.D.	DATE: 7/15/90	ACTUAL RANGE
W.O.	SAMPLE SIZE: 25	SPECIFICATION: ROCKWELL "C"	32 - 39	34 -37.5
P.O.	AQL %:	PART NUMBER: 1234567		MEAN 35.7

ROCKWELL "C"	FREQUENCY 1	2	3	4	5	6	7	8	9	10	11	12	13	14	15	16	17	18	19	20	21	22	f	ACCUM. f	ACCUM %
32																									
32.5																									
33																									
33.5																									
34	X																						1	1	4
34.5	X	X																					2	3	12
35	X	X	X	X																			4	7	28
35.5	X	X																					2	9	36
36	X	X	X	X	X	X	X	X															8	17	68
36.5	X																						1	18	72
37	X	X	X	X	X																		5	23	88
37.5	X	X																					2	25	100
38																									
38.5																									
39																									

Fig. 4 Final hardness distribution analysis for a typical quench and temper operation. Source: Ref 1

Choosing SPC or SQC

It is necessary to make a distinction between SPC and its relative, SQC. Because SQC is an after-the-fact tool, its best use is in the control of

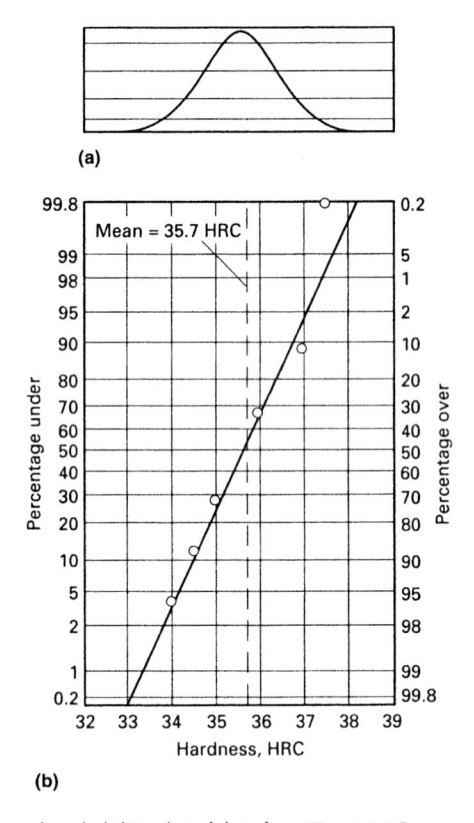

Fig. 5 Normal probability plot of data from Fig. 4. (a) Frequency distribution. (b) Distribution analysis sheet. Specification, mean = 35 HRC; range = 7 HRC. Results, mean = 35.7 HRC, 6σ = 5.5 HRC. Action, adjust temper to adjust mean to 35 HRC. Source: Ref 1

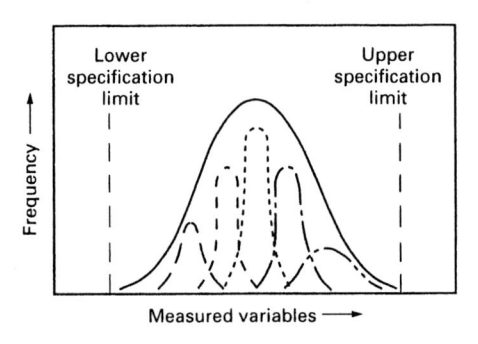

Fig. 6 Factors contributing to overall heat treating process result variations. Source: Ref 1

continuous processing where trends can be noticed and corrected before significant damage occurs.

In batch processing, however, SQC is of little value in preventing problems because at least one entire load of parts will be adversely affected before a problem can even be noticed. Even if the problem is caught after one load, the proposed solution cannot be tested without committing yet another load. Statistical quality control can be very helpful in batch or short run type (setup dominated) processes by using it to analyze setup variables. If the process is then set up to optimal setup parameters (as determined by experimentation or evaluation of part outputs), meeting parts specifications will necessarily result.

The idea behind true SPC is that the results of a process can be guaranteed if none of the relevant process parameters are allowed to stray outside of previously established control limits.

The longstanding problem in applying SPC to heat treatment has been finding methods to quantify and measure process parameters that are of known importance (outside of the obvious ones). Many SPC programs are based on charting controlled parameters such as temperature, atmosphere carbon potential, quenchant temperature, and so forth. While this approach is certainly not incorrect, it does often lead to a situation where a deviation in an SQC chart (output results) commonly cannot be attributed to any special cause deviation in a corresponding SPC chart (processing parameters) because all the things being charted are controlled variables that by design will not normally change.

With induction and flame heat treating, parts are typically processed one at a time. Using part evaluation techniques (SQC) to predict negative results becomes difficult and impractical. Thus, the focus must shift to SPC and the identification, monitoring, and controlling of the process variables to ensure repeatability of the results. Electric power, flame temperature, scan speed, coil dimension, part positioning, and quenchant pressure and temperature are some variables that need to be considered.

A fact of life in any heat treating process is that the equipment gradually succumbs to the wear and tear of constant operation, thus the process inevitably gets worse with time. The challenge is to counter this natural deterioration with corrective action before out-of-specification parts are produced. Statistical process control techniques can be used to measure furnace performance and address process deterioration in heat treating. By monitoring key process variables and/or key process outputs, preferably in online fashion, trends can be spotted and action taken before nonconforming product is produced.

Other Quality Assurance Tools

In-Process Quality Control. In-process inspection and process audits are important to ensure all systems are operating as designed and the material being processed will meet final specifications. The most important aspect of in-process control is the monitoring of system and equipment performance. Tests need to be made on the work being processed as well as critical parameters of the equipment used in processing.

A process control audit form used to verify heat treating systems and furnace control systems is shown in Fig. 7. Verifying the proper operation of major systems such as the cooling water system, generator operation, air compressors, nitrogen purge system, and ammonia system that were each critical to the operation of the furnace equipment was mandatory.

In-process tests should be selected to produce critical information, but this information must be rapidly and economically analyzed to allow for rapid response to problems that may be identified. For these reasons, hardness checks, ultrasonic or eddy current scanning, and dimensional checks are widely used. A sampling procedure should be established according to sound SQC principles. This will give confidence that all steps in the process were followed properly and consistent quality assured.

A control chart is a frequency distribution of observed values plotted as a function of time. This enables any unusual occurrence to be related to the time it occurred. The chart has upper and lower limits established, which may be specific values of the property being measured, or may be some function of the standard deviation.

Each data point on the control chart is the average of a small number of samples taken at a particular time (typically five). These averages are used rather than single values because averages are more normally distributed than single values. Two kinds of control charts may be used: an x-bar chart for averages and an R chart for ranges.

Figure 8(a) shows an x-bar chart indicating a process well under control. Figure 8(b) shows a systematic trend indicating a problem developing that needs attention. It may be a gradual increase in the temperature of the quenching bath, or some other factor identified with the property being monitored. In these charts, limits have been set in terms of standard deviation. Figure 8(c) shows a sudden out-of-control point for which corrective action was taken and control restored.

Final Inspection

Final inspection is to ensure that the specified properties have been met. If proper attention has been paid to the control of incoming material and equipment and processing variables, there should be no surprises at this stage. Final inspection may involve much more extensive tests than those used during in-process testing (physical, mechanical, or destructive). Generally, the customer determines what properties should be evaluated and what final tests should be carried out.

Hardness generally can be verified by the use of SQC through the use of statistical sampling techniques. Table 2 shows a typical sampling plan

TUESDAY / FRIDAY PROCESS CONTROL AUDIT

SYSTEM CHECKS	DATE	TIME		BY			REVIEWED BY					PROPANE		
												Full	Empties	
COOLING WATER	E Tower	C Tower	Supply	Fce Sys	Ind Sys	STATUS	AIR		NITROGEN			Tank	Tank	System	STATUS
F	psi	psi	psi	psi	psi							psi		psi	
(66-87)	(15-20)	(15-20)	(40-65)	(40-65)	(40-60)							(100-160)	(100 Min.)	(70-76)	
AMMONIA SYSTEM	Tank	Tank	Line	System	STATUS		Dew Pt.		#1 Comp	#2 Comp		Comp	Comp	System	STATUS
	psi	%	psi	psi								psi	psi	psi	
	(100-120)	(40 Min.)	(40-50)	(25-45)								(105-115)	(105-115)	(110-115)	
ATM. GENERATORS	#1 Gen	#2 Gen	#3 Gen	Out-cfh	Air-cfh	Dew Pt.	Temp.		Filters			STATUS			
	oz	oz	oz	cfh	cfh	%	F								
	(8-12)	(8-12)	(8-12)	(76-160)	(76-160)	(53-68)	(1940-1960)								
Primary Furnaces	1 Pac	2 Pac	3 Pac	4 Pac	5 Pac	Lind	Holc		Pusher			Bert		John	5X8
X - If Down															
1. Temperature Systems															
2. Heat Recovery Time															
3. Charts/Computer															
4. Carbon Control												n/a	n/a	n/a	n/a
5. Probe Air Flow												n/a	n/a	n/a	n/a
6. Quenchant Temp.												n/a	n/a	n/a	n/a
7. Gas Flows												n/a	n/a	n/a	n/a
8. Flowmeters												n/a	n/a	n/a	n/a
9. Nitrogen Purge "ON"															
Tempering Furnaces	2 Pac	3 Pac	4 Pac	5 Pac	6 Lind.	5 X 5	4 X 4		3 X 3			2Z Ipsen	1 Holc	2 Holc
X - If Down															
10. Temperature Systems															
11. Heat Recovery Time															
12. Charts															
13. Cooling Water						n/a	n/a		n/a			n/a	n/a	n/a
Other Equipment	Pc. Wash	Hc. Wash	Psh Wash	Psh Draw	1 Belt	2 Belt	50 RF		60 RF			100 KW	150 KW	150 RF
X - If Down															
14. Temperature Systems															
15. Concentration				n/a		n/a	n/a		n/a			n/a	n/a	n/a
15. Charts	n/a	n/a	n/a												
16. Quenchant Temp.	n/a	n/a	n/a	n/a											
(Circle Problem Areas)															

Fig. 7 Process audit form for heat operation. Source: J.L. Dossett, unpublished work

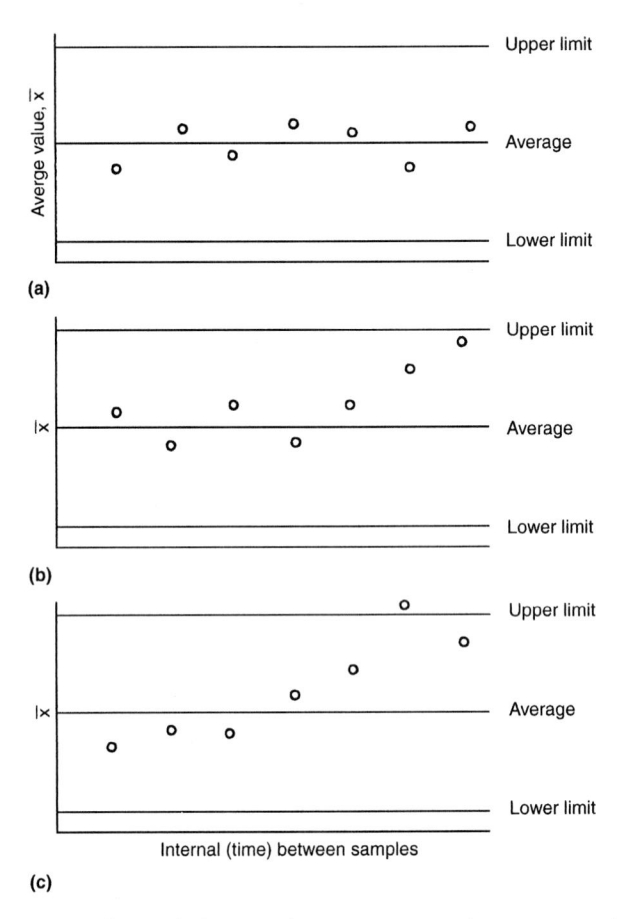

Fig. 8 Use of control charts. Each point represents the average results of a specific number of tests (usually five or ten) taken after given time intervals. Sample values may be plotted as averages as here ($\bar{x}$ chart) or against the value of the range (R chart), or the difference between the sample average and the overall average. Source: Ref 2

Table 2 Increased final inspection for hardness

Batch size	Sample size	Reject if # defects ≥
2–8	5	1
9–15	5	1
16–25	5	1
26–50	5	1
51–90	7	1
91–150	11	1
151–280	13	1
281–500	16	1
501–1200	19	1
1201–3200	23	1
3201–10000	29	1
10001–35000	35	1

Source: Mil Std 105-D (Single Sampling Plan Normal AQL 2.5%, C = 0)

to ensure an average quality level of 2.5% when checking to the specified range and allowing no readings to be outside the range ($C = 0$).

It is important that good records be maintained during processing so that operations are well documented and the sources of any problems can be identified. Of course, in-process inspections should make these situations occur rarely. So long as equipment has been functioning properly, inspection techniques are reliable, and important variables are measured, meeting high-quality standards should not be a problem.

REFERENCES

1. J.L. Dossett, G.M. Baker, T.D. Brown, and D.W. McCurdy, Statistical Process Control of Heat-Treating Operations, *Heat Treating,* Vol 4, *ASM Handbook,* ASM International, 1991, p 620–637
2. R.W. Bohl, Testing and Quality Control of Metals and Alloys, *Practical Heat Treating,* Course 42, Lesson 13, Materials Engineering Institute, ASM International, 1995
3. W.E. Dowling and N. Palle, Design for Heat Treatment, *Materials Selection and Design,* Vol 20, *ASM Handbook,* ASM International, 1997, p 774–780

APPENDIX A

Glossary of Heat Treating Terms

THE MOST COMMONLY USED heat treating terms are defined and/or described below. For a complete listing of metallurgical terms, see the "Glossary of Metallurgical and Metalworking Terms" in the *Metals Handbook Desk Edition,* 2nd ed., ASM International, 1998, p 3 to 63.

A

Ac$_{cm}$, Ac$_1$, Ac$_3$, Ac$_4$. Defined under *transformation temperature.*

Ac$_{cm}$, A$_1$, A$_3$, A$_4$. Same as Ae$_{cm}$, Ae$_1$, Ae$_3$, and Ae$_4$.

acicular ferrite. A highly substructured nonequiaxed ferrite that forms upon continuous cooling by a mixed diffusion and shear mode of transformation that begins at a temperature slightly higher than the temperature transformation range for upper bainite. It is distinguished from bainite in that it has a limited amount of carbon available; thus, there is only a small amount of carbide present.

Ae$_{cm}$, Ae$_1$, Ae$_3$, Ae$_4$. Defined under *transformation temperature.*

age hardening. Hardening by aging, usually after rapid cooling or cold working. See *aging.*

aging. A change in the properties of certain metals and alloys that occurs at ambient or moderately elevated temperatures after hot working or a heat treatment (quench aging in ferrous alloys, natural or artificial aging in ferrous and nonferrous alloys) or after a cold working operation (strain aging). The change in properties is often, but not always, due to a phase change (precipitation), but never involves a change in chemical composition of the metal or alloy. See also *age hardening, artificial aging, natural aging, precipitation hardening,* and *precipitation heat treatment.*

allotropy. A near synonym for *polymorphism.* Allotropy is generally restricted to describing polymorphic behavior in elements, terminal phases, and alloys whose behavior closely parallels that of the predominant constituent element.

alloy steel. Steel containing specified quantities of alloying elements (other than carbon and the commonly accepted amounts of manganese, copper, silicon, sulfur, and phosphorous) within the limits recognized for constructional alloy steels, added to effect changes in mechanical or physical properties.

artificial aging. Aging above room temperature. Compare with *natural aging*. See also *aging*.

austempering. A heat treatment for ferrous alloys in which a part is quenched from the austenitizing temperature at a rate fast enough to avoid formation of ferrite or pearlite and then held at a temperature just above M_s until transformation to bainite is complete. Although designated as bainite in both austempered steel and austempered ductile iron (ADI), austempered steel consists of two phase mixtures containing ferrite and carbide, while austempered ductile iron consists of two phase mixtures containing ferrite and austenite.

austenite. A solid solution of one or more elements in face-centered cubic iron. Unless otherwise designated (such as nickel austenite), the solute is generally assumed to be carbon.

austenitic steel. An alloy steel whose structure is normally austenitic at room temperature.

B

bainite. A metastable aggregate of ferrite and cementite resulting from the transformation of austenite at temperatures below the pearlite range but above M_s. Its appearance is feathery if formed in the upper part of the bainite transformation range; acicular.

banded structure. A segregated structure consisting of alternating nearly parallel bands of different composition, typically aligned in the direction of primary hot working.

bright annealing. Annealing in a protective medium to prevent discoloration of the bright surface.

Brinell hardness test. A test for determining the hardness of a material by forcing a hard steel or carbide, 10 mm, or 0.4 in., in diameter, into it under a specified load. The result is expressed as the Brinell hardness number, which is the value obtained by dividing the applied load in kilograms by the surface area of the resulting impression in square millimeters.

C

carbon equivalent. For cast iron, an empirical relationship of the total carbon, silicon, and phosphorus contents expressed by the formula:

$$CE = TC + 1/3(Si + P)$$

carbonitriding. A case-hardening process in which a suitable ferrous material is heated above the lower transformation temperature in a gaseous atmosphere of such composition as to cause simultaneous absorption of carbon and nitrogen by the surface and by diffusion create a concentration gradient. The heat treating process is completed by cooling at a rate that produces the desired properties in the workprice.

carbonization. Conversion of an organic substance into elemental carbon. (Should not be confused with carburization.)

carbon potential. A measure of the ability of an environment containing active carbon to alter or maintain, under prescribed conditions, the carbon level of the steel. Note: In any particular environment, the carbon level attained will depend on such factors as temperature, time, and steel composition.

carbon restoration. Replacing the carbon lost in the surface layer from previous processing by carburizing this layer to substantially the original carbon level. Sometimes called recarburizing.

carbon steel. Steel having no specified minimum quantity for any alloying element (other than the commonly accepted amounts of manganese, silicon, and copper) and that contains only an incidental amount of any element other than carbon, silicon, manganese, copper, sulfur, and phosphorus.

carburizing. Absorption and diffusion of carbon into solid ferrous alloys by heating to a temperature usually above Ac_3 in contact with a suitable carbonaceous material. A form of case hardening that produces a carbon gradient extending inward from the surface enabling the surface layer to be hardened either by quenching directly from the carburizing temperature or by cooling to room temperature then reaustenitizing and quenching.

case. That portion of a ferrous alloy, extending inward from the surface, whose composition has been altered so that it can be case hardened. Typically considered to be the portion of the alloy (a) whose composition has been measurably altered from the original composition, (b) that appears dark on an etched cross section, or (c) that has a hardness, after hardening, equal to or greater than a specified value. Contrast with *core*.

case hardening. A generic term covering several processes applicable to steel that change the chemical composition of the surface by absorption of carbon or nitrogen, or both, and by diffusion create a concentration gradient. See *carburizing, carbonitriding, nitriding,* and *nitrocarburizing*.

cast iron. A generic term for a large family of cast ferrous alloys in which the carbon content exceeds the solubility of carbon in austenite at the eutectic temperature. Most cast irons contain at least 2% C plus silicon and sulfur, and may or may not contain other alloying elements. For the various forms *gray cast iron, white cast iron, malleable cast iron,* and *ductile cast iron,* the word "cast" is often left out, resulting in "gray iron," "white iron," "malleable iron," and "ductile iron," respectively.

cementite. A compound of iron and carbon, known chemically as iron carbide and having the approximate chemical formula Fe_3C. It is characterized by an orthorhombic crystal structure. When it occurs as a phase in steel, the chemical composition will be altered by the presence of manganese and other carbide-forming elements.

cold treatment. Exposing to suitable subzero temperatures for the purpose of obtaining desired conditions or properties such as dimensional or microstructural stability. When the treatment involves the transformation of retained austenite, it is usually followed by tempering. May be referred to as cryogenic treatment.

combined carbon. The part of the total carbon in steel or cast iron that is present as other than *free carbon, graphitic carbon,* or *temper carbon.*

constitution diagram. A graphical representation of the temperature and composition limits or phase fields in an alloy system as they actually exist under the specific conditions of heating or cooling (synonymous with phase diagram). A constitution diagram may be an equilibrium diagram, an approximation to an equilibrium diagram, or a representation of metastable conditions or phases. Compare with *equilibrium diagram.*

core. In a ferrous alloy prepared for case hardening that portion of the alloy that is not part of the case. Typically considered to be the portion that (a) appears dark (with certain etchants) on an etched cross section. (b) has an essentially unaltered chemical composition or (c) has a hardness after hardening less than a specified value.

critical point. (1) The temperature or pressure at which a change in crystal structure, phase, or physical properties occurs. Also termed *transformation temperature.* (2) In an equilibrium diagram that combination of composition temperature and pressure at which the phases of an inhomogeneous system in equilibrium

critical temperature. (1) Synonymous with *critical point* if the pressure is constant. (2) The temperature above which the vapor phase cannot be condensed to liquid by an increase in pressure.

critical temperature ranges. Synonymous with *transformation ranges,* which is the preferred term.

cryogenic treatment. See *cold treatment.*

D

dead soft. A temper of nonferrous alloys and some ferrous alloys corresponding to the condition of minimum hardness and tensile strength produced by full annealing.

decarburization. Loss of carbon from the surface layer of a carbon-containing alloy due to reaction with one or more chemical substances in a medium that contacts the surface.

direct quenching. (1) Quenching carburized parts directly from the carburizing operation. (2) Also used for quenching pearlitic malleable parts directly from the malleablizing operation.

distortion. An irreversible change in size or shape (warpage) of a component that occurs during heat treat processing. The dimensional change and/or warpage can be the result of one or more of the following: changes in metallurgical structure (phases present), nonuniform heating or cooling (quenching), or prior residual stresses.

double aging. Employment of two different aging treatments to control the type of precipitate formed from a supersaturated matrix in order to obtain the desired properties. The first aging treatment, sometimes referred to as intermediate or stabilizing, is usually carried out at higher temperature than the second.

ductile cast iron. A *cast iron* that has been treated while molten with an element such as magnesium or cerium to induce the formation of free graphite as nodules or spherulites, which imparts a measurable degree of ductility to the cast metal. Also known as nodular cast iron, spherulitic graphite cast iron, and SG iron.

ductility. The ability of a material to deform plastically without fracturing.

E

electron-beam heat treating. A selective surface hardening process that rapidly heats a surface by direct bombardment with an accelerated stream of electrons.

end-quench hardenability test. A laboratory procedure for determining the hardenability of a steel or other ferrous alloy; widely referred to as the Jominy test. Hardenability is determined by heating a standard specimen above the upper critical temperature; placing the hot specimen in a fixture so that a stream of cold water impinges on one end; and, after cooling to room temperature is completed, measuring the hardness near the surface of the specimen at regularly spaced intervals along its length. The data are normally plotted as hardness versus distance from the quenched end.

equilibrium. A dynamic condition of physical, chemical, mechanical, or atomic balance, where the condition appears to be one of rest rather than change.

equilibrium diagram. A graphical representation of the temperature, pressure, and composition limits of phase fields in an alloy system as they exist under conditions of complete equilibrium. In metal systems, pressure is usually considered constant.

eutectic. (1) An alloy having the composition indicated by the eutectic point on a phase diagram. (2) An alloy structure of intermixed solid constituents formed by a eutectic reaction.

eutectic carbide. Carbide formed during freezing as one of the mutually insoluble phases participating in the eutectic reaction of ferrous alloys.

eutectic melting. Melting of localized microscopic areas whose composition corresponds to that of the eutectic in the system.

eutectoid. (1) An alloy having the composition indicated by the eutectoid point on a phase diagram. (2) An alloy structure of intermixed solid constituents formed by a eutectoid reaction.

F

fatigue strength. The maximum stress that can be sustained for a specified number of cycles without failure, the stress being completely reversed within each cycle unless otherwise stated.

ferrite. A solid solution of one or more elements in body-centered cubic iron. Unless otherwise designated (for instance, as chromium ferrite), the solute is generally assumed to be carbon. On some equilibrium diagrams, there are two ferrite regions separated by an austenite area. The lower area is alpha ferrite; the upper, delta ferrite. If there is no designation, alpha ferrite is assumed.

ferritizing anneal. A treatment given as-cast gray or ductile (nodular) iron to produce an essentially ferritic matrix. For the term to be meaningful, the final microstructure desired or the time-temperature cycle used must be specified.

fixturing. See *racking*

flame annealing. Annealing in which the heat is applied directly by a flame.

flame hardening. A process for hardening the surfaces of hardenable ferrous alloys in which an intense flame is used to heat the surface layers above the upper transformation temperature, whereupon the workpiece is immediately quenched.

flame straightening. Correcting distortion in metal structures by localized heating with a gas flame.

fluidized-bed heating. Heating carried out in a medium of solid particles suspended in a flow of gas.

fog quenching. Quenching in a fine vapor or mist.

free carbon. The part of the total carbon in steel or cast iron that is present in elemental form as graphite or temper carbon. Contrast with *combined carbon*.

free ferrite. Ferrite that is formed directly form the decomposition of hypoeutectoid austenite during cooling, without the simultaneous formation of cementite. Also, proeutectoid ferrite.

full hard. A *temper* of nonferrous alloys and some ferrous alloys corresponding approximately to a cold-worked state beyond which the material can no longer be formed by bending. In specifications, a full hard temper is commonly defined in terms of minimum hardness or minimum tensile strength (or, alternatively, a range of hardness or strength) corresponding to a specific percentage of cold reduction following a full anneal. For aluminum, a full hard temper is equivalent to a reduction of 75% from *dead soft*; for austenitic stainless steels, a reduction of about 50 to 55%.

G

gamma iron. The face-centered cubic form of pure iron, stable from 910 to 1400 °C (1670 to 2550 °F).

grain size. For metals, a measure of the areas or volumes of grains in a polycrystalline material, usually expressed as an average when the individual sizes are fairly uniform. In metals containing two or more phases, the grain size refers to that of the matrix unless otherwise specified. Grain sizes are reported in terms of number of grains per unit area or volume, average diameter, or as a grain-size number derived from area measurements.

graphitic carbon. Free carbon in steel or cast iron

graphitic steel. Alloy steel made so that part of the carbon is present as graphite.

graphitization. Formation of graphite in iron or steel. Where graphite is formed during solidification, the phenomenon is called primary graphitization; where formed later by heat treatment, secondary graphitization.

graphitizing. Annealing a ferrous alloy in such a way that some or all of the carbon is precipitated as graphite.

gray cast iron. A *cast iron* that gives a gray fracture due to the presence of flake graphite. Often called gray iron.

Grossmann chart. A chart describing the ability of a quenching medium to extract heat from a hot steel workpiece in comparison to still water.

H

half hard. A *temper* of nonferrous alloys and some ferrous alloys characterized by tensile strength about midway between that of *dead soft* and *full hard* tempers.

hardenability. The relative ability of a ferrous alloy to form martensite when quenched from a temperature above the upper critical temperature. Hardenablity is commonly measured as the distance below a quenched surface where the metal exhibits a specific hardness (50 HRC, for example) or a specific percentage of martensite in the microstructure.

hardening. Increasing hardness by suitable treatment, usually involving heating and cooling. When applicable, the following more specific terms should be used: *age hardening, flame hardening, induction hardening, precipitation hardening,* and *quench hardening*.

hardness. Resistance of metal to plastic deformation, usually by indentation. However, the term may also refer to stiffness or temper, or to resistance to scratching, abrasion, or cutting. Indentation hardness may be measured by various hardness tests, such as *Brinell, Knoop, Rockwell,* and *Vickers.*

hardness profile. Hardness as a function of distance from a fixed reference point (usually form the surface).

homogeneous carburizing. Use of a carburizing process to convert a low-carbon ferrous alloy to one of uniform and higher carbon content throughout the section.

homogenizing. Holding at high temperature to eliminate or decrease chemical segregation by diffusion.

hypereutectic alloy. In an alloy system exhibiting a *eutectic,* any alloy whose composition has an excess of alloying element compared with the eutectic composition, and whose equilibrium microstructure contains some eutectic structure.

hypereutectoid alloy. In an alloy system exhibiting a *eutectoid,* any alloy whose composition has an excess of alloying element compared with the eutectoid composition, and whose equilibrium microstructure contains some eutectoid structure.

hypoeutectic alloy. In an alloy system exhibiting a *eutectic,* any alloy whose composition has an excess of alloying element compared with the eutectic composition, and whose equilibrium microstructure contains some eutectic structure.

hypoeutectoid alloy. In an alloy system exhibiting a *eutectoid,* any alloy whose composition has an excess of alloying element compared with the eutectoid composition, and whose equilibrium microstructure contains some eutectoid structure.

I

ideal critical diameter (D_I). The diameter of a bar (in inches), which, if heated to above the upper critical temperature (A_{c3}) and quenched at an infinite cooling rate ($H = \infty$, where, H is the quench severity factor), the center of the bar would quench to a structure of 50% martensite.

induction hardening. A surface-hardening process in which only the surface layer of a suitable ferrous workpiece is heated by electromagnetic induction to above the upper critical temperature and immediately quenched.

induction heating. Heating by combined electrical resistance and hysteresis losses induced by subjecting a metal to the varying magnetic field surrounding a coil carrying alternating current.

induction tempering. Tempering of steel using low-frequency electrical *induction heating.*

intensive quenching. Quenching in which the quenching medium is cooling the part at a rate at least two and a half times faster than still water. See *Grossmann chart.*

interrupted quenching. A quenching procedure in which the workpiece is removed from the first quench at a temperature substantially higher than that of the quenchant and is then subjected to a second quenching system having a different cooling rate.

interstitial solid solution. A solid solution in which the solute atoms occupy positions that do not correspond to lattice points of the solvent. Contrast with *substitutional solid solution.*

ion carburizing. A method of surface hardening in which carbon ions are diffused into a workpiece in a vacuum through the use of high-voltage electrical energy. Synonymous with plasma carburizing or glow-discharge carburizing.

ion nitriding. A method of surface hardening in which nitrogen ions are diffused into a workpiece in a vacuum through the use of high-voltage electrical energy. Synonymous with plasma nitriding or glow-discharge nitriding.

isothermal annealing. Austenitizing a ferrous alloy and then cooling to and holding at a temperature at which austenite transforms to a relatively soft ferrite carbide aggregate.

isothermal transformation. A change in phase that takes place at a constant temperature. The time required for transformation to be completed, and in some instances the time delay before transformation begins, depends on the amount of supercooling below (or superheating above) the equilibrium temperature for the same transformation.

K

Knoop hardness. Microhardness determined from the resistance of metal to indentation by a pyramidal diamond indenter, having edge angles of 172° 30′ and 130°, making a rhombohedral impression with one long and one short diagonal.

M

maraging. A precipitation-hardening treatment applied to a special group of iron-base alloys to precipitate one or more intermetallic compounds in a matrix of essentially carbon-free martensite. Note: the first developed series of maraging steels contained, in addition to iron, more than 10% Ni and one or more supplemental hardening elements. In this series, aging is done at 480 °C (900 °F).

malleable cast iron. A cast iron made by a prolonged anneal of *white cast iron* in which decarburization or graphitization, or both, take place to eliminate some or all of the cementite. The graphite is in the form of temper carbon. If decarburization is the predominant reaction, the product will have a light fracture, hence, "whiteheart malleable;" otherwise, the fracture will be dark, hence, "blackheart malleable." Ferritic malleable has a predominantly ferritic matrix; pearlitic malleable may contain pearlite, spheroidite, or tempered martensite depending on heat treatment and desired hardness.

malleablizing. Annealing *white cast iron* in such a way that some or all of the combined carbon is transformed to graphite or, in some instances, part of the carbon is removed completely.

martempering. (1) A hardening procedure in which an austenitized ferrous workpeice is quenched into an appropriate medium whose temperature is maintained substantially at the M_s of the workpeice, held in the medium until its temperature is uniform throughout—but not long enough to permit bainite to form—and then cooled in air. The treatment is frequently followed by tempering. (2) When the process is applied to carburized material, the controlling M_s temperature is that of the case. This variation of the process is frequently called marquenching.

martensite. (1) In an alloy, a metastable transitional structure intermediate between two allotropic modifications whose abilities to dissolve a given solute differ considerably, the high-temperature phase having the greater solubility. The amount of the high-temperature phase transformed to martensite depends to a large extend on the temperature attained in cooling, there being a rather distinct beginning temperature. (2) A metastable phase of steel, formed by a transformation of austenite below the M_s (or Ar'') temperature. It is an interstitial supersaturated solid solution of carbon in iron having a body-centered tetragonal lattice. Its microstructure is characterized by an acicular, or needlelike, pattern.

martensite range. The temperature interval between M_s and M_f.

martensitic transformation. A reaction that takes place in some metals on cooling, with the formation of an acicular structure called *martensite*.

McQuaid-Ehn test. A test to reveal grain size after heating into the austenitic temperature range. Eight standard McQuaid-Ehn grain sizes rate the structure, No. 8 being finest, No. 1 coarsest.

metastable. Refers to a state of pseudoequilibrium that has a higher free energy than the true equilibrium state.

M_f temperature. For any alloy system, the temperature at which martensite formation on cooling is essentially finished. See *transformation temperature* for the definition applicable to ferrous alloys.

microhardness. The hardness of a material as determined by forcing an indenter such as a Vickers or Knoop indenter into the surface of a material under very light load; usually, the indentations are so small they must be measured with a microscope. Capable of determining hardnesses of different microconstituents within a structure, or of measuring steep hardness gradients such as those encountered in case hardening.

microsegregation. Segregation within a grain, crystal, or small particle.

M_s temperature. For any alloy system, the temperature at which martensite starts to form on cooling. See *transformation temperature* for the definition applicable to ferrous alloys.

N

natural aging. Spontaneous aging of a supersaturated solid solution at room temperature. See *aging,* and compare with *artificial aging* (aging above room temperature).

nitriding. Introducing nitrogen into the surface layer of a solid ferrous alloy by holding at a suitable temperature (below Ac_1 for ferritic steels) in contact with a nitrogenous material usually ammonia or molten cyanide of appropriate composition. Quenching is not required to produce a hard case.

nitrocarburizing. Any of several processes in which both nitrogen and carbon are absorbed into the surface layers of a ferrous material at temperatures below the lower critical temperature and, by diffusion, create a concentration gradient Nitrocarburizing is performed primarily to provide an antiscuffing surface layer and to improve fatigue resistance. Compare with *carbonitriding.*

O

overaging. Aging under conditions of time and temperature greater than those required to obtain maximum change in a certain property so that the property is altered in the direction of the initial value. See *aging.*

overheating. Heating a metal or alloy to such a high temperature that its properties are impaired. When the original properties cannot be restored by further heat treating, by mechanical working, or by a combination of working and heat treating, the overheating is known as burning.

oxygen probe. An atmosphere-monitoring device that electronically measures the difference between the partial pressure of oxygen in a furnace or furnace supply atmosphere and the external air.

P

pack carburizing. A method of surface hardening of steel in which parts are packed in a steel box with the carburizing compound and heated to elevated temperatures.

pack nitriding. A method of surface hardening of steel in which parts are packed in a steel box with the nitriding compound and heated to elevated temperatures.

partial annealing. An imprecise term used to denote a treatment given cold-worked material to reduce the strength to a controlled level or to effect stress relief. To be meaningful, the type of material, the degree of cold work, and the time-temperature schedule must be stated.

pearlite. A metastable lamellar aggregate of ferrite and cementite resulting from the transformation of austenite at temperatures above the bainite range.

phase. A physically homogeneous and distinct portion of a materials system.

phase diagram. A graphical representation of the temperature and composition limits of phase fields in an alloy system as they actually exist under the specific conditions of heating or cooling (synonymous with constitution diagram). A phase diagram may be an equilibrium diagram, an approximation to an equilibrium diagram, or a representation of metastable conditions or phases. Compare with *equilibrium diagram.*

physical metallurgy. The science and technology dealing with the properties of metals and alloys, and of the effects of composition, processing, and environment on those properties.

plasma carburizing. Same as *ion carburizing.*

plasma nitriding. Same as *ion nitriding.*

polymorphism. A general term for the ability of a solid to exist in more than one form. In metals alloys and similar substances, this usually means the ability to exist in two or more crystal structures, or in an amorphous state and at least one crystal structure See also *allotropy.*

precipitation hardening. Hardening caused by the precipitation of a constituent from a supersaturated solid solution. See also *age hardening* and *aging.*

precipitation heat treatment. *Artificial aging* (aging above room temperature) in which a constituent precipitates from a supersaturated solid solution.

preheating. Heating before some further thermal or mechanical treatment. For tool steel, heating to an intermediate temperature immediately before final austenitizing. For some nonferrous alloys, heating to a high temperature for a long time, in order to homogenize the structure before working. In welding and related processes, heating to an intermediate temperature for a short time immediately before welding, brazing, soldering, cutting, or thermal spraying.

process annealing. An imprecise term denoting various treatments used to improve workability. For the term to be meaningful, the condition of the material and the time-temperature cycle used must be stated.

pusher furnace. A type of continuous furnace in which parts to be heated are periodically charged into the furnace in containers, which are pushed along the hearth against a line of previously charged containers thus advancing the containers toward the discharge end of the furnace, where they are removed.

Q

quench annealing. Annealing an austenitic ferrous alloy by *solution heat treatment* followed by rapid quenching.

quench cracking. Fracture of a metal during quenching from elevated temperature. Most frequently observed in hardened carbon steel, alloy steel, or tool steel parts of high hardness and low toughness. Cracks often emanate from fillets, holes, corners, or other stress raisers and result form high stresses due to the volume changes accompanying transformation to martensite.

quench hardening. (1) Hardening suitable alpha-beta alloys (most often certain copper or titanium alloys) by solution treating and quenching to develop a martensite-like structure. (2) In ferrous alloys, hardening by austenitizing and then cooling at a rate such that a substantial amount of austenite transforms to martensite.

R

racking. A term used to describe the placing of parts to be heat treated on a rack or tray. This is done to keep parts in a proper position to avoid heat-related distortions and to keep the parts separated. Also called fixturing.

recrystallization. (1) The formation of a new, strain-free grain structure from that existing in cold-worked metal, usually accomplished by heating. (2) The change from one crystal structure to another, as occurs on heating or cooling through a critical temperature.

recrystallization annealing. Annealing cold-worked metal to produce a new grain structure without phase change.

recrystallization temperature. The approximate minimum temperature at which complete recrystallization of a cold-worked metal occurs within a specified time.

residual stress. An internal stress not depending on external forces resulting from such factors as cold working, phase changes, or temperature gradients.

Rockwell hardness test. An indentation hardness test based on the depth of penetration of a specified penetrator into the specimen under certain arbitrarily fixed conditions.

rotary furnace. A circular furnace constructed so that the hearth and workpieces rotate around the axis of the furnace during heating.

S

Scleroscope test. A hardness test in which the loss in kinetic energy of a falling metal "tup," absorbed by indentation upon impact of the tup on the metal being tested, is indicated by the height of rebound.

selective quenching. Quenching only certain portions of an object.

shell hardening. A surface-hardening process in which a suitable steel workpiece, when heated through and quench hardened, develops a martensitic layer or shell that closely follows the contour of the piece and surrounds a core of essentially pearlitic transformation product. This result is accomplished by a proper balance among section size, steel hardenability, and severity of quench.

sigma phase. A hard, brittle, nonmagnetic intermediate phase with a tetragonal crystal structure, containing 30 atoms per unit cell, space group $P4_2/mnm$, occurring in many binary and ternary alloys of the transition elements. The composition of this phase in the various systems is not the same, and the phase usually exhibits a wide range in homogeneity. Alloying with a third transition element usually enlarges the field of homogeneity and extends it deep into the ternary section.

sigma-phase embrittlement. Embrittlement of iron-chromium alloys (most notably austenitic stainless steels) caused by precipitation at grain boundaries of the hard, brittle intermetallic *sigma phase* during long periods of exposure to temperatures between approximately 565 and 980 °C (1050 and 1800 °F). Sigma-phase embrittlement results in severe loss in *toughness* and *ductility*.

siliconizing. Diffusing silicon into solid metal, usually steel, at an elevated temperature.

sintering. The bonding of adjacent surfaces in a mass of particles by molecular or atomic attraction on heating at high temperatures below the melting temperature of any constituent in the material. Sintering strengthens a powder mass and normally produces densification and, in powdered metals, recrystallization.

slack quenching. The incomplete hardening of steel due to quenching from the austenitizing temperature at a rate slower than the critical cooling rate for the particular steel, resulting in the formation of one or more transformation products in addition to martensite.

snap temper. A precautionary interim stress-relieving treatment applied to high-hardenability steels immediately after quenching to prevent cracking because of delay in tempering them at the prescribed higher temperature.

solution heat treatment. Heating an alloy to a suitable temperature, holding at that temperature long enough to cause one or more constituents to enter into solid solution, and then cooling rapidly enough to hold these constituents in solution.

spheroidizing. Heating and cooling to produce a spheroidal or globular form of carbide in steel. Spheroidizing methods frequently used are (1) prolonged holding at a temperature just below Ae_1; (2) heating and cooling alternately between temperatures that are just above and just below Ae_1; (3) heating to a temperature above Ae_1 or Ae_3 and then cooling very slowly in the furnace or holding at a temperature just below Ae_1.

steel. An iron-base alloy, malleable in some temperature ranges as initially cast, containing manganese, usually carbon, and often other alloying elements. In carbon steel and low-alloy steel, the maximum carbon is about 2.0%; in high-alloy steel, about 2.5%. The dividing line between low-alloy and high-alloy steels is generally regarded as being at about 5% metallic alloying elements.

Steel is to be differentiated from two general classes of irons: the cast irons, on the high-carbon side, and the relatively pure irons such as ingot iron, carbonyl iron, and electrolytic iron, on the low-carbon side. In some steels containing extremely low carbon, the manganese content is the principal differentiating factor, steel usually containing at least 0.25%; ingot iron, considerably less.

substitutional solid solution. A solid solution in which the solvent and solute atoms are located randomly at the atom sites in the crystal structure of the solution.

surface hardening. A generic term covering several processes applicable to a suitable ferrous alloy that produces, by quench hardening only, a surface layer that is harder or more wear resistant than the core. There is no significant alteration of the chemical composition of the surface layer. The processes commonly used are *carbonitriding, carburizing, induction hardening, flame hardening, nitriding,* and *nitrocarburizing.* Use of the applicable specific process name is preferred.

T

temper. (1) In heat treatment, reheating hardened steel or hardened cast iron to some temperature below the eutectoid temperature for the purpose of decreasing hardness and increasing toughness. The process also is sometimes applied to normalized steel. (2) In nonferrous alloys and in some ferrous alloys (steels that cannot be hardened by heat treatment). the hardness and strength produced by mechanical or thermal treatment or both and characterized by a certain structure mechanical properties or reduction in area during cold working.

temper carbon. Clusters of finely divided graphite, such as that found in malleable iron that are formed as a result of decomposition of cementite, for example by heating white cast iron above the ferrite-austenite transformation temperature and holding at these temperatures for a considerable period of time. Also known as annealing carbon.

thermocouple. A device for measuring temperatures, consisting of lengths of two dissimilar metals or alloys that are electrically joined at one end and connected to a voltage-measuring instrument at the other end. When one junction is hotter than the other, a thermal electromotive force is produced that is roughly proportional to the difference in temperature between the hot and cold junctions.

time quenching. A term used to describe a quench in which the cooling rate of the part being quenched must be changed abruptly at some time during the cooling cycle.

total carbon. The sum of the free and combined carbon (including carbon in solution) in a ferrous alloy.

toughness. Ability of a material to absorb energy and deform plastically before fracturing.

transformation ranges. Those ranges of temperature within which austenite forms during heating and transforms during cooling. The two ranges are distinct, sometimes overlapping but never coinciding. The limiting temperatures of the ranges depend on the composition of the alloy and on the rate of change of temperature, particularly during cooling. See also *transformation temperature.*

transformation temperature. The temperature at which a change in phase occurs. The term is sometimes used to denote the limiting temperature of a transformation range. The following symbols are used for iron and steels:

Ac_{cm}. In hypereutectoid steel, the temperature at which the solution of cementite in austenite is completed during heating.

Ac_1. The temperature at which austenite begins to form during heating.

Ac_3. The temperature at which transformation of ferrite to austenite is completed during heating.

Ac_4. The temperature at which austenite transforms to delta ferrite during heating.

Ae_{cm}, Ae_1, Ae_3, Ae_4. The temperatures of phase changes at equilibrium.

Ar_{cm}, In hypereutectoid steel, the temperature at which precipitation of cementite starts during cooling.

Ar_1. The temperature at which transformation of austenite to ferrite or to ferrite plus cementite is completed during cooling.

Ar_3. The temperature at which austenite begins to transform to ferrite during cooling.

Ar_4. The temperature at which delta ferrite transforms to austenite during cooling.

Ar'. The temperature at which transformation of austenite to pearlite starts during cooling.

M_f. The temperature at which transformation of austenite to martensite finishes during cooling.

M_s (or Ar''). The temperature at which transformation of austenite to martensite starts during cooling.

Note: All these changes except the formation of martensite occur at lower temperatures during cooling than during heating and depend on the rate of change of temperature.

V

vacuum carburizing. A high-temperature gas carburizing process using furnace pressures between 7 and 55 kPa during the carburizing portion of the cycle.

vacuum furnace. A furnace using low atmospheric pressures instead of a protective gas atmosphere like most heat treating furnaces. Vacuum furnaces are categorized as hot wall or cold wall, depending on the location of the heating and insulating components.

vacuum nitrocarburizing. A subatmospheric nitrocarburizing process using a basic atmosphere of 50% ammonia/50% methane, containing controlled oxygen additions of up to 2%.

Vickers hardness test. An indentation hardness test employing a 136° diamond pyramid indenter (Vickers) and variable loads enabling the use of one harness scale for all ranges of hardness from very soft lead to tungsten carbide.

W

white iron. A cast iron that is essentially free of graphite, and most of the carbon content is present as separate grains of hard Fe_3C. White iron exhibits a white, crystalline fracture surface because fracture occurs along the iron carbide platelets.

APPENDIX B

Decarburization of Steels

DECARBURIZATION is defined as "a loss of carbon atoms from the surface of a ferrous material, thereby producing a surface with a lower carbon gradient than at some short distance below the surface" (Ref 1). This is a serious problem that can be either the result of a preexisting condition in the material being heat treated or the result of a heat treating atmosphere that allows the surface carbon to be depleted. Decarburization not only results in a loss of surface hardness in any untreated or hardened steel surface but also causes a degradation of mechanical properties, or even cracking (during hardening)—under certain conditions.

Decarburization can result in the total loss of base material carbon, where the surface structure is reduced to ferrite (0.02% C), or only a partial loss of base carbon, or a combination of both.

Figure 1(b) shows an increase in ferrite content as one moves from the arrow to the surface. Figure 1(a) shows decarburization extending from the arrow to the surface. This microstructure becomes a totally ferritic structure at the surface; the ferritic structure extends in approximately 0.125 mm (0.005 in.) from the surface, with the remainder being only a partial reduction in base carbon.

Figure 2(b) shows surface decarburization and the decrease in surface hardness nearer to the surface for a 1.3% C hypereutectoid steel in the quench and tempered condition. The microstructure of that steel is shown in Fig. 2(a). Figure 2(c) shows the completely ferritic surface layer in a commercial 1.0% C strip in the austenitized and quenched condition.

Sources of Decarburization

Steel Originated. Decarburization is a surface condition that is common to all hot-rolled steel products to some extent. It is produced during heating and rolling operations when oxygen in the atmosphere at rolling or forging temperature reacts with and removes carbon from the hot surface. The depth of decarburization depends on temperature, time at temperature, nature of the furnace atmosphere (if any), reduction of area between the billet and the finished size, and the type of steel. To ensure the removal of decarburization and other steel surface defects, the following guidelines have been established:

- *Stock removal for cold-finished steel.* Standard quality cold-finished bars are produced from hot-rolled steel. Therefore, the original decarburization that was present on the hot-rolled stock is still present on the steel, although it has been reduced in thickness as the bar diameter was reduced by cold drawing. Guidelines provide recommendations for stock removal of surface decarburization and any other surface defects that might be present. The recommended stock removal for all nonresulfurized grades is 0.025 mm per 1.59 mm (0.001 in. per 1/16 in.) of cross section, or 0.254 mm (0.010 in.), whichever is greater.

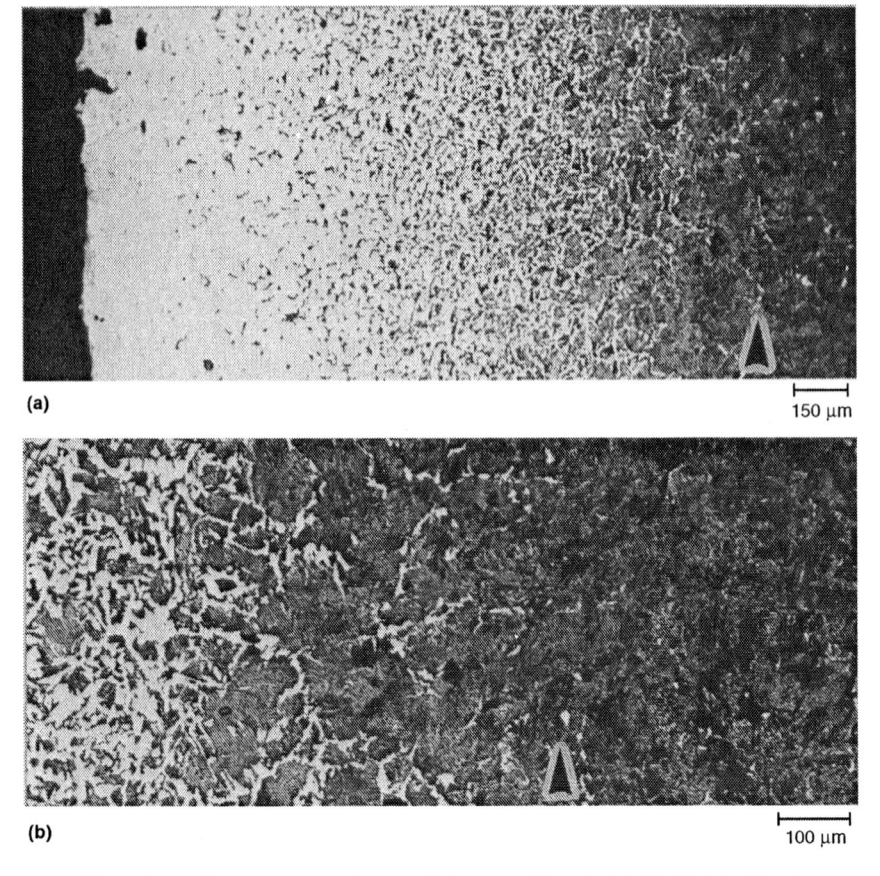

Fig. 1 (a) Decarburization in a 0.8% C eutectoid steel, 0.78C-0.30Mn. Picral etch, 50×. (b) Transverse section of a hot-rolled bar; normalized. Arrows indicate total depth of decarburization. Picral etch, 100×. Source: Ref 2

For example, for a 25 mm (1 in.) bar, recommended stock removal is 0.254 mm (0.016 in.) per side. For the resulfurized grades, recommended stock removal is 0.038 mm per 1.59 mm (0.0015 in. per 1/16 in.) of cross section or 0.38 mm (0.015 in.), whichever is greater.
- *Stock removal for hot-rolled stock.* Stock removal of 3.18 mm (1/8 in.) is recommended for diameters of 38 through 76 mm (1½ through 3 in.) and 6.35 mm (1/4 in.) of larger diameter bars.

Castings. Decarburization is a normal condition found on the surface of steel and iron castings. Even investment castings, with the close finished tolerances, always have surface decarburization due to the metal-mold reaction. Thus, investment castings requiring hardening are generally given a carbon restoration treatment prior to any final heat treatment.

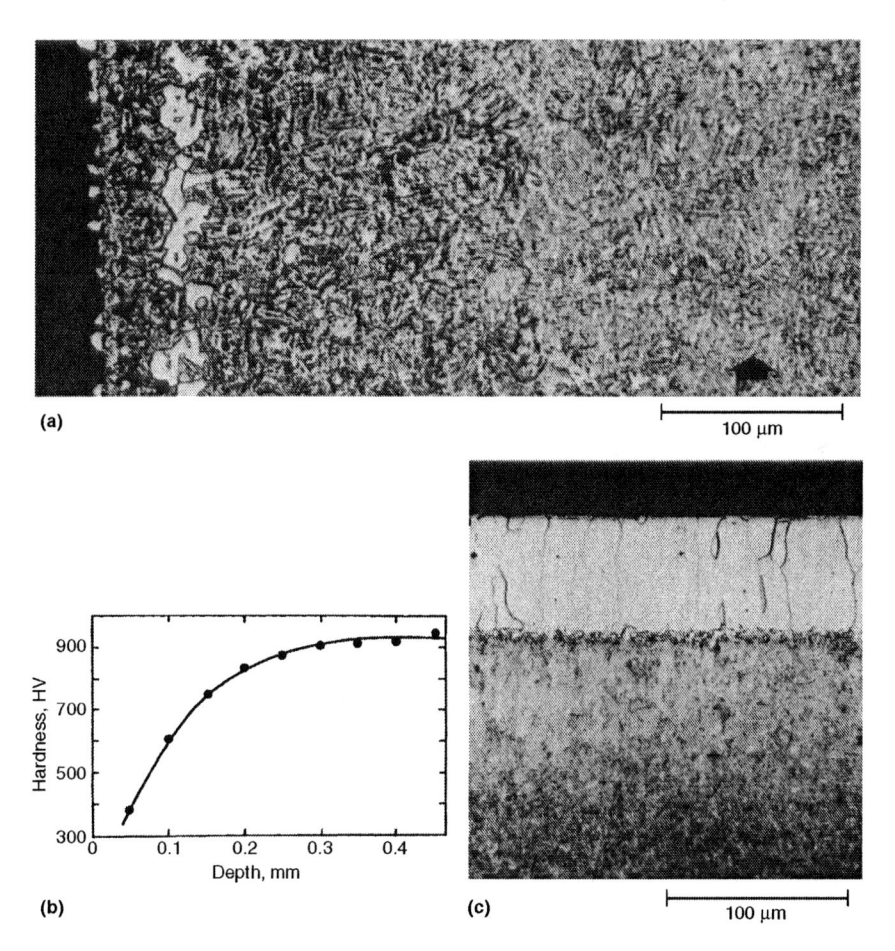

Fig. 2 Decarburization in a 1.3% C hypereutectoid steel (1.20C-0.17Si-0.40Mn) in the quench-hardened condition. Austenitized at 850 °C (1560 °F), water quenched, tempered at 175 °C (350 °F). (a) 1% nital etch, 250×. (b) Variation of hardness with depth in the quenched and tempered hypereutectoid steel. (c) Completely ferritic surface layer in a commercial 1% C strip in the quench-hardened condition. 3% nital etch, 250×. Source: Ref 2

Forgings. Steel used for forging process is generally heated to forging temperatures without the benefit of protective atmospheres. Therefore, unless a post-carbon-restoration process is performed or the surface is adequately machined, forgings will have decarburization.

Effect of Heat Treatment

Furnace Treatments. The goal of any furnace hardening treatment of finished parts is to maintain the surface and base carbon condition. Depending on the furnace atmosphere conditions and the surface condition of the workpiece, the heat treatment can either maintain the same carbon as the workpiece, restore the surface carbon, or cause decarburization. See Chapter 5, "Instrumentation and Control of Heat Treating Processes," for information about the control devices necessary to maintain the proper surface carbon conditions in furnace atmospheres.

Induction Hardening. Because induction hardening involves heating to above the critical temperature in air, the treatment can be a source of decarburization. For a typical 5 s heat cycle to 950 °C (1750 °F), the decarburization depth is calculated to be 0.00197 cm (0.00078 in.). This depth is so shallow that decarburization caused during induction heat treating generally is not a problem. The more common problem is decarburization that has not been removed prior to the induction hardening process.

Fluidized Bed Processing. The depth of decarburization of cold-worked steel in a fluidized bed at various temperatures is shown in Fig. 3.

REFERENCES

1. G. Parrish, *Carburizing: Microstructures and Properties,* ASM International, 1999
2. L. Samuels, *Light Microscopy of Carbon Steels,* ASM International, 1999

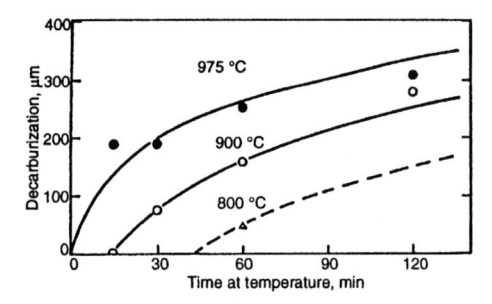

Fig. 3 Depth of decarburization of a cold-worked steel in a fluidized bed in air. Source: Ref 1

APPENDIX C

Boost/Diffuse Cycles for Carburizing

AFTER THE WORKPIECE is heated to above the critical temperature in a protective environment (atmosphere or vacuum), the carbon level in the surrounding environment is elevated above the material base carbon in order to cause the adsorption of carbon and diffusion from the surface inward. A plot of carbon content starting from the surface into the piece is called a "carbon penetration profile." With increasing time, the total depth of carbon diffusion (total case depth, TDC) increases according to the formula:

$$\text{TDC} = K \times \sqrt{t}$$

where t is time and K is a constant based on temperature, the base material, and the carbon potential of the carburizing environment. The shape of the carbon potential profile when a single carbon level is maintained during the carburizing cycle is shown in Fig. 1, curve 3.

As the carburizing process has become more sophisticated and controllable, it also has become easy to change the carbon potential during the carburizing process. Boost/diffuse carburizing cycles involve changing the carbon potential from a higher value of 1.05 to 1.20% C (boost) to a somewhat lower value generally in the range of 0.80 to 0.90% C (diffuse) later in the cycle in order to flatten and extend the initial portion of the curve as shown in Fig. 1, curve 2. If the potential is lowered too early in the cycle, a penetration profile curve such as that shown in Fig. 1, curve 3 results, and if the change is "too late" in the cycle, a carbon penetration profile such as the curve in Fig. 1, curve 1 is the result. Thus, in order to achieve the desired benefits, Fig. 1, curve 2, it is very important that the change in the carbon potential be made at the right time in the overall cycle.

The "boost/diffuse" carburizing process offers two distinct advantages over single potential carburizing:

- The cycle time needed to achieve the same total depth of carbon penetration is decreased by up to 25%.
- The case quality (residual compressive stress depth) level is significantly improved.

Boost/diffuse cycles generally are not used for case depths of less than 1 mm (0.040 in.) because of cycle duration limitation.

Table 1 lists typical carburizing constants and boost/diffuse time ratios to obtain a surface carbon content of 0.80 to 0.90% with a low-carbon, low-alloy steel.

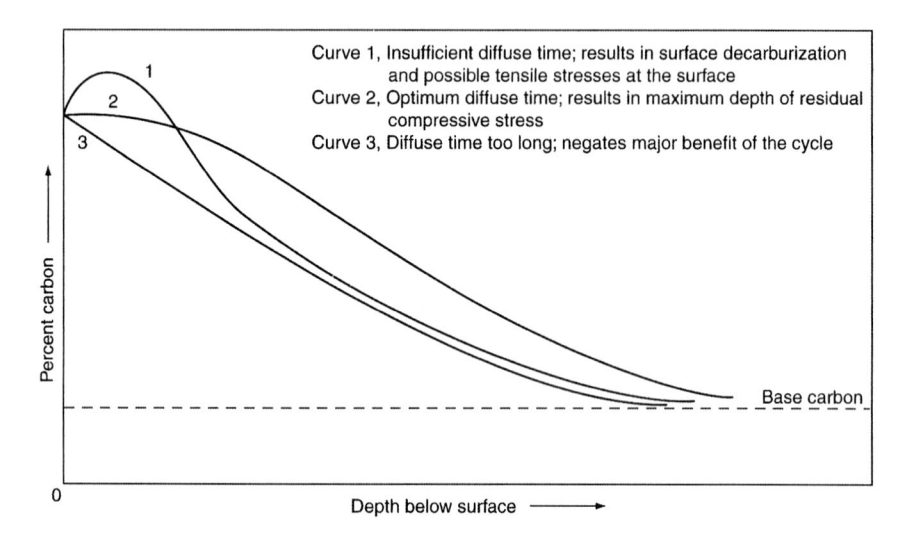

Fig. 1 Possible carbon penetration profiles from boost/diffuse cycles. Source: Ref 1

Curve 1, Insufficient diffuse time; results in surface decarburization and possible tensile stresses at the surface
Curve 2, Optimum diffuse time; results in maximum depth of residual compressive stress
Curve 3, Diffuse time too long; negates major benefit of the cycle

Table 1 Typical carburizing constants and boost/diffusion ratios needed to obtain a 0.80 to 0.90% surface carbon content in a low-alloy, low-carbon steel

Temperature		Carburizing constant		Boost/diffusion ratio, r
°C	°F	K(a)	K(b)	
840	1550	0.25	0.010	0.75
870	1600	0.33	0.013	0.65
900	1650	0.41	0.016	0.55
925	1700	0.51	0.020	0.50
950	1750	0.64	0.025	0.45
980	1800	0.76	0.030	0.40
1010	1850	0.89	0.035	0.35
1040	1900	1.02	0.0400	0.30

(a) To obtain effective case depth (50 HRC hardness), D, in millimeters when $D = K\sqrt{t}$ and t is in hours. (b) To obtain effective case depth (50 HRC hardness), D, in inches when $D = K\sqrt{t}$ and t is in hours. Source: Ref 2

REFERENCES

1. J.L. Dossett, *Furnace Atmospheres Presentation,* Satellite Seminar, ASM International, 1995
2. J. St. Pierre, Vacuum Carburizing, *Heat Treating,* Vol 4, *ASM Handbook,* ASM International, 1991, p 348–351

APPENDIX D

Use of Test Coupons for Process Verification

THE RESULTS of certain heat treating processes must be verified for case quality and case depth by destructively sectioning a part or parts that were subjected to the process. Test coupons or test pins are often used for diffusion processes such as carburizing, carbonitriding, nitriding, and ferritic nitrocarburizing to provide an accurate heat treating process evaluation. Their use has a twofold benefit:

- The cost is significantly lower than for processes that involve destruction of actual parts.
- The material and section size variables are removed to provide a measurement of actual process variation.

Test coupons must be carefully designed to be an effective statistical process control (SPC) tool. They must be properly selected for size, shape, and material that can be directly correlated to the material and parts configuration being processed. The coupons should be prepared in sufficient quantities—enough for a year or more of use—from the same heat of steel to eliminate or minimize the material uniformity variable from processing variation.

By using SPC with test coupons in conjunction with statistical quality control (SQC) on heat treated parts, critical process variations may be identified and controlled.

Typical Example of Use: Monitoring Carburizing and Hardening of Gears

Test pins were used to monitor carburizing and hardening processing for 5 to 8-pitch gears made mostly from 8620H steel. This procedure was used to monitor the process variation in carburizing for surface hardness, effective case depth, and core hardness. The test pin diameter chosen was based on the gear tooth thickness and the fact that the test pin center cooling rate would be on the steeper portion of the Jominy hardenability curve. This meant that monitoring the center core hardness of the test pins would respond to quench system performance.

Purchase of Test Pins. The quantity of test pins purchased was approximately 10,000 pieces made from a single heat of 8620H steel. The size tolerance of the round pins was ±0.13 mm (±0.005 in.) with the absolute size being 15.6 mm (0.625 in.). The length was 64 ± 1.6 mm (2½ ± ¹⁄₁₆ in.). A groove was added to the pin for attaching the pin by wire to the load.

Processing of Test Pins. The following procedures were used:

- At least one test pin was processed with each batch load or one pin was run every 4 h on each row of all continuous furnaces.
- Test pins were hung in all furnace loads in a location where the processing was typical of the parts processed.
- The test pins were evaluated in the "as-quenched" condition. Test pins evaluated for purposes of SPC control were from cycles with no noted changes in times, atmosphere conditions, temperatures, or quenching procedure.

Testing Procedures for Test Pins: Surface Hardness. File check the surface of the pin to check for file hardness per SAE J534 and to make a smooth surface. Three hardness readings were checked by Rockwell C hardness tests, and the average was recorded on a form such as that shown in Fig. 1. Only flat or spot anvils were used.

Core Hardness. A parallel section 6.4 mm (¼ in.) thick was cut from the center of the test pin. For hardness testing, the diamond and anvil were set by checking the hardness at mid-radius. Then the center hardness were checked by Rockwell C hardness test and recorded. The core hardness values were used to monitor the effectiveness of the oil quench media.

Effective Case Depth. Using the section cut from above, the cross section to be checked was polished on a 120-grit or finer paper. Rockwell 15N tests were made from the surface in to the point where the hardness is 85.5 HR15N (50 HRC). A measurement from the surface to the center of that mark was made using a Brinell glass. The reading was then recorded as effective case depth in thousandths of an inch.

The referee method for checking effective case depth was by Knoop 500 g microhardness testing to a depth of 50 HRC equivalent. At least one of every ten checks and/or any check of effective case depth not within the specified limits was verified by the microhardness method.

The results are to be plotted by standard cycle number and furnace on the form shown in Fig. 1.

Setting Upper and Lower Control Limits for Each Variable. This method can be started and used on a monitoring basis for a short time

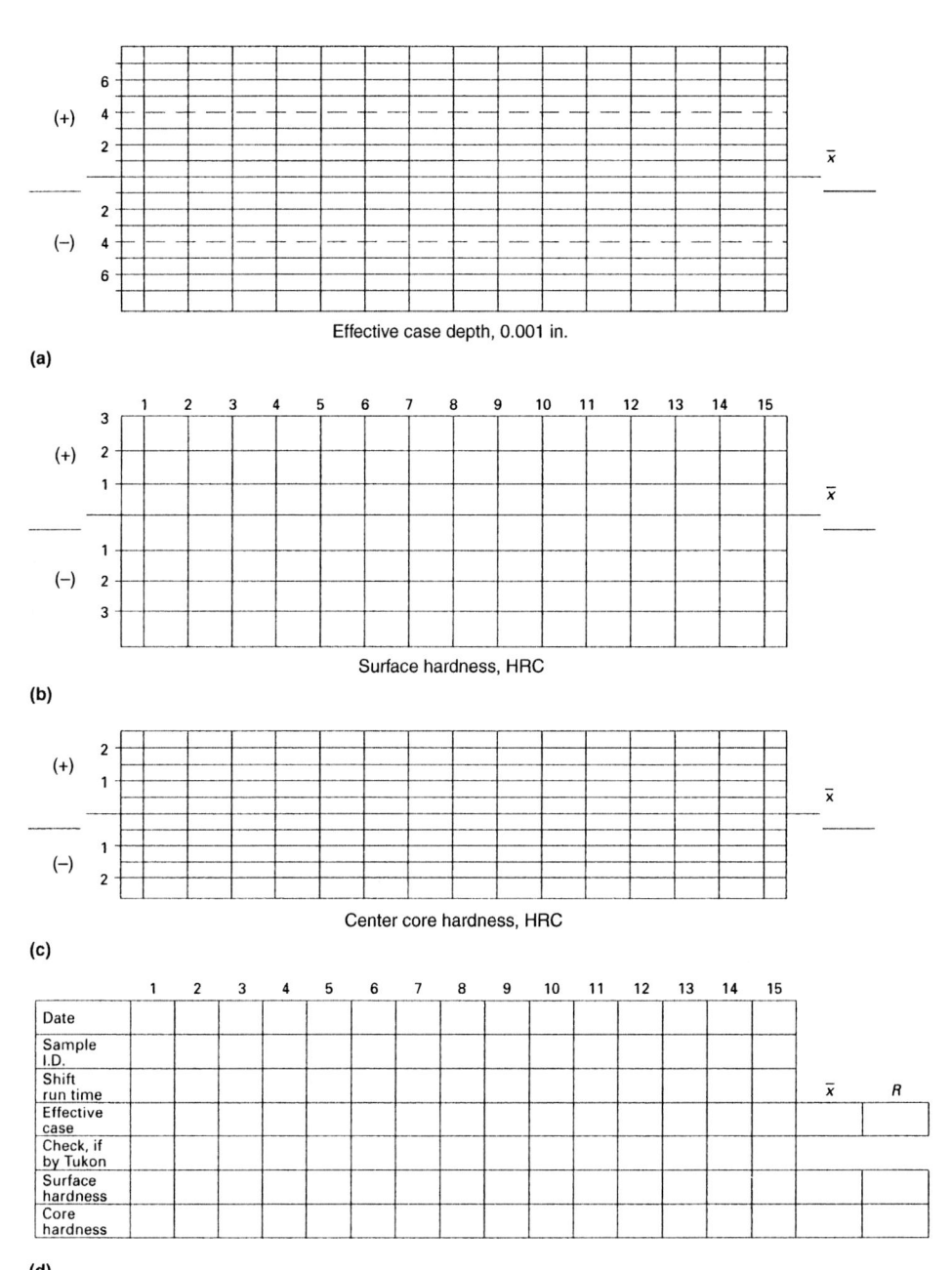

Fig. 1 Chart for plotting 8620 steel test pin variation by characteristic. (a) Effective case depth characteristic. (b) Surface hardness characteristic. (c) Center core hardness characteristic. (d) Chart to plot data from (a), (b), and (c) by cycle and furnace. Source: Ref 1

until mean values with upper control limits (UCL) and lower control limits (LCL) can be established based on the appropriate statistical formulas.

REFERENCE

1. J.L. Dossett, G.M. Baker, T.D. Brown, and D.W. McCurdy, Statistical Process Control of Heat-Treating Operations, *Heat Treating,* Vol 4, *ASM Handbook,* ASM International, 1991, p 620–637

Index

A

Alloy steels
AISI-SAE compositions, 127, 128(T), 129(T)
alloying elements, purposes served by, 126(T), 127–130(T)
austempering, 136
classification of, 125–126
composition of standard alloy steels, 128(T)
compositions of standard boron (alloy) steels, 129(T)
defined, 125
designation of, 126–127(T), 128(T), 129(T)
effects of alloying on hardenability, 130–132(F,T)
free-machining alloy steels, 127
H-steels, 127
hardenability curves for several alloy steels, 52(F)
hardenability, effect of boron, 129(T), 132, 133(F)
hardenability for alloy grades, 131–132(F,T)
heat treating, 125–140(F,T)
heat treating procedures for specific grades, 132–134
martempering, 137–138(F)
numerical designations of AISI and SAE grades of alloy steels, 126(T)
restricted hardenability (RH) steels, 127
tempering, effects of, 135(F), 136–138
Alloying
alloying mechanisms, 13–14(F)
atom exchange, 13
cementite, 9–10
interstitial mechanism, 13(F)
phase, defined, 9
Alloying elements
boron-treated steels, 46
chromium, 130
effect on hardenability, 45–47(F,T)
hardenabilities for various steels, 48(T)
manganese, 129
molybdenum, 130
nickel, 129–130
overview, 127
silicon, 128(T), 129
vanadium, 126(T), 130
Alloying mechanisms
austenite, transformation of, 18–19(F)
effect of carbon on the constitution of iron, 14–15(F)
iron-cementite phase diagram, 15–17(F)
martensite, forming of, 19
phase, defined, 16
phase diagram, defined, 16–17
solubility of carbon in iron, 17–18(F)
Alpha iron
defined, 13
maximum solubility of carbon in, 18
nine-atom lattice, 12(F)
Aluminum-copper alloys
aging, 237
overaging, 237–238
overview, 236–237(F)
part of the aluminum-copper phase diagram, 236
schematic aging curve and microstructure, 237(F)
Aluminum oxide sensor, 94(F), 94–95
American Iron and Steel Institute (AISI), 10
Annealing
overview, 2
specific process names, 3
stress-relief annealing (stress relieving), 3
Argon, 83
Atmosphere control
overview, 90–91
purpose of, 90
Atmosphere sensors and control systems
aluminum oxide sensor, 94(F)
dew point instrument, 94–95(F)

Atmosphere sensors and control systems (continued)
  infrared analyzers, 95
  overview, 91
  oxygen probe, 91–92(F)
  oxygen probe systems, 92–94(F)
  typical oxygen probe construction, 92(F)
Austempering
  advantages, 121
  alloy steels, 136
  applications, 122
  compositions and characteristics of salts used for austempering, 121(T)
  difference between conventional quenching/tempering, 112(F), 120–121(F)
  overview, 6
  procedure, 120
  quenching mediums, 121(T)
  selection of grade for, 112(F), 121–122
Austenite
  formation of, 9–10
  overview, 17–18(F)
  transformation of, 18–19(F)
Austenitic grades (stainless steels), 176–182(F,T)
  corrosion resistance of, 182
  effects of nickel and/or manganese, 178–180(F), 181(F)
  full annealing temperatures for austenitic stainless steels, 181(T)
  heat treating of austenitic grades, 180–182(T)

B

Bainite microstructure, 23(F), 24(F)
Banding, 244–245(F)
Batch-type furnaces, 59–61(F)
  batch-type multiple purpose furnace, 60(F)
  bell-type furnaces, 61
  box-type batch furnace, 59–61(F)
  car-bottom batch furnace, 56(F)
  car-bottom furnaces, 59–60
  elevator car-bottom furnace, 60
  heating mode, principally radiation, 60–61
  low-temperature heat processing furnace, 57(F)
  overview, 59
  small batch-type furnace, 59(F)
  temperature range, 60–61
  top loader type, 57(F)
Belt-type continuous furnaces, 62
Body-centered cubic (bcc)
  defined, 12
  lattice structure, 12(F)
Boost/diffuse cycles for carburizing, 279–282(F,T)

Boron-treated carbon steels, 102–103(F)
  compositions for six standard grades, 99(T)
  hardenability, effect of manganese on, 99(T)
Brinell hardness number (HB), 31
Brinell testing
  automatic Brinell hardness tester system with digital readout, 33(F)
  Brinell hardness number (HB), 31, 32(T)
  Brinell indention process, 30(F)

C

Carbon
  effect on annealed steels, 40–42
  effect on the constitution of iron, 14–15(F)
  role in hardened steels, 42–44(F)
Carbon equivalent (CE), 209–210
Carbon H-steels
  compositions of standard carbon H-steels and standard carbon boron H-steels, 99(T)
  effect of composition on hardenability, 103(F)
  effect of manganese on hardenability, 103(F)
  overview, 102
Carbon penetration profile, 279, 280(F)
Carbon steel compositions
  compositions of selected standard nonresulfurized carbon steels, 98(T)
  compositions of standard carbon H-steels and standard carbon boron H-steels, 99(T)
  compositions of standard nonresulfurized carbon steels (over 1.0% Mn), 98(T)
  compositions of standard rephosphorized and resulfurized carbon steels, 99(T)
  compositions of standard resulfurized carbon steels, 99(T)
  development/standardization of, 97
  types and approximate percentages of identifying elements in standard carbon steels, 97(T)
Carbon steels
  AISI classification definition, 100
  AISI-SAE coding system, 100–101(T)
  austempering, 120–122(F,T)
  boron-treated carbon steels, 102, 103(F)
  carbon H-steels, 102, 103(F)
  classification for heat treatment, 106–109(F)
  compositions, 97(T), 98(T), 99(T)
  free-machining additives/free cutting additives, effects of, 104–105
  free-machining carbon steels, 104
  heat treating, 97–124(F,T)
  heat treating procedures for specific grades, 109–118(F)
  higher manganese carbon steels, 101

lead, effects of, 105–106
martensitic and nonmartensitic structures, 112(F), 118–119
overview, 106
phosphorus, effects of, 105
precautions concerning free-machining additives, 105–106
sulfur, effects of, 105(F)
tempering of quenched carbon steels, 111(F), 115(F), 119–120
Carbon steels (classification for heat treatment)
group I (0.08 to 0.25% C), 106–107(F)
group II (0.30 to 0.50% C), 107–108
group III (0.55 to 0.95% C), 108–109
Carbonitriding
applications, 152–153(F), 154(F)
furnace atmospheres, 153
furnaces, 153
overview, 4, 152
steels for, 153
tempering, 153, 154(F)
Carburized case
effect of temperature, 145–146(F,T)
effect of time, 145(T), 146–147(F)
overview, 144–145(F,T)
relationship of carbon concentration with distance from the surface, 144(F)
total case versus effective case, 144(F), 145
values of total case depth for various times/temperatures, 145(T)
Carburizing
batch-graphite integral oil-quench vacuum furnace, 71(F)
boost/diffuse cycles, 275–278(F)
gas carburizing, 143–147(F,T)
liquid carburizing, 147–150(T)
overview, 4, 141–142
pack carburizing, 150–151
vacuum carburizing, 147
Case hardening
carbonitriding, 152–156(F,T)
carburizing processes, 141–151(F,T)
classification of, 141, 142(T)
gas nitriding, 153–156(F,T)
gaseous nitrocarburizing, 156–157
liquid carburizing, 147–150(T)
overview, 141
pack carburizing, 150–151
plasma (ion) nitriding, 155–156
plasma nitrocarburizing, 157
salt bath nitriding, 156
typical characteristics of case hardening treatments, 142(T)
vacuum carburizing, 147
Cast irons
carbon equivalent (CE), 209–210
differences among types, 208–209(F,T)

ductile irons, 222–228(F,T)
gray irons, 212–221(F,T)
hardenability of gray and ductile irons, 212–228(F,T)
heat treating of, 207–230(F,T)
heat treatments for gray irons, 212–213
iron-graphite phase diagram at 2½% Si, 207(F)
measuring hardness of, 210–212(F,T)
overview, 207–208(F)
typical compositions of major classes of cast irons, 209(T)
typical microstructures of four types of cast iron, 210(F)
Cementite (iron carbide), 9–10
Conduction, 56–57
Continuous furnaces, 61–63(F)
Convection
low-temperature heat processing furnace, 57(F)
overview, 57
tempering, common application of, 57(F)
Copper-beryllium alloys
attaining maximum hardness, 240–241(F)
beryllium-copper phase diagram, 239(F)
dispersion hardening, 241
overview, 238
phase diagrams, 238–239(F)
precipitation-hardening curves of beryllium-copper binary alloys, 240(F)
solution treating and aging, 239–240(F)

D

Decarburization, 275–278(F)
castings, 277
defined, 275
forgings, 277
furnace treatments, 278(F)
overview, 78
removal of, guidelines for, 275–277
sources of, 275–278
steel originated, 275–277
stock removal for cold-finished steel, 276–277
stock removal for-hot-rolled stock, 277
Dew point instrument, 94–95(F)
aluminum oxide sensor, 94(F)
Diamond indenter, 37(F)
Dispersion hardening, 241
Dry-hydrogen atmospheres, 83
Ductile irons
annealing, 222–223
annealing practice for casting with differing alloy content, 223
austempering, 225–226(F), 227(F)
cooling, 222

Ductile irons (continued)
 hardenability of, 212–228(F,T)
 heat treating, 222
 laser/plasma torch process, 228
 nitriding, 228
 normalizing of, 223
 quenching and tempering, overview, 224–225(F)
 response to induction hardening, 227
 stress relieving of, 222
 stress-relieving temperatures for different types, 222
 surface hardening, 226–228
 surface hardness, 228
Duplex grades (stainless steels)
 heat treating, 184(T)
 overview, 184
 recommended annealing temperatures for selected duplex stainless steels, 184(T)

E

Electromotive force (emf), 92
Emissivity, defined, 58
End-quench test (Jominy), 49–50(F), 51(F)
Endothermic gas generator, 81(F)
Equilibrium transformations, defined, 14–15
Eutectoid steels, 20–21(F)
 spheroidite, 20–21

F

Ferrite (or delta iron), 12(F)
 alpha ferrite, defined, 13
 delta ferrite, defined, 13
Ferritic grades (stainless steels)
 annealing, 183–184(T)
 annealing treatments of ferritic stainless steels, 184(T)
 heat treating, 180(F), 183–184(T)
 overview, 182–183
Flame hardening
 combination of progressive and spinning methods, 163–164(F)
 combination progressive-spinning method of flame hardening, 164(F)
 equipment, 161
 materials, 161
 methods of flame heating, 162(F)
 overview, 160–161
 progressive method, 161–162(F)
 quenching, 162(F), 163(F), 164(F)
 spinning method, 162–163(F)
 spot (stationary) method, 161, 162(F)
 tempering, 164
Fluidized bed furnaces, 66–67(F)
 applications, 67
 externally fired (low-temperature application), 66, 67(F)
 externally gas-fired fluidized-bed furnace, 68(F)
 internally fired (high-temperature applications), 66–67(F)
Free-machining carbon steels
 AISI 11XX grade, 104
 AISI 12XX grade, 104
 compositions of 12XX series, 99(T)
 leaded steels, 99(T), 104
Furnace atmospheres, 77–83(F)
 commercial nitrogen-based atmospheres, 81–82(F)
 decarburization, 77–78
 dissociated ammonia, 82–83
 endothermic atmospheres, 80, 81(F)
 exothermic atmospheres, 80
 gaseous atmospheres, 79
 industrial-gas nitrogen-base atmosphere processes, 82(F)
 natural atmospheres (air), 79
 overview, 77–79
 products of combustion, 79–80
 purpose of, 79
 schematic diagram of an endothermic gas generator, 81(F)
 steam atmospheres, 83
 vacuum atmospheres, 82
Furnace parts and fixtures
 bar frame type basket, 73(F)
 baskets and fixtures, 71–74(F)
 castings, 71
 compositions of selected heat-resistant alloys, 72(T)
 heat-resistant materials, 70, 72(T)
 overview, 70–71
 tray/fixture for carburizing ring gears, 73(F)
 trays and grids, 71, 72(F)
Furnaces and related equipment, 55–84(F,T)
 batch-graphite integral oil-quench vacuum furnace, 71(F)
 batch-type versus continuous-type furnaces, 59–61(F)
 bell-type furnaces, 61
 belt-type continuous furnaces, 62
 box-type batch furnace, 59(F)
 car-bottom batch furnace, 56(F)
 car-bottom furnaces, 59–60
 classification by heat transfer medium, 58–59
 continuous furnaces, 61–63(F)
 elevator car-bottom furnace, 60
 externally gas-fired fluidized-bed furnace, 68(F)
 fluidized bed furnaces, 66–67(F)
 gas-fired fluidized-bed furnace with internal combustion, 66(F)

Index / 291

heat transmission, modes of, 56–58(F)
liquid bath furnaces, 63–65(F)
low-temperature heat processing furnace, 57(F)
ovens/furnaces, difference between, 55–56
overview, 55–56(F)
parts and fixtures, 70–74(F,T)
principal types of salt bath furnaces, 65(F)
pusher-type continuous furnaces, 62–63
roller-hearth continuous furnaces, 61–62(F)
roller-hearth furnace, 62(F)
rotary-hearth continuous furnaces, 62(F)
rotary hearth forging or heat treating furnace, 63(F)
single-chamber batch-type pressure-quench vacuum furnace, 70(F)
small batch-type furnace, 60(F)
three-chamber cold-wall vacuum oil-quench furnace, 69(F)
top loader type, batch-type furnace, 57(F)
vacuum furnaces, 67–70(F)

G

Gamma iron, 12
Gas carburizing
carburized case, characteristics of, 144–147(F)
carburizing gases, 143
drip carburizing, 143
furnaces, 143
overview, 143
values of total case depth for various times/temperatures, 145(T)
Gas nitriding
case characteristics, 155
furnaces, 155
overview, 153–154
processing, 155
steels for, 154–155(T)
Gaseous nitrocarburizing
aerated salt bath nitriding (salt bath nitrocarburizing), 156–157
furnaces, 157
overview, 156–157
primary objective, 157
Glossary of heat treating terms, 257–274(F)
Gray irons
annealing of, 214–216
austempering, 219–220(F)
austempering and martempering, comparing, 218–219
austenitizing and tempering, 217–218
effect of stress relieving temperature on hardness of gray irons, 213(F)
ferritizing anneal, 214–215
flame hardening, 220–221

full annealing, 215
graphitizing anneal, 215–216
hardenability of, 212–228(F,T)
hardening and tempering of, 217–218, 219(F)
heat treatments for, 212–213
induction hardening, 221–222(T)
martempering, 220
normalizing of, 216–217(F)
quenching, 218
room-temperature hardness of gray iron normalizing, 217(F)
stress relieving of, 213–214(F)
tempering, 218, 219(F)

H

H-steels, 52–54(F)
chemistry only specification, 53(F)
hardenability bands, 53–54(F)
Hardenability, 44–54(F,T)
austenitic grain size of steels, determining, 45, 46(F)
effect of alloying elements on, 45–47(F,T)
effect of composition on hardenability, 103(F)
effects of alloying on, 130–132(F,T)
end-quench test (Jominy), 49–50(F), 51(F)
evaluating, methods of, 48–52(F)
H-steels, 52–54(F)
hardenabilities for various steels, 48(T)
hardenability bands, 53(F), 54(F)
hardness penetration diagram, 49(F)
hardness penetration diagram test, 48, 49(F)
ideal critical number D_I, defined, 45, 46(F)
meaning of, 44–45(F)
quantifying, 45
S-A-C test (Rockwell inch test), 48–49
variations in, 51–52(F)
Hardenability bands, 53(F), 54(F)
Hardened steels
example of hardness versus carbon content, 43(F)
role of carbon, 42–44(F)
Hardened zones (low-hardenability steels), 159, 160(F)
Hardening
electron-beam surface hardening, 172
hardened zones in low-hardenability steels, 159, 160(F)
induction hardening, 165–171(F,T)
laser-surface hardening, 172
shell hardening in molten metal or salt, 160
Hardness, 27–44(F,T)
apparent hardness, defined, 42
apparent hardness versus actual hardness, 195–196(F)

Hardness (continued)
    effects of carbon on annealed steels, 40–42
    hardness testing, 28–40(F,T)
    role of carbon in hardened steels, 42–44(F)
    true hardness, defined, 42

Hardness testing
    Brinell indention process, 30(F)
    Brinell testing, 30–31(F,T)
    Brinell's method, 29
    diamond indenters, applications, 34, 35(T)
    Foeppl's method, 28–29(F)
    hardness test, 27
    history of, 28–29(F)
    Knoop indenter, 38(F)
    Leeb tester (Equotip testers), 36, 37
    microhardness testing, 37–40(F)
    Reaumur's method, 28, 29(F)
    Rockwell testing, 31–36(F,T)
    Scleroscope testing, 36
    steel ball indenters, applications, 34, 35(T)
    systems, 30–37(F,T)
    ultrasonic microhardness testing, 36–37
    Vickers indenter, 36, 37(F)
    Vickers testing, 36

Heat transmission
    conduction, 56–57
    convection, 57(F)
    low-temperature heat processing furnace, 57(F)
    modes, 56–58(F)
    radiation, 57–58, 59(F)

Heat treating
    alloy steels, 125–140(F,T)
    carbon steels, 97–124(F,T)
    cast irons, 207–230(F,T)
    constitution of iron, 11–13(F)
    furnaces and related equipment, 55–84(F,T)
    importance of, 1–2
    metallurgical phenomena, 11–19(F)
    nonferrous alloys, 231–242(F,T)
    quality assurance program, 243–256(F,T)
    stainless steels, 175–190(F,T)
    steel, 9–26(F)
    tool steels, 191–206(F,T)

Heat treating procedures (alloy steels)
    4037, 4037H, 134, 135(F)
    4140, 4140H, 134, 135(F)
    4340, 4340H, 134–136(F)
    E52100, 135(F), 136

Heat treating procedures (carbon steels)
    1008 through 1019, 107(F), 109
    1020, 110
    1035, 110, 111(F)
    1045, 1045H, 110–112(F)
    1050, 111(F), 112
    1060, 111(F), 113
    1070, 111(F), 113
    1080, 111(F), 113
    1095, 111(F), 113–114
    1137, 114, 115(F)
    1141, 115(F), 116
    1144, 99(T), 115(F), 116
    1151, 115(F), 116
    12L14, 114
    1522, 1522H, 117
    1566, 115(F), 118
    15B41H, 115(F), 117

Heat treating processes
    annealing, 2–3
    atmosphere control, 90–95(F)
    classification of, 2–7
    cold and cryogenic treatment, 7
    instrumentation and control of, 85–96(F,T)
    integrated control systems, 95–96(F)
    normalizing, 2
    quenching/quenchants, 5–7
    stress-corrosion cracking (SCC), 3
    stress-relief annealing (stress relieving), 3
    surface hardening, 4–5
    temperature-control systems, 85–86
    tempering, 5–6
    tool steels, 193–194(T)

Hydrogen
    dry hydrogen, applications, 83
    production methods, 83

Hypereutectoid steels, 21–22(F)

Hypoeutectoid steels, 19–20(F), 21(F)

Hysteresis
    arret, defined, 22
    chauffage, defined, 22
    overview, 22
    refroidissement, defined, 22

I

Ideal critical number D_I, 45, 46(F)

Induction hardening, 165–171(F,T)
    equipment for, 167–172(F,T)
    fluidized bed processing, 278(F)
    frequency, selection of, 170–171(T)
    furnace tempering, 171
    furnace treatments, 278(F)
    induction hardening of a gear, 167(F)
    induction heating, 165–167(F)
    induction tempering, 171–172(F)
    inductors, 168–170(F)
    minimum hardness depths for production work in surface hardening, 171(T)
    overview, 165, 167–168(F)
    power supplies, 168, 169(T)
    process development, 171

Induction heating
    advantages of, 165
    characteristics of the four major power sources, 169(T)

induction hardening of a gear, 167(F)
magnetic fields and induced currents produced by various induction coils, 166(F)
principles of, 165–167(F)
Inductors (induction hardening)
inductor with separate internal chambers for flow of quenchant and cooling water, 169(F)
typical work coils for induction heating, 170(F)
Industrial Heating Equipment Association (IHEA), 56
Integrated control systems
overview, 95
programmable logic controller (PLC), 95–96(F)
Iron
allotropy of, 13(F), 14(F)
constitution of, 11–13(F)
constitutional diagram for commercially pure iron, 12(F)
effect of carbon on the constitution of iron, 14–15(F)
ferrite (or delta iron), 12(F)
gamma iron, 12
solubility of carbon in, 17–18(F)
Iron-cementite phase diagram, 15–17(F)

J

Jominy end-quench hardenability test, 49–50(F), 51(F)

K

Knoop hardness number (HK), 39
Knoop indenter, 38(F)
comparing to Vickers indenter, 39(F)
Knoop indention testing, 38–40(F)
Knoop hardness number (HK), 39
Vickers indenter, comparing to, 39(F)

L

Leaded steels, 104
Leeb tester (Equotip testers), 36, 37
Liquid bath furnaces
disadvantages, 63
molten salt baths, 64–65(F)
overview, 63–64
principal types of salt bath furnaces, 65(F)
Liquid carburizing
advantages and limitations of, 149–150

control of case depth and carbon concentration, 149, 150(F)
cyanide-type baths, 148–149(T)
furnaces, 149
noncyanide liquid carburizing, 149
operating compositions of liquid carburizing baths, 148(T)
overview, 147–148
Low-hardenability steels, 159, 160(F)

M

Maraging steels, 6
Martempering
alloy steels, 137–138(F)
overview, 6
Martensite, forming, 19
Martensitic grades (stainless steels)
annealing practice, 185–187(T)
austenitizing and quenching, 186(T), 187
hardenability, 185, 186(F)
heat treating, 185–188(T)
overview, 178(T), 185
temperatures for heat treating several martensitic stainless steels, 186(T)
tempering, 187–188(F)
Metallurgical phenomena, 11–19(F)
Methylacetylene propadiene (MAPP), 160
Microhardness, defined, 38
Microhardness testers, 40
components of a typical microhardness tester, 41(F)
Microhardness testing, 37–40(F)
applications, examples, 38
components of a typical microhardness tester, 41(F)
Knoop indenter, 38(F)
Knoop indention testing, 38–40(F)
microhardness, defined, 38
microhardness testers, 40, 41(F)
Vickers indenter, 37, 38(F)
Molten salt baths
advantages/disadvantages, 64–65
equipment required, 64–65
overview, 64
principal types of salt bath furnaces, 65(F)
salt bath equipment, 64–65
temperature range, 64
Mottled cast irons, 209

N

Nitriding, overview, 4
Nitrocarburizing
gaseous nitrocarburizing, 156–157
overview, 4–5, 156–157
plasma nitrocarburizing, 157

Nonferrous alloys
  aging cycles, 235
  aging (precipitation hardening), 234–235
  aluminum-copper alloys, 236–238(F)
  annealing of nonferrous metals and alloys, 232–233(F)
  basic requirements for hardening by heat treatment, 233–234
  cooling from the solutionizing temperature, 234
  copper-beryllium alloys, 238–241(F)
  examples of some precipitation-hardenable alloys, 235(T)
  heat treating, 231–242(F,T)
  heat treating procedures, overview, 234–235
  part of the aluminum-copper phase diagram, 236(F)
  solubility and insolubility, 233–234
  work hardening, 231–232(F)
Normalizing
  ductile irons, 223
  gray irons, 216–217(F)
  overview, 2
  tool steels, 193

O

Ovens/furnaces, difference between, 55–56
Oxygen probe
  overview, 91–92(F)
  typical oxygen probe construction, 92(F), 93(F)
Oxygen probe systems
  applications, 92
  endothermic generators, controlled by, 93
  programmable digital carbon controller, 93(F)
  shim analysis procedure, 94

P

Pack carburizing
  carbon potential and gradient, 151
  furnaces, 151
  overview, 150–151
  process limitations, 151
  processing, 151
Phase, defined, 16
Phase diagram, defined, 16–17
Plasma (ion) nitriding, 155–156
Plasma nitrocarburizing, 157
Power supplies (induction hardening)
  characteristics of the four major power sources, 169(T)
  frequency inverters, 168
  motor generator, 168
  solid-state frequency converters, 168, 169(T)
  vacuum tube oscillator, 168
Precipitation-hardening grades (stainless steels)
  chemical compositions of precipitation-hardening stainless steels, 179(T)
  heat treating, recommended practice, 188–189
  overview, 188–189
Process capability study, 247–249(F), 250(F)
  base materials contributions, 248
  contributing factors, 248
  evaluation method contribution, 248
  example of a capability study, 249(F), 250(F)
  metallurgical requirements, establishing, 248
  part-related contributions, 248
  process-related contributions, 248
  samples, minimum number required per variable, 248
  SPC or SQC, choosing, 250–251
  three steps to use in conjunction with, 249
Programmable logic controller (PLC), 95–96(F)
  batch atmosphere furnace programmable logic controller (PLC) control system, 96(F)
Pusher-type continuous furnaces, 62–63

Q

Quality assurance program, 243–256(F,T)
  banding, 244–245(F)
  components of, 243–244
  control charts, 252, 254(F)
  decarburization, 245
  final inspection, 252–255(F,T)
  hypothetical hardness values for two batches of parts, 247(T)
  in-process quality control, 251–252, 253(F)
  in-process tests, 252
  increased final inspection for hardness, 254(T)
  material considerations, 244–245(F)
  normal distribution curve, 246(F)
  overview, 243
  process audit form, 253(F)
  process capability study, 247–249, 250(F)
  process control audit form, 252, 253(F)
  statistical control, 245–246(F), 247(F,T)
  statistical process control (SPC), 243–244
  statistical quality control (SQC), 243–244
Quenching, defined, 74
Quenching media
  gaseous quenchants, 5
  liquid quenchants, 5
  overview, 5

Index / 295

Quenching systems
  agitation, effects of, 75–76(T)
  applications, 74
  cooling power, 75–76
  cooling rates, 75
  effect of agitation on the effectiveness of quenching, 76(T)
  examples, 75
  metallurgical aspects, 74–75(F)
  oil-quenching system, 77, 78(F)
  quenching, defined, 74
  tank system for water quenching, 77(F)
Quenching/quenchants, 5–7
  overview, 5
  quenching media, 5
  temperature range, 5

R

Radiation, 57–58, 59(F)
  emissivity, defined, 58
Radiation pyrometers, 90(F)
Resistance temperature detectors (RTDs), 86
Rockwell testing, 31–36(F,T)
  automatic Rockwell testing system, 37(F)
  Brale indenter, 31, 33(F)
  components of a Rockwell testing machine, 35(F)
  indention made on a Brale indenter, 33(F)
  Rockwell standard hardness scales, 34(F)
Roller-hearth continuous furnaces, 61–62(F)
Rotary-hearth continuous furnaces, 62(F)
  rotary hearth forging or heat treating furnace, 63(F)

S

Salt bath nitriding
  furnaces, 156
  overview, 156
  proprietary salt bath processes, 156
  types of salts, 156
Scleroscope testing, 36
Shell hardening, 160
Society of Automotive Engineers (SAE), 10
Special-purpose steels
Spheroidite, 20–21
Stainless steels
  annealing treatments of ferritic stainless steels, 184(T)
  austenitic grades, 176–182(F,T)
  chemical compositions of precipitation-hardening stainless steels, 179(T)
  chemical compositions of selected wrought duplex stainless steels, 177(T)
  chemical compositions of standard AISI austenitic stainless steels, 177(T), 178(T)
  classification of wrought stainless steels, 176–189(F,T)
  corrosion resistance of austenitic grades, 182
  defined, 175–176
  duplex grades, 184
  ferritic grades, 180(F), 182–184(T)
  heat treating, 175–190(F,T)
  heat treating of austenitic grades, 180–182(T)
  heat treatment conditions, 189
  martensitic grades, 178(T), 185–188(F)
  nominal chemical composition of standard AISI 400-series ferritic, 177(T)
  overview, 175
  precipitation-hardening grades (PH), 179(T), 188–189
  recommended annealing temperatures for selected duplex stainless steels, 184(T)
Statistical process control (SPC)
  overview, 250–251
  quality assurance program, 244
Statistical quality control (SQC)
  overview, 250–251
  quality assurance program, 244
Steam atmospheres, 83
Steel
  case hardening, 141–158(F,T)
  chemistry only specification, 53(F)
  classification by carbon content, 19–22(F)
  classification of, 10
  cold and cryogenic treatment, 6
  constitution of iron, 11–13(F)
  defined, 9–10
  eutectoid steels, 20–21(F)
  H-steels, 52–54(F)
  heat treating, 9–26(F)
  hypereutectoid steels, 21–22(F)
  hypoeutectoid steels, 19–20(F)
  importance of, 10–11
  martempering, 6
  metallurgical phenomena, 11
  phases of, defined, 9–10
  role of carbon in hardened steels, 42–44(F)
Stress-corrosion cracking (SCC), 3
Stress-relieving (or stress-relief annealing), 3
Surface hardening
  case, defined, 4
  core defined, 4
  minimum hardness depths for production work in surface hardening, 171(T)
  overview, 4–5

T

Temperature-control systems
  limits of error for thermocouples, 88(T)
  noncontact temperature instruments, 89–90(F)
  optical pyrometer, 90(F)

Temperature-control systems (continued)
   overview, 85–86
   process temperatures, controlling, 85
   radiation pyrometers, 90(F)
   ramp-soak, programmable, single-point temperature controller, 89(F)
   temperature control, 88–89(F)
   temperature measurement, 87–88(T)
   temperature sensors, 86–90(F)
   thermocouple types, nominal ranges, and material combinations, 87(T)
Temperature sensors, 86–90(F)
   resistance temperature detectors (RTDs), 86
   thermocouples, 86–87(F,T)
Tempering
   austempering of steel, 6
   maraging steels, 6
   martempering of steel, 6
   overview, 5–6
Test coupons (or test pins), 283–286(F)
Test pins, monitoring carburizing/hardening of gears
   processing of, 284
   purchase of, 284
   setting upper/lower control limits for variables, 285–286
   testing procedures: core hardness, 284
   testing procedures: effective case depth, 284–285(F)
   testing procedures: surface hardness, 284–285(F)
Thermocouples, 86–87
   limits of error for thermocouples, 88(T)
   simple thermocouple, 87(F)
   thermocouple types, nominal ranges, and material combinations, 87(T)
Time-temperature-transformation (TTT), 23
Tool steels
   apparent hardness versus actual hardness, 195–196(F)
   classification and approximate compositions of some principal types of tool steels, 194(T)
   classification of, 191–193, 194(T)
   double and multiple tempering, 204(F), 205–206(F)
   grouping by AISI, 193
   heat treating of, 191–206(F,T)
   heat treating practice for specific groups, 196–202(F,T)
   heat treating processes, general, 193–194(T)
   overview, 191
   tempering temperature versus hardness, 204–205(F), 206(F)
Tool steels, heat treating practice for
   air-hardening, medium-alloy cold-work tool steels (A grades), 197–198(T)
   grades D2 and D5 tool steels, 199(T)
   grades P2 and P4 mold steels, 199(T)
   high-carbon, high-chromium cold-work tool steels (D grades), 198, 199(T)
   high-speed steels (T, M, and ultrahigh-speed grades), 200–202, 203(T)
   hot-work tool steels, 201(T)
   hot-work tool steels (H grades), 200, 201(T)
   low-alloy, special-purpose tool steels (L grades), 199
   mold steels (P grades), 199–200(T)
   molybdenum high-speed steels, 203(T)
   O-and A-grade tool steels, 197(T)
   oil-hardening cold-work tool steels (O grades), 197, 198(T)
   shock-resisting tool steels (S grades), 196–197(T)
   tungsten high-speed steels, 202(T), 202–206(F)
   W-and S-grade tool steels, 197(T)
   water-hardening tool steels (W grades), 196, 197(T)
Total case depth (TCD), 279
Transformation
   bainite microstructure, 24(F)
   defined, 15
   effect of time on, 23–25(F)
   equilibrium transformations, 14–15
   time-temperature-transformation (TTT) diagram, 23(F)

U

Ultrasonic microhardness testing, 36–37
Unified Numbering System (UNS), 10, 98–99

V

Vacuum carburizing, 147
Vacuum furnaces
   applications, 69
   batch-graphite integral oil-quench vacuum furnace, 71(F)
   cold-wall vacuum furnace, 69(F)
   overview, 67
   single-chamber batch-type pressure-quench vacuum furnace, 70(F)
   vacuum conditions, 69
Vacuum heat treating, 67–68
Vickers hardness number (HV), 36, 37(F)
Vickers testing, 36
   diamond pyramid indenter, 37(F)
   Vickers hardness number (HV), 36, 37(F)
   Vickers indenter, 36

W

White cast irons, 209(T)
Wrought stainless steels, classification of, 176–189(F,T)